Antarctic Lakes

Johanna Laybourn-Parry

and

Jemma L. Wadham

OXFORD
UNIVERSITY PRESS

Antarctic Lakes. Johanna Laybourn-Parry and Jemma L. Wadham.
© Johanna Laybourn-Parry and Jemma L. Wadham 2014. Published 2014 by Oxford University Press.

UNIVERSITY PRESS

Great Clarendon Street, Oxford, OX2 6DP,
United Kingdom

Oxford University Press is a department of the University of Oxford.
It furthers the University's objective of excellence in research, scholarship,
and education by publishing worldwide. Oxford is a registered trade mark of
Oxford University Press in the UK and in certain other countries

© Johanna Laybourn-Parry and Jemma L. Wadham 2014

The moral rights of the authors have been asserted

First Edition published in 2014

Impression: 1

All rights reserved. No part of this publication may be reproduced, stored in
a retrieval system, or transmitted, in any form or by any means, without the
prior permission in writing of Oxford University Press, or as expressly permitted
by law, by licence or under terms agreed with the appropriate reprographics
rights organization. Enquiries concerning reproduction outside the scope of the
above should be sent to the Rights Department, Oxford University Press, at the
address above

You must not circulate this work in any other form
and you must impose this same condition on any acquirer

Published in the United States of America by Oxford University Press
198 Madison Avenue, New York, NY 10016, United States of America

British Library Cataloguing in Publication Data
Data available

Library of Congress Control Number: 2014932546

ISBN 978–0–19–967049–9 (hbk.)
ISBN 978–0–19–967050–5 (pbk.)

Printed and bound by
CPI Group (UK) Ltd, Croydon, CR0 4YY

Links to third party websites are provided by Oxford in good faith and
for information only. Oxford disclaims any responsibility for the materials
contained in any third party website referenced in this work.

DUDLEY LIBRARIES	
000001021141	
Askews & Holts	29-Feb-2016
	£34.99
DU	

Preface

"Several tributary glaciers hung down these slopes and reached the 1500 feet level. The water from these glaciers drained into Lake Bonney".

"Lake Bonney evidently occupies an area of internal drainage in which ablation just about balances inflow from the glaciers. As far as I could judge the bottom of the lake was practically at sea level but the head of the defile near Suess Glacier was 100 to 200 feet above sea level."

Griffith Taylor 1880—British Antarctic ('Terra Nova') Expedition (1910–1913).
The Physiography of the McMurdo Sound and Granite Harbour Region.

The above comments were among the first limnological observations made in Antarctica. The earlier British Discovery Expedition, in 1901–1903, also led by Robert Falcon Scott, made the first measurements of lake level in Lake Bonney and measured the distance across the narrows between its east and west lobes. These were important data against which present water levels can be compared. In fact there has been a very significant increase in the depth of the McMurdo Dry Valleys lakes since 1903, indicating that ablation does not balance inflows from the glaciers. These observations were made during the age of Antarctic exploration. It was not until the International Geophysical Year (1957–1958), the establishment of the International Committee on Antarctic Research (SCAR), and the establishment of permanent research stations in the 1950s and 1960s, that Antarctic limnology really started to gain momentum.

Antarctica contains the most diverse range of lakes on the planet. There are many freshwater and brackish to hypersaline lakes in the ice-free areas, but there are also freshwater epishelf lakes that either overlie seawater or have a connection to the sea and are therefore tidal, cryolakes on glaciers, ice shelf ponds and lakes, and most remarkable of all, a vast network of subglacial lakes under the continental ice sheet, of which lake Vostok, Whillans, and Ellsworth are the best known. The latter confront scientists endeavouring to unravel their secrets with major challenges, and effectively represent the modern equivalent of the age of exploration.

There are a number of excellent books that deal with the limnology of specific areas of Antarctica, for example *Ecosystem Dynamics in a Polar Desert* edited by John Priscu and The *Schirmacher Oasis* edited by Peter Bormann and Diedrich Fritzsche, but there is no single volume that pulls together the data for the entire continent. Our aim was to produce a book that would be of general interest to those with a limited knowledge of Antarctic lakes, as well as a reference book for experienced researchers in the field. The first chapter is intended as an introduction to Antarctic lakes, while subsequent chapters provide an in-depth consideration of specific lake types. The final chapter considers future directions.

There are still major gaps in our knowledge of Antarctic limnology, as this volume will show, but nonetheless the expanding database provides us with a clear picture of the formation and ecology of some of the most extreme water bodies on our planet. Antarctic lakes are usually depauperate systems and are characterized by truncated microbially dominated food webs. Moreover, unlike many lakes at lower latitudes that suffer the direct impacts of Man's industrial and agricultural activities, Antarctic lakes are pristine. However, they are subject to the indirect anthropogenic effects of global climate change and enhanced UV radiation. Polar lakes, both in the Arctic and Antarctic, are widely recognized as sentinels of local and global climate change. We have not included a specific chapter on this important issue, but embedded information throughout the book.

Ice is an important factor in polar limnology. Lakes are either covered by it, in the case of subglacial lakes by up to a 4 km thickness, or they are located on glaciers or ice shelves. Within Antarctic lakes research traditional limnologists have worked with glaciologists, and the boundary between what were traditionally two distinct disciplines has blurred. The study of Antarctic lakes exemplifies the need for a multi-disciplinary approach which is particularly well illustrated by the McMurdo Long Term Ecosystem Research Program.

We are indebted to colleagues and friends worldwide who have kindly given us access to their photographs. We thank our editors Ian Sherman and Lucy Nash and the many other people who have contributed to the production of this volume. Particular thanks go to Simon Powell at Bristol University for his excellent work on the illustrations. Lastly we would like to thank the US, Australian, New Zealand, and British Antarctic programmes and acknowledge funding from a wide range of bodies, both national and international, that has supported our own research in Antarctica.

Johanna Laybourn-Parry
Jemma L. Wadham
University of Bristol January 2014

Contents

1 An introduction to Antarctic lakes — 1

1.1 Introduction — 1
1.2 History of Antarctic limnology and logistics — 3
1.3 Climatic conditions in Antarctica — 4
1.4 Glaciological history of Antarctica — 7
1.5 Diversity of lakes — 11
1.6 Lake types and geochemical conditions — 15
 1.6.1 Salinity — 15
 1.6.2 Redox conditions — 17
 1.6.3 Nutrient and organic carbon supply — 19
 1.6.4 Geochemical indicators of lake history — 22
1.7 Geomorphology of Antarctic lakes — 23
1.8 Antarctic lake biota — 27
 1.8.1 Archaea and Bacteria — 29
 1.8.2 Viruses — 30
 1.8.3 Protozoa — 31
 1.8.4 Algae — 34
 1.8.5 Rotifers — 35
 1.8.6 Crustaceans — 36
 1.8.7 Other invertebrates — 37
1.9 Habitats in Antarctic lakes — 38

2 Freshwater lakes — 41

2.1 Introduction — 41
2.2 Formation of freshwater lakes — 49
2.3 Temperature and stratification — 60
2.4 Water chemistry — 62
2.5 The planktonic biota of freshwater lakes — 64
 2.5.1 Heterotrophic bacteria — 64
 2.5.2 Viruses — 66
 2.5.3 Protozoa — 68
 2.5.4 The phytoplankton — 72
 2.5.5 The zooplankton — 73

	2.6 Carbon cycling in the planktonic environment	75
	2.6.1 Primary production	75
	2.6.2 Bacterial production	78
	2.6.3 Heterotrophic grazing	80
	2.7 The benthic communities	83
	2.7.1 Phototrophic benthic communities	83
	2.7.2 Heterotrophic benthic communities	87
	2.7.3 Carbon cycling in the benthos	88

3 Saline lakes   91

	3.1 Introduction	91
	3.2 Distribution of saline lakes in Antarctica	94
	3.3 Formation of saline lakes	97
	3.4 Patterns of stratification and temperature	101
	3.5 Water chemistry	102
	3.6 The planktonic biota of saline lakes	105
	3.6.1 Heterotrophic Bacteria and Archaea	106
	3.6.2 Photosynthetic bacteria	110
	3.6.3 Viruses	112
	3.6.4 Protozoa	114
	3.6.5 Algae	121
	3.6.6 Zooplankton	122
	3.7 Carbon cycling in the plankton	123
	3.7.1 Primary production	123
	3.7.2 Bacterial production	125
	3.7.3 Heterotrophic grazing and carbon cycling	125
	3.8 The biota of saline lake ice covers	129
	3.9 The benthic community	130
	3.10 Carbon cycling in the benthos	131
	3.11 A unique Antarctic lake—Lake Vida	132

4 Epishelf lakes   134

	4.1 Introduction	134
	4.2 Formation and physico/chemical characteristics of epishelf lakes	135
	4.2.1 Geomorphology	135
	4.2.2 Physico/chemical characteristics	137
	4.3 The planktonic biota of epishelf lakes	140
	4.4 Carbon cycling in the plankton of epishelf lakes	143
	4.5 The benthic communities of epishelf lakes	144

5 Lakes and ponds on glaciers and ice shelves   147

	5.1 Introduction	147
	5.2 Supraglacial lakes	148
	5.2.1 Types of cryolakes	148
	5.2.2 The physical/chemical environment and biology of cryolakes	150
	5.3 Ice shelf ponds and lakes	152

6 Subglacial lakes — **156**

    6.1 Introduction — 156
    6.2 Distribution and physiographic characteristics of subglacial lakes in Antarctica — 157
    6.3 Detailed studies of subglacial lakes — 159
        6.3.1 Lake Vostok — 159
        6.3.2 Lake Ellsworth — 159
        6.3.3 Lake Whillans — 162
        6.3.4 Hodgson Lake — 163
    6.4 Formation of subglacial lakes and hydrological conditions — 163
    6.5 Geochemical conditions in subglacial lakes — 166
    6.6 The biota of subglacial lakes — 170

7 Conclusions and future directions — **174**

    7.1 Antarctic lakes in a global context — 174
    7.2 Inter-annual variations and longer-term trends — 175
    7.3 The gaps in the data—the way forward — 176
    7.4 Future directions — 177

Glossary — 181
References — 187
Index — 211

CHAPTER 1

An introduction to Antarctic lakes

1.1 Introduction

Around 98% of the Antarctic landmass is covered by an ice sheet up to 4 km thick. Small ice-free areas occur in the coastal margins (Figure 1.1) and are often referred to as oases, because they carry lakes in which life is supported and they may also have sparse pockets of lichens and mosses, where there is sufficient seasonal melt to support them. Examples are the Vestfold Hills, the Schirmacher Oasis, Syowa Oasis, the Bunger Hills, and the Larsemann Hills. The McMurdo Dry Valleys of Southern Victoria Land (Figure 1.1) represent a large inland ice-free area in which there are numerous large lakes and ponds. The Antarctic Peninsula and the nearby groups of islands that constitute what is known as the Maritime Antarctic also have ice-free areas that carry lakes and ponds. Beyond these ice-free zones, the basal regions of the ice sheet are also now known to harbour over 380 subglacial lakes (Wright and Siegert 2011, 2012); their potential to support life is currently of great topical interest. Antarctica is a desolate, isolated continent that cannot support human life. Effectively, it is a continent dominated by microorganisms, with some invertebrates. The seals, penguins, and other birds that occur around its margins are part of the marine food chain. They use the continent to nest, as in the case of penguins, or as a place to give birth (Weddell seals) and to bask in the austral sun. In contrast, the Arctic has vegetation with land mammals and birds. Human life can be supported and there are numerous groups of indigenous peoples who have successfully made a living from the sea and land. While the Arctic is the northern extension of continental land masses that has allowed species to colonize during interglacial periods, something that is still happening today, the isolation of the Antarctic continent renders its colonization challenging. This is an important issue we shall explore elsewhere in this volume.

Antarctica is not a sovereign nation, it 'belongs' to no one. It is administered through the Antarctic Treaty and is effectively a huge nature reserve where there is promotion of international scientific cooperation. The Antarctic Treaty was signed in Washington in December 1959 and came into force in June 1961. It involved the 12 nations who engaged in Antarctic research during the Geophysical Year, 1957–1958. These were Australia, Argentina, Belgium, Chile, France, Japan, New Zealand, Norway, USSR, South Africa, the United Kingdom, and the United States of America. Subsequently the membership has grown so that today 38 other countries have acceded to the Treaty, bringing the membership to 50. Among these nations, 28 have consultative status in that they attend consultative meetings and participate in decision making. These are nations that are 'conducting substantial research activity' in Antarctica. The remaining 22 signatories are non-consultative parties who may attend meetings, but do not participate in decision making. The Treaty establishes Protocols and Conventions that the signatory nations must adhere to. For example the Environment Protocol has a series of annexes that provide clear guidelines on issues such as the protection of flora and fauna, marine pollution, and waste disposal. Further information on the Antarctic Treaty can be obtained from the website: <http://www.ats.aq/e/ats.htm>.

Thus scientific research in Antarctica is very different from that elsewhere in the world. Firstly, there is a very strong collaborative network among nations. One only needs to look at the literature to see that papers are often co-authored by scientists

Antarctic Lakes. Johanna Laybourn-Parry and Jemma L. Wadham.
© Johanna Laybourn-Parry and Jemma L. Wadham 2014. Published 2014 by Oxford University Press.

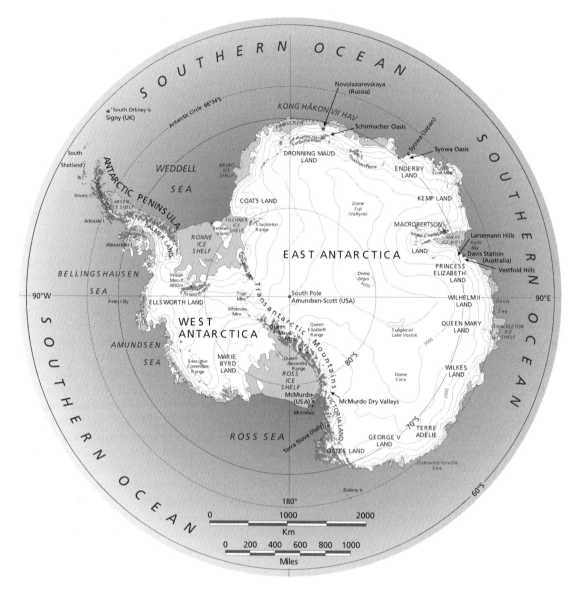

Figure 1.1 Map of Antarctica showing the position of coastal oases, the Dry Valleys and the Maritime Antarctic. (See Plate 1 for better resolution)

from a range of countries. While this does occur elsewhere in other scientific disciplines, it is very strong among the Antarctic community. Secondly, one cannot simply go to Antarctica to conduct research, as one can when one works in the Arctic. While there are some constraints among the various Arctic nations about working in their specific Arctic territories, it is much easier and often less logistically difficult to get to field locations. To be able to work in the Antarctic, a scientist has to be part of the Antarctic organization of a nation, in order to gain access to logistics (transport, research stations, field camps, training, medical support, etc.) and approval for their research project under the protocols and conventions of the Antarctic Treaty. The scientist must usually also be in receipt of research grant funding from their government, or other sources such as a charitable foundation. The logistical complexity of doing this kind of science begs the question as to why a scientist should want to do limnological research in Antarctica. The answer is that the varied lakes of Antarctica are extraordinary, and in many cases one has the opportunity to see organisms and processes that have never before been seen or described. Moreover, these pristine water bodies appear to respond rapidly to climatic variability and may prove to be valuable sentinels of wider climate change.

Unlike many lakes at lower latitudes that have been subject to the direct effects of Man's activity, Antarctic lakes are pristine. At lower latitudes the development of agriculture, resulting in the clearing of forests and urbanization, have changed the catchment hydrology of lakes leading to greater sedimentation and changes in nutrient status. Polar lakes are, however, subject to the indirect effects of Man's activities that have resulted in ozone depletion over Antarctica and increases in greenhouse gases and climate change. Thus in Antarctica we are able to see these indirect effects in isolation, without them being superimposed on the direct anthropogenic influences.

This volume is intended for the specialist and non-specialist. This first chapter provides a background understanding of Antarctic lakes, laying the foundation for more detailed treatment in subsequent chapters. Chapter 1 also includes introductory information on basic biology, biogeochemistry, and the geomorphology of Antarctic lakes for the non-specialist.

1.2 History of Antarctic limnology and logistics

During the early years of the twentieth century there were numerous national expeditions by the Belgians, the Germans, the French, and the British to the Maritime Antarctic islands, the Peninsula, and the continent during which lakes were sampled and some information on the biota published (for a review see Heywood 1972). The Dry Valleys were discovered by Scott during the British Antarctic Expedition of 1903 and the first detailed description of the Dry Valley lakes was given by Taylor (1922). The early biological focus was on algal mats, crustaceans, and larger protozoans, as well as the characteristics of the physical and chemical environment. At that time, the modern understanding of the functioning of the plankton and the importance of microorganisms were many decades in the future.

As Heywood (1972) points out, biological studies in Antarctica really only came of age after the International Geophysical Year (IGY) of 1957–1958. The establishment of the Scientific Committee on Antarctic Research (SCAR) by the Antarctic Treaty occurred in 1958. Its remit was to coordinate Antarctic research and at its third meeting in 1959, guidelines for marine and freshwater biology were drawn up. Research conducted by New Zealand and the United States in the McMurdo Dry Valleys has been ongoing since the IGY, and in 1993 the Taylor Valley became the site of a National Science Foundation (NSF) Long-Term Ecosystem Research (LTER) Program (Figure 1.1). This programme has provided invaluable long-term data sets on all aspects of the limnology in the Taylor Valley lakes. The severity of the climate and the distance from McMurdo Station make winter studies logistically challenging, so data mostly pertain to the summer. In recent years, work has started in September and extended into late autumn (March/April), but as yet we have no detailed biological data for the winter months.

The New Zealanders established a station on the shore of Lake Vanda in the Wright Valley (McMurdo Dry Valleys) in 1968 which was occupied during the summers until 1991. Wintering parties were maintained in 1969, 1970, and 1974. Rising water levels in the lake and changes in research priorities saw the removal of the station, and by 1994/1995 all traces of it had been successfully eradicated. Today there are two refuge huts on the shore of Lake Vanda.

Various research programmes by US and New Zealand scientists have focused on the McMurdo Ice Shelf and lakes on Ross Island. McMurdo Station (USA) and Scott Base (NZ) are close by. In 1985, the Italians set up a station at Terra Nova Bay in the Ross Sea, an area with some lakes and ponds that have been the focus of limnological research.

Davis Station in the Vestfold Hills (Australian Antarctic Territory) was established in the IGY of 1957, with an overwintering complement of five (Figure 1.1). The Vestfold Hills, along with the Dry Valleys, has some of the most exciting suites of surface freshwater and saline lakes in Antarctica. The Vestfold Hills carries some 300 lakes and ponds that range from freshwater to hypersaline. The Soviets had already undertaken some brief investigations in the Vestfold Hills in preparation for the IGY. Their early presence is indicated by some of the lake names, such as Lake Druzhby and Lake Zvezda. The Soviets later established Mirny Station, some 350 km east of Davis Station, and a small station in the Larsemann Hills. Davis Station has grown significantly over the years and now provides excellent laboratory facilities and accommodation. Easy access to the lakes by helicopter in summer and overland using various vehicles in winter has allowed year-round studies on a number of the lakes, notably the saline meromictic Ace Lake, brackish Pendant and Highway Lakes, and the freshwater Crooked Lake and Lake Druzhby. Some 80 km distant from the Vestfold Hills are the Larsemann Hills, where a summer station was established in 1987. The Larsemann Hills contains a suite of small lakes and ponds (Figure 1.1). At various times the Australians have mounted expeditions to the Bunger Hills, which also possesses an interesting suite of lakes including epishelf lakes. The first of these expeditions was mounted in 1977.

Early geographical and geological observations in the Schirmacher Oasis were initiated by a Soviet expedition in February 1959. However, it was a Norwegian expedition that mapped the region during 1958–1959. In 1960/1961, Novolazarevskaya Station was established in the eastern part of the Oasis, in Queen Maud Land (Figure 1.1). A series of observations on the lakes commenced during the 1960s. Between 1976 and 1978 the German Democratic Republic built a research station a few kilometres from the Soviet station, initially to study the ionosphere, and later extending their studies to other areas including limnology (Bormann and Fritzsche 1995).

Japanese involvement in Antarctic limnological research developed in 1957 with the building of Syowa Station (Figure 1.1) on East Ongul Island, Lutzow-Holm Bay, Queen Maud Land. The Syowa Oasis contains suites of freshwater and marine-derived saline lakes. Some of the saline lakes are hypersaline with salinities around seven times that of seawater.

The Maritime Antarctic includes the Antarctic Peninsula, the South Shetland Islands, the South Orkney Islands, the South Sandwich Islands, and Bouvet Island. Signy Island in the South Orkney Islands was a site of the whaling industry in the past. In 1947, the British Antarctic Survey established a base on the site of a previous whaling station that hosted a range of research including limnology. This included year-long studies of a number of the lakes. The South Shetland Islands, which were also known for the whaling industry, have been the focus of numerous recent studies. A number of nations, including Argentina, Spain, and Bulgaria maintain seasonal stations on Livingston Island (South Shetlands). The Antarctic Peninsula has numerous lakes that have been researched by scientists from a range of nations with stations in the region. The climate is less severe in the Maritime Antarctic and the vegetation better developed than on the main continent, with mosses, lichens, and the Antarctic hair grass *Deschampsia antarctica*.

Most limnological databases cover the summer only. It is only where logistics allow, and/or where stations are manned throughout the year, that long-term annual studies are possible. Relatively few nations maintain wintering facilities.

1.3 Climatic conditions in Antarctica

Antarctica is the coldest continent on Earth and in common with the Arctic, experiences the lowest annual levels of Photosynthetically Active Radiation (PAR). While there are many hours of daylight in summer, the winter is characterized by darkness, when the sun does not rise above the horizon. The length of the winter darkness depends on latitude across the continent. At the South Pole there are six months of darkness, the sun disappears in March and does not reappear until September. In summer there are many months of 24-hour daylight. In contrast, on the Antarctic coast at Davis Station in the Vestfold Hills (68°S) the sun disappears below the horizon from late May through June to early July, and in summer, 24-hour daylight occurs during late November through December to early January. Thus the crucial process of photosynthesis, the first stage in the carbon cycle where inorganic carbon (CO_2) is transformed into plant biomass, is severely curtailed through lack of light energy for a significant part of the year. As we shall see in later chapters, the autotrophic organisms in Antarctic lakes have evolved adaptations that allow them to survive and thrive under these light-limited cold conditions.

Until the International Geophysical Year of 1957–1958 there were little continuous climate data for

Antarctica. Today there are many continuous data sets from the various stations across the continent. The examples shown in Figure 1.2 are for the McMurdo Dry Valleys and the Vestfold Hills, both areas of intense limnological research. Davis Station in the Vestfold Hills is occupied all year round and data are collected each day by a team of observers from the Australian Meteorological Bureau. The McMurdo Dry Valleys are uninhabited during winter, but an automated weather station logs basic meteorological data. What is clear from Figure 1.2 is that over short time spans there are considerable inter-annual variations, with occasional very warm years that can lead to longer phases of ice loss, or colder years when ice does not break out or only partially breaks out on the coastal lakes. Note that temperatures at the Lake Hoare site in the Dry Valleys are very much lower than the Vestfold Hills, which is coastal and further north. Ice cover is important in the temperature dynamics of the water column. Ice is effectively a thermal blanket that protects the water column from the cooling impact of the overlying cold atmosphere. Where lake ice is missing, as is the case in extreme hypersaline lakes like Deep Lake (Vestfold Hills) with a salinity ten times that of seawater, summer water temperatures range from 7 °C to 11.5 °C, and in winter the temperature plummets to between –17 °C and –18 °C (Ferris and Burton 1988). Saline lakes that have winter ice covers maintain temperatures of only a few degrees below zero. The Dry Valleys experienced a particularly warm year in 2001/2002, when there was significant stream flow from the glaciers that resulted in an increase in lake levels, cancelling out the previous 14 years of lowering water levels (Doran et al. 2008). These short-term changes should not be confused with long-term trends in climate change.

Considerable effort is being devoted to assessing long-term trends in climate in Antarctica. An

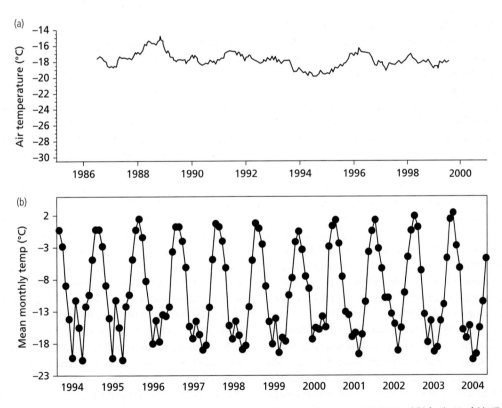

Figure 1.2 Air temperature data for (a) Lake Hoare (data from Doran et al. 2002a) over the years 1986–2000 and (b) for the Vestfold Hills for the years 1994–2004, courtesy of the Australian Meteorological Bureau.

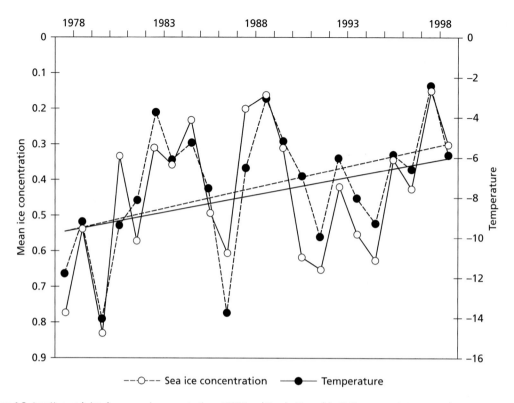

Figure 1.3 June/August (winter) mean sea ice concentration at 70°W and Faraday/Vernadsky Station mean winter near-surface temperature. Regression lines are shown for both time series. From Turner et al. (2005), with the permission of John Wiley & Sons.

analysis of 50-year data sets from 19 stations from the Peninsula and coastal stations, as well as Scott–Amundsen Base (South Pole) and Vostok, reveals that the Antarctic Peninsula has experienced a statistically significant increase in temperature. At the Faraday/Vernadsky Station, annual temperature has increased by 0.56 °C per decade, and by 1.09 °C per decade over winter (Figure 1.3) (Turner et al. 2005). These long-term warming trends have contributed to the collapse of the Larsen A and part of the Larsen B ice shelves. Today only 1670 km² of Larsen B remains of the original 11,500 km² (Vaughan and Doake 1996). Other parts of Antarctica are cooling, and at Vostok Station within the East Antarctic interior, there has been no significant change in the records extending back over 40 years. Changes at coastal stations around the continent are more variable, some locations showing warming while others showing no change (Turner et al. 2005). While the collapse of ice shelves is often rapid and spectacular, long-term climate warming on the Peninsula is also contributing to the gradual retreat of glaciers. Data covering the period 1940 to 2001 indicate a clear transition between mean advance and mean retreat, a southerly migration of that transition as ice shelves are retreating, and progressive atmospheric warming. Over this period 146 glaciers have retreated less than 500 m, while 27 have advanced less than 500 m. Fifty-two have retreated between 500 and 2000 m compared with five that have advanced that distance. A further 13 have retreated between 2000 and 5000 m (Cook et al. 2005).

We are currently in an interglacial period (see Section 1.4). The ice sheet of Antarctica contains entrapped air inclusions throughout its depth that provide a direct record of past changes in the atmosphere. The deuterium content of the ice (δD_{ice}) is a proxy for local temperature change. The Vostok ice core provides 3623 m of such palaeoclimate records covering the past 420,000 years and four glacial–interglacial cycles during which the climate has been in a state of change. During these interglacial cycles, temperature and atmospheric concentrations of carbon dioxide (CO_2) and methane (CH_4) have risen steadily. Both CO_2 and CH_4 correlate strongly with temperature, indicating that these greenhouse gases have contributed significantly to glacial–interglacial change. Today's concentrations of these gases are uniquely elevated compared with levels during the last 420,000 years (Petit et al. 1999). As we shall see in succeeding chapters, there is clear evidence that Antarctic lakes respond rapidly to short-term climatic perturbations and longer-term changes. For example, the lakes of the McMurdo Dry Valleys have risen significantly during the twentieth century (Chinn 1993). A number of the Dry Valley lakes (Vanda, Fryxell, and Bonney) had lost their ice covers and evaporated to small hypersaline ponds by about 1000–1200 years before present (B.P.), indicating that there was a major decrease in lake levels. Subsequent climate warming resulted in lakes refilling to the size we see today (Lyons et al. 1998a).

1.4 Glaciological history of Antarctica

The large body of permanent glacial ice in Antarctica comprises two separate ice sheets; the West Antarctic Ice Sheet (WAIS) and the much larger East Antarctic Ice Sheet, separated by the Transantarctic Mountains (Figure 1.1). A number of smaller glaciers and ice caps also exist along the Antarctic Peninsula and in mountain ranges at the margins of the ice sheet, such as in the McMurdo Dry Valleys. The two ice sheets have quite contrasting histories, thus influencing the age and genesis of lakes beneath the ice sheet and around its periphery.

Much of the history of the West and East Antarctic Ice Sheets has been inferred from mid-latitude ocean core (Shackleton 1987; Zachos et al. 1992, 2001) and ice core records (Jouzel et al. 2007; Petit et al. 1999; Watanabe et al. 2003). Both types of study have relied quite heavily upon the analysis of oxygen isotope ratios in order to infer global temperature and ice volume changes. In marine cores, this involves analysis of the oxygen isotope composition of calcium carbonate shells of Foraminifera (Protozoa) that inhabit surface ocean waters until they die and sink to the ocean bottom and become incorporated into sediments. For the ice core records, scientists analyse the oxygen isotope composition of water molecules locked up in continental ice sheets (Emiliani 1955; Shackleton 1987). The marine records go back some tens of millions of years (Zachos et al. 1992, 2001), whereas ice core records are limited to the last 800 kyr at most (Jouzel et al. 2007). Box 1.1 gives a description of how these records are compiled.

Box 1.1 Stable isotopes

Isotopes refer to atoms of the same element, but with a slightly different mass, as a result of a varying number of neutrons being present in the atom. It is common in elements for multiple isotopes to exist, and the ratio between a heavier and lighter isotope in a pair gives the isotopic signature of a sample. Isotopic signatures may vary between different source materials. In addition, the isotopic composition of a substance can be influenced by a range of physical (e.g. phase changes) and chemical (e.g. photosynthesis) processes. These have what is known as a 'fractionation' effect, where one isotope becomes progressively enriched or depleted with respect to the other. An isotope pair that is commonly used in palaeoclimatic studies is the ratio of ^{16}O to ^{18}O, where the mass of the oxygen atom is 16 and 18 respectively, a reflection of the presence of 8 neutrons and 8 protons in the former 'light' isotope and 10 neutrons and 8 protons in the latter 'heavy' isotope. Hence, the ratio $^{18}O/^{16}O$ may be determined, and expressed as a difference (δ) in parts per thousand (or ‰) relative to this ratio in a known standard, most commonly the Standard Mean Ocean Water (SMOW) standard. Hence, the following equation would apply:

$$\delta x = \left[\left(R_{sample} / R_{standard} \right) - 1 \right] \times 10^3, \quad (1)$$

where x is the isotope of interest and R represents the ratio of the isotope of interest (e.g. $^{18}O:^{16}O$).

In ocean sediment cores, the calcium carbonate ($CaCO_3$) shells of Foraminifera have different ratios of the isotope pair ^{18}O and ^{16}O, depending upon the isotopic composition of the ocean water at their time of formation (Emiliani 1955; Shackleton 1987). While these organisms die, their carbonate shells sink to the ocean floor and

continued

Box 1.1 *Continued*

form layers of ocean sediment, which may be recovered by coring, and subsequently dated. The variable $\delta^{18}O$ of the shells of these Foraminifera generally reflects changes in local temperature and global ice volume (Shackleton 1987). In a similar manner, the $\delta^{18}O$ of water molecules deposited upon ice sheets during different snowfall events also reflects changes in global temperatures and recovery of these data from ice cores reveals variations in climate during the latter half of the Quaternary.

In general, ice locked up in the Earth's ice sheets is enriched in the lighter ^{16}O isotope, since it is more difficult to evaporate water molecules containing the heavier isotope from the oceans. In addition, it is also easier for the heavier isotope to be lost in deposition as rain at lower latitudes, resulting in the deposition of isotopically light water molecules on ice sheets. However, the degree to which water molecules are isotopically depleted in ^{18}O also depends upon ocean temperatures, so that as ocean temperatures decrease it becomes increasingly difficult for heavy water molecules to form vapour. Hence, during cold periods the water locked up in ice sheets becomes isotopically lighter, and the calcium carbonate shells of foraminifera in the oceans become enriched in ^{18}O (Shackleton 1987). In the oceans, there is an additional effect which reflects the fact that progressively more ^{16}O becomes stored in ice sheets during cold periods as they grow. Hence, the ocean $\delta^{18}O$ records reflect local seawater temperatures and global ice volumes (Shackleton 1987).

Ice core records currently go back 800,000 years, with the well-known Vostok ice core covering the last 420 kyr (Petit et al. 1999; Jouzel et al. 2007). These ice core records show good correspondence with benthic ocean core records (Figure 1.4) (Lisiecki and Raymo 2005). In general, both of these records show cyclically varying climate during the Quaternary period, whereby cold 'glacial' periods lasting approximately 100 kyr are separated by shorter (10–20 kyr) 'interglacial' warm periods. Data from deep ocean cores derived from multiple locations can be stacked to produce very long time series of $\delta^{18}O$, and hence temperature change, with some records extending to 65 Ma (Zachos et al. 2001) (Figure 1.5).

In both West and East Antarctica, rapid cooling at the Eocene–Oligocene boundary at approximately 34 million years before present (hereafter, Myr B.P.) triggered a transition from a 'greenhouse' to 'icehouse' climate during the late Cenozoic (Zachos et al. 2001). This fast transition is believed to have led to the initiation of permanent ice cover in Antarctica (Barrett 1996; Flower 1999). During the early Eocene, and prior to glaciation, the Antarctic continent exhibited positive mean annual air temperatures and was ice-free and vegetated, where recent core data from the Wilkes Land margin show the presence of near-tropical forest (Pross et al. 2012), later believed to revert to tundra (DeConto et al. 2012). As the ice sheet grew, isostatic depression

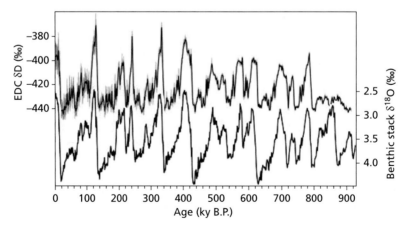

Figure 1.4 Comparison of the EPICA Dome C δD record with the benthic $\delta^{18}O$ record over the last 800 years. Adapted from Jouzel et al. (2007), with the permission of Highwire Press.

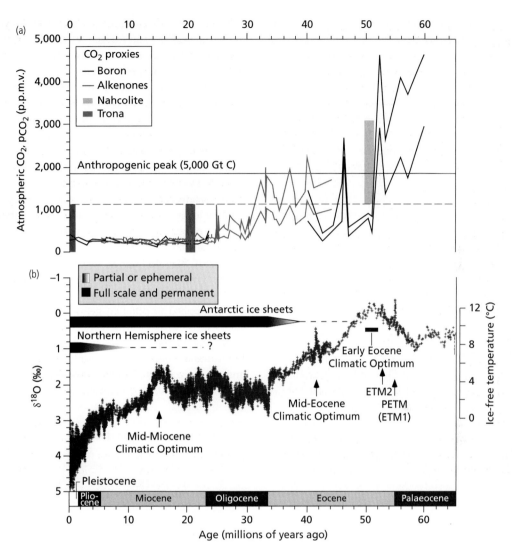

Figure 1.5 Evolution of atmospheric CO_2 and global climate over the last 65 million years (adapted from Zachos et al. 2008). (a) Cenozoic PCO_2 for the period 0 to 65 Myr B.P. Data are from a compilation of marine and lacustrine sources (full details in Zachos et al. 2008). (b) Climatic conditions for the same period derived from stacked benthic Foraminifera oxygen-isotope curves from marine cores (smoothed with a 5-point running mean). The $\delta^{18}O$ temperature scale, on the right axis, assumes an ice-free ocean; it therefore applies only to the time preceding the onset of large-scale glaciation in Antarctica (about 35 million years ago). Major climatic events during this period are shown (2-million-year-long Early Eocene Climatic Optimum and the more transient Mid-Eocene Climatic Optimum, and the very short-lived early Eocene hyperthermals, such as the PETM (also known as Eocene Thermal Maximum 1, ETM1) and Eocene Thermal Maximum 2 (ETM2; also known as ELMO). ‰ parts per thousand. With the permission of the Nature Publishing Group.

of the basement rocks would have caused a relative rise in sea level around the continent and the ingress of marine waters and deposition of marine sediments on formerly sub-aerial parts of the continent. This, together with continual erosion of the underlying bedrock by glacial ice, has produced some very substantial subglacial overdeepened basins beneath the ice sheet (Young et al. 2011), containing up to 14 km of sedimentary sequences (Ferraccioli et al. 2009). These points are significant for an understanding of the substrata of subglacial lakes, since thick pre-existing sediments (either of permafrost or marine origin) that become overridden by advancing ice have the potential to provide an important source of carbon and nutrients to lake ecosystems and other aquatic ecosystems beneath the ice sheet.

The establishment of ice sheets of a size close to that of the present day occurred at around 30 Myr B.P. in East Antarctica, and while there is some evidence for ice marginal fluctuations, the general view is of stability and similar ice volumes since 30 Myr B.P. (Huybers and Denton 2008; Pollard and DeConto 2009). In contrast, the West Antarctic Ice Sheet has fluctuated considerably in volume and spatial extent during the last 10 Myr (Naish et al. 2009; Ravelo et al. 2004; Scherer 1991). The grounding of large areas of the West Antarctic Ice Sheet below sea level (Young et al. 2011) leaves it vulnerable to rapid retreat (Alley and Whillans 1991; Macayeal 1992; Mercer 1978). This may occur during phases of sea level change or ocean warming, when fast retreat of the ice shelves that buttress this ice sheet are thought to cause ice acceleration and ice thinning. In contrast, the East Antarctic Ice Sheet is largely grounded above sea level (Young et al. 2011). There is debate as to the timing and frequency of collapses of the West Antarctic Ice Sheet, which are indicated largely by changes in global ice volume, inferred from a range of data sources including sea level records, marine oxygen isotope data, and the presence of open ocean diatoms in subglacial sediments (Scherer et al. 2008). There is strong evidence for a collapse at approximately 3 Myr B.P. during the warmth of the early Pliocene (Scherer 1991), during Marine Isotope Stage 31 at 1 Myr B.P. (Naish et al. 2009), and some records also support more recent collapses during Pleistocene interglacials (Scherer et al. 1998).

The Quaternary period, the most recent of the three periods of the Cenezoic Era (Figure 1.5) and corresponding to approximately the last 2.5 Myr is perhaps the most relevant to understanding the genesis and evolution of lakes around the Antarctic periphery. This period includes two geological epochs: the Pleistocene refers to the time period from approximately 2.6 Myr B.P. until 12 kyr B.P., during which there was a series of repeated glaciations. This was followed by the Holocene, which is the current warm period. During this time period, the East Antarctic Ice Sheet margin exhibited relative stability, with the exception of small variations along the Wilkes Land margin (Hill et al. 2007) and around other major marine inlets such as Prydz Bay (O'Brien et al. 2007). As previously discussed, the West Antarctic Ice Sheet margin is believed to have fluctuated dramatically during this time including several major or partial collapses. The extent of the Antarctic Ice Sheet was at its maximum at the 'Last Glacial Maximum', at approximately 21 kyr B.P., when sub-marine geomorphology shows that ice extended to the continental shelf edge in some locations (Cofaigh et al. 2008; Evans et al. 2006).

Knowledge of the spatial patterns and timing of local ice margin advance and retreat during the Quaternary is inferred from the dating of glacial geomorphological features and accumulated sediments in lake and marine cores. These data are most abundant for the Holocene period, when radiocarbon dating can be applied to different geomorphological indicators of ice extent. These variations are important for an understanding of the evolution of lake ecosystems around the Antarctic periphery (see Box 1.2), and are discussed in detail in subsequent chapters.

Box 1.2 Radiocarbon dating

Radiocarbon dating is a radiometric dating method that is based upon analysis of the radiogenic ^{14}C isotope to estimate the age of carbon-containing materials up to approximately 50,000 yrs B.P. The uncalibrated radiocarbon values are quoted in year before present (B.P.), where present is 1950 AD. In order to apply this method to sediments, there must be a sufficient amount of organic carbon present. Application of this method to dating organic carbon in sedimentary materials in order to arrive at an age for those materials relies on the following principles. When plants or other phototrophic organisms fix carbon dioxide from the atmosphere as organic matter, ^{14}C is incorporated in a concentration that approximates that in the atmosphere at the time. Once these organisms die, ^{14}C accumulation ceases and the ^{14}C that is present at the time of death decays at a known exponential rate, where the half-life usually assumed in radiocarbon studies (the time taken to reduce the amount of the radiogenic isotope by half) is ~5570 years. Knowledge of this decay and the present ^{14}C content enables us to calculate the age of the organic carbon, and hence sediments.

During the Holocene, a number of climatic periods are noteworthy towards understanding the evolution of lakes located around the ice sheet margin. Broadly, the Antarctic Ice Sheet began to retreat late during the last deglaciation and perhaps as late as 10 kyr B.P. The retreat of the West Antarctic Ice Sheet is believed to have occurred rapidly in some places via the collapse of ice shelves (Domack et al. 1999), potentially forced by sea level rise. Elsewhere in Antarctica, the most rapid retreat and thinning occurred after 12 kyr B.P. up until 6 kyr B.P. (Siegert 2001). Complicating the pattern of ice sheet history during the Holocene is the Mid-Holocene Hypsithermal in the Southern Hemisphere that attained its maximum at approximately 9 kyr B.P. These climate changes during the Holocene have dramatically affected lakes in many ice marginal areas and a more detailed discussion of their impacts on lake geomorphology can be found in Section 1.6.

1.5 Diversity of lakes

The Antarctic continent possesses the greatest diversity of lake types on the planet. The surface lakes are mainly located on the coastal oases of the continent, the McMurdo Dry Valleys, and the Maritime Antarctic. These range from freshwater to extremely hypersaline and from relatively small water bodies including ponds to large deep systems such as Crooked Lake (Vestfold Hills), which is around 9 km^2 and has a maximum depth of over 150 m (Laybourn-Parry et al. 1992) and Untersee in Queen Maud Land with an area of 11.4 km^2 and a maximum depth of 169 m (Richter and Bormann 1995).

Freshwater lakes occur around the periphery of the continent, in areas where meltwaters provide inflows to lakes. They also occur on the surface of ice sheets and glaciers. These are often known as supraglacial lakes or cryolakes (Figure 1.6). They can contain a high sediment content which, like the sediment or cryoconite contained in cryoconite holes on glacier surfaces, is likely to support significant biological activity (Laybourn-Parry et al. 2012). While some supraglacial lakes are long-lived others are transitory, they appear during summer and then drain through a moulin and disappear. Shallow ponds and small lakes occur on ice shelves and range in conductivity from close to freshwater to hypersaline. A large network of such water bodies is found on the McMurdo Ice Shelf, the most extensive surface ablation area in Antarctica. Basal freezing incorporates marine-derived debris into the ice, which is transported upwards though the ice shelf as the surface ablates. Ablation is the process whereby there is a loss of surface ice through melting, sublimation, and wind erosion. Areas of

Figure 1.6 Cryolake on Canada Glacier, Taylor Valley. Photo courtesy of M. Tranter. (See Plate 2)

Figure 1.7 Ice shelf lakes on the McMurdo Ice Shelf. Photo courtesy of W.F. Vincent. (See Plate 3)

dark sediment promote surface melt, forming shallow water bodies (Figure 1.7). Being shallow, these ponds and small lakes freeze to their bases in winter providing challenging conditions to their biota (Howard-Williams et al. 1989; Hawes et al. 1999, 2011c).

Many of the coastal lakes lose all or part of their ice covers for a few weeks to a few months each summer, whereas those further south in the Dry Valleys possess thick perennial ice covers around which moats often form in the summer. Where coastally located lakes abut glaciers or the ice sheet they also carry thick perennial ice covers, for example Chelnok Lake which abuts the Sørsdal Glacier and a range of unnamed lakes that abut the ice sheet in the Vestfold Hills (Figure 1.8). In the Maritime Antarctic, for example the South Orkney Islands, a range of small lakes are located in the ice-free regions. These are freshwater systems and differ from continental lakes in that they have vegetated catchments of mostly mosses, lichens, and grasses.

Epishelf lakes are freshwater lakes that may overlie colder, denser seawater adjacent to ice shelves (Figure 1.9), or be connected to the sea via a conduit

Figure 1.8 Lake Nottingham (unofficial name), Vestfold Hills. Perennially ice covered freshwater lake that abuts the continental ice sheet. Photo J. Laybourn-Parry. (See Plate 4)

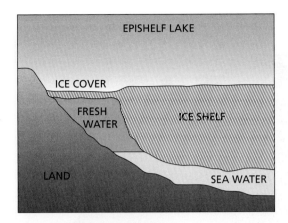

Figure 1.9 Diagrammatic representation of an epishelf lake.

under a glacier or ice shelf (see Chapter 4). Epishelf lakes are almost unique to Antarctica as climate warming and the collapse of ice shelves is leading to their disappearance in the Arctic (Veillette et al. 2008). Epishelf lakes have been described from the Schirmacher Oasis, the Bunger Hills, MacRobertson Land, and Alexander Island. The largest epishelf lake is Beaver Lake in MacRobertson Land, so called because Beaver aircraft used to land on its perennial ice cover (Figure 1.10a, b).

Beneath the polar ice sheet there is an extraordinary array of over 380 subglacial lakes (Wright and Siegert 2011) and a complex hydrological system that connects groups of them (Figure 1.11

Figure 1.10 (a) Rafted freshwater ice caused by tidal action on the shore of Beaver Lake, Amery Oasis. Beaver Lake camp in the background. (b) The perennial ice surface of Beaver Lake with twin otter used for deployment from Davis Station in the Vestfold Hills. Photos J. Laybourn-Parry. (See Plate 5)

14 ANTARCTIC LAKES

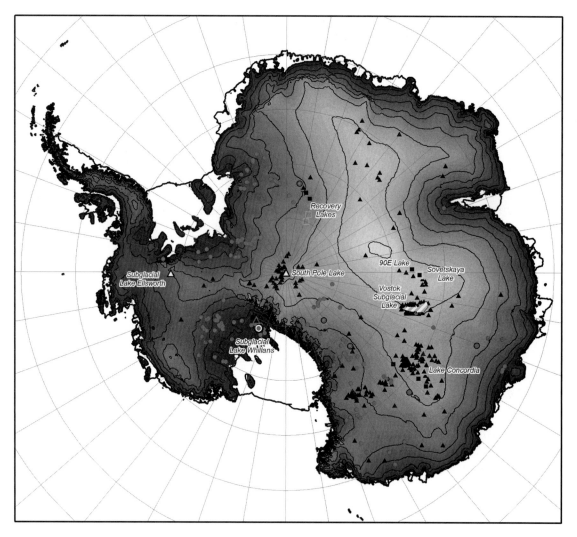

Figure 1.11 The position of subglacial lakes under the Antarctic Ice Sheet, from Wright and Siegert (2012) with the permission of Cambridge University Press. Colours and shapes indicate the type of investigation for each site: black triangle = radio-echo sounding, yellow = seismic sounding, green = gravitational field mapping, red circles = surface height change measurement, square = shape identified from ice surface feature. Lake Vostok is shown in outline. Original illustration provided by M. Siegert. (See Plate 6)

(Siegert et al. 2001, 2007). These lakes have been located by airborne radio-echo sounding (RES) since the 1970s (Oswald and Robin 1973), with recent additions to the inventory inferred from laser altimetry, which detects changes in the ice sheet surface elevation as subglacial lakes fill and drain (Fricker et al. 2007; Smith et al. 2009; Wingham et al. 2006). The best known lake is Lake Vostok beneath the East Antarctic Ice Sheet. It is at least 240 km long and 50 km wide and lies below around 3750–4150 m of ice. The depth of the water column is estimated to be at least 1000 m, lying in a crescent-shaped basin with steep sided walls. It is argued that Lake Vostok has existed since the formation of the stable East Antarctic Ice Sheet some 20 million years ago (Siegert et al. 2001). Thus if subglacial lakes contain life, some of the communities have been separated from the rest of the living world since at least the

Mid-Cenozoic. Lake Whillans and Lake Vostok were sampled in 2013 (see Chapter 6). Prior to this accretion ice from above Lake Vostok was shown to contain bacterial life (Priscu et al. 1999a; Christner et al. 2006). Accretion ice is formed when the underlying lake waters freeze onto the underside of the ice sheet in zones of net freezing.

Wetlands and seepages occur in the deglaciated areas of Antarctica for a few weeks during the austral summer when areas become saturated with liquid water from melting permafrost, snowfields, and glaciers. As yet we have relatively little information about them. They are particularly common in the Maritime Antarctic. At this time, air temperatures can rise to around 8 °C and rarely falls below –5 °C. These wetlands possess microbial communities dominated by filamentous Cyanobacteria. Such organisms have to be tolerant of freezing and desiccation, because for most of the year they are frozen and non-functional (Sabacká and Elster 2006). Seepage communities display a distinct vertical structure and species succession during summer (Komárek and Komárek 2003).

Ponds and associated wetlands are relatively common in the McMurdo Dry Valleys. They are small and spatially isolated and like the maritime wetlands, have only short phases of activity in summer. One of the interesting aspects of the McMurdo wetlands that occur at around 300 m above sea level is their contribution to organic carbon in adjacent soils. One pond studied in 2001 contributed 1388 kg of organic matter to its surrounding soil. The seepage or wetted soils close to the ponds support cyanobacterial mats and a range of invertebrates including tardigrades, rotifers, and nematodes. These ponds and the surrounding saturated soils can vary significantly in area from season to season in relation to the degree of glacial melt. For example after the 'flood' of the summer of 2001–2002, the ponds were much larger and deeper and the areas of wetted soils more extensive (Moorhead et al. 2003; Moorhead 2007).

1.6 Lake types and geochemical conditions

Crucial to an understanding of the abundance and diversity of microorganisms within Antarctic lakes is knowledge of a number of geochemical attributes which can be summarized under the headings of (1) salinity, (2) redox conditions, (3) nutrient and organic carbon supply, and specifically the abundance of different electron acceptors available to fuel respiratory processes and chemotrophic redox transformations. As for any lake system, in situ geochemical conditions in lake waters reflect the chemistry of the hydrological inputs to the lake and outputs from the lake, together with any in situ chemical transformations of inorganic and organic solute species in lake waters. The latter may include processes such as cryo- and evapo-concentration, which serve to concentrate certain solutes within the lake as water is removed by evaporation or frozen to form ice. They may also include biogeochemical uptake (e.g. of key nutrients) or transformations of one form of solute to another. The latter may include important microbially mediated redox processes such as sulphate reduction, methanogenesis, etc.

1.6.1 Salinity

Antarctic lakes may display a range of salinities, forming four broad categories: (a) freshwater lakes, (b) brackish or hyposaline lakes, (c) saline lakes (similar to seawater), and (d) hyper-saline lakes. Saline lakes such as those found in the Vestfold Hills are usually coastal endhoreic or closed basins, containing seawater from when the sea level was at a previous high stand. Some exhibit hyper-salinity due to evapo-concentration. In the McMurdo Dry Valleys, perennially ice-covered saline lakes are thought to be formed by the gradual shrinkage of larger lakes formed during warming after the Last Glacial Maximum (Lyons et al. 1998a). Hence, these saline waters may become increasingly concentrated over time due to evaporation and freeze-concentration.

Antarctic freshwater lakes usually have inflows from glacial and snow melt in addition to outflows, whereas, as indicted above, saline lakes are usually closed or endorheic basins, with inflow but no outflow. Examples of freshwater lake systems include those located around the Antarctic periphery, those on the surface of glaciers, ice sheets (supraglacial lakes and cryolakes, Figure 1.6), and epishelf lakes.

Figure 1.12 Lake Fryxell, Taylor Valley with the Canada Glacier in the background. Note polar haven erected over the sampling hole to stop it from freezing up. Photo J. Laybourn-Parry. (See Plate 7)

In some cases the inflow is glacial melt, for example, Lakes Miers and Hoare in the McMurdo Dry Valleys (Figure 1.12) or snow melt as in the case of the saline lakes of the Vestfold Hills (see Figure 1.13). Beneath the ice sheet itself, subglacial lakes are believed to be freshwater systems, receiving meltwater inputs via the melting of the underside of the ice sheet due to geothermal heating. Up to 2012/2013, only the accretion ice of Lake Vostok has been analysed for its chemical composition, and estimates of the solute composition of the lake waters has indicated a freshwater system with concentrations of dissolved solutes in line with those found in other glacial meltwaters worldwide (Christner et al. 2006).

Saline lakes may be permanently stratified meromictic systems, where the latter defines lakes where a proportion of the water remains unmixed with the remainder of the lake. They have strong physical and chemical gradients marking the transition between an upper mixed layer, termed the mixolimnion, and a lower anoxic unmixed layer called the monimolimnion (Figure 1.14a, b). The

Figure 1.13 Arial photograph of a suite of saline lakes in the Vestfold Hills. From the left bottom seaward: Lake Jabs, Club Lake, Deep Lake (salinity X 10 seawater), Lake Stinear, and Lake Dingle. Photo J. Laybourn-Parry. (See Plate 8)

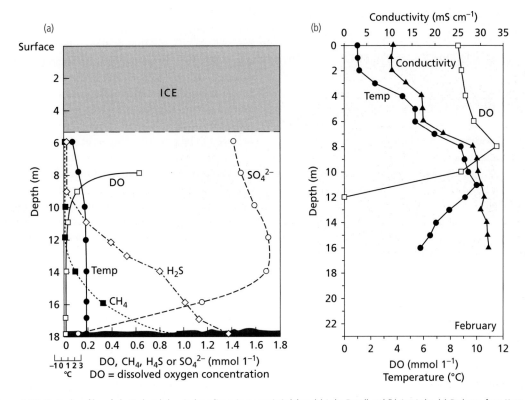

Figure 1.14 Vertical profiles of physical and chemical gradients in meromictic lakes, (a) Lake Fryxell and (b) Ace Lake. (a) Redrawn from Karr et al. (2003), (b) from Bell 1998, unpublished PhD thesis. Note that the data for Ace Lake were collected in February in a year when the ice cover had completely melted.

transition zone is termed the chemocline. As Figure 1.14 shows, the upper waters are less saline than the lower waters, which are lacking in oxygen and are the domain of anaerobic organisms. The upper, less saline waters often reflect seasonal meltwater inputs. Temperature also shows a marked change, with the lower waters possessing a higher temperature than the upper mixed mixolimnion. Other saline lakes may show no stratification over an annual cycle, while the extreme hypersaline lakes undergo seasonal thermal stratification in summer; for example, Deep Lake (Vestfold Hills) which is around ten times the salinity of seawater. Here, warmer water reaching maxima between 7 °C and 11.5 °C overlies colder water (Ferris and Burton 1988). Lakes that are mixed throughout their water columns at some time in an annual cycle are classified as holomictic. Deep Lake is monomictic, because it experiences one phase of stratification in an annual cycle. Epishelf lakes form between the ice shelf and the land as a result of drainage from a glacier or another lake. Because these lakes very often overlie seawater, they are tidal and their edges are characterized by rafted ice produced by daily tidal movement (Figure 1.10a). In epishelf lakes, freshwater and the underlying seawater layers remain discrete, with negligible mixing.

1.6.2 Redox conditions

Antarctic lake waters display a range of dissolved oxygen contents, from full oxic conditions to anoxia (indicating the absence of molecular oxygen in waters). In general, anoxic conditions evolve when processes removing oxygen from waters (e.g. the oxidation of organic carbon during respiration, Equation 2) exceed those that supply it (e.g.

diffusion from the atmosphere and photosynthesis, Equation 3),

$$CH_2O + O_2 \rightarrow CO_2 + H_2O \qquad (2)$$

$$6CO_2 + 6H_2O \underset{light}{\rightarrow} \underset{glucose}{C_6H_{12}O_6} + 6O_2 \qquad (3)$$

Oxic conditions are commonly observed in surface lake waters and around the ice margin, with hyperoxia (i.e. dissolved oxygen concentrations in excess of saturation) present in some perennially ice covered lakes, indicating oxygen generation by photosynthesis beneath an impermeable ice cover (Priscu et al. 1996; Vincent 1981). Similar oversaturation with respect to dissolved oxygen is observed in closed system cryoconite holes in the McMurdo Dry Valleys, for similar reasons (Tranter et al. 2004). The upper oxic zone in meromictic lakes is termed the mixolimnion. Dissolved oxygen concentrations generally decline with water depth, as oxygen consumption by microbial respiration starts to exceed oxygen additions from diffusion from the atmosphere/melting ice or phototrophic production (Equation 3). This gives rise to a dissolved oxygen profile such as that presented in Figure 1.14b, for meromictic Ace Lake (Vestfold Hills). The point at which the dissolved oxygen concentration of the lake displays a steep gradient towards zero concentrations is termed the oxycline. Lake waters below this gradient are termed the monimolimnion. The depth of the oxycline and the steepness of the gradient depend upon the rate of oxygen consumption within lake waters. Where very high concentrations of dissolved or particulate organic carbon are found in the upper water column, respiratory processes quickly remove both organic carbon and oxygen from lake waters generating a relatively steep gradient in shallow waters (e.g. as observed in Lake Fryxell, where dissolved oxygen concentrations fall to zero at approximately 50% of the lake depth) (Karr et al. 2006). Anoxic or hypoxic (waters depleted in dissolved oxygen) to anoxic bottom waters are found in a great many meromictic saline lakes, including Lake Bonney (Ward et al. 2003), Lake Fryxell (Lawrence and Hendy 1985), Lake Vanda (Canfield and Green 1985), and a number of meromictic lakes in the Vestfold Hills (Bowman et al. 2000). Some freshwater systems in the Maritime Antarctic, which are oxic in summer, evolve towards anoxic conditions during the polar winter when the activity of oxygen-producing phototrophs falls to zero (Ellis-Evans 1996). Anoxia has yet to be demonstrated in Antarctic cryoconite holes, cryolakes, and subglacial lakes. Certainly in the latter case, anoxic conditions with lake sediments are believed to be possible (Siegert et al. 2003).

The oxygen saturation of lake waters imposes a fundamental control on the biogeochemical processes that dominate in different parts of Antarctic lakes and smaller water bodies. Once dissolved oxygen has been depleted to zero concentrations in lake waters, facultative and obligate anaerobic microbes resort to a range of lower energy yielding electron acceptors to either oxidize organic matter (by heterotrophy) or to oxidize reduced forms of elements (during chemoautotrophy), thus generating energy. Geochemical profiles in lakes such as those shown in Figure 1.14 for Ace Lake and Lake Fryxell provide clues to microbial community zonation and redox progression.

Denitrification and dissimilatory nitrate reduction (Figure 1.15) have been reported in many lakes, often attaining a maximum at the top of the anoxic layer where the supply of nitrate and N_2O are provided from the upper water column

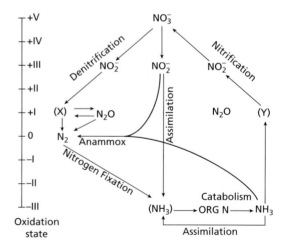

Figure 1.15 Schematic of the nitrogen cycle from Codispoti et al. (2001), with the permission of Institut de Ciències del mar de Barcelona (CSIC).

for denitrification. An example is in the west lobe of Lake Bonney (Priscu 1997). Sulphate reduction (Equation 4) is reported in some brackish and meromictic saline lakes, where there is a sulphate pool, for example in Ace Lake and neighbouring lakes in the Vestfold Hills (Franzmann et al. 1988),

$$SO_4^{2-} + 2CH_2O \rightarrow H_2S + HCO_3 . \qquad (4)$$

Iron reduction has been inferred in some of the Maritime Antarctic lakes, particularly in winter where greater anoxia enables iron reducers to outcompete other microbial groups. The result is the accumulation of Fe(II), which then diffuses upwards in the water column where it is oxidized above the oxycline to Fe(III) oxyhydroxides. These precipitates subsequently sink and are either scavenged to sediments or reduced again (Ellis-Evans and Lemon 1989). This cycle is termed the 'ferrous wheel' and is common in temperate lakes (Campbell and Torgersen 1980).

The process of methanogenesis (Equations 5 and 6) is reported in only the most anoxic lake waters, since this process yields the least amount of energy per mole of carbon consumed, and can only occur once high energy yielding electron acceptors such as nitrate, Fe(III), and sulphate have been removed,

$$CH_3COOH \rightarrow CH_4 + CO_2 \qquad (5)$$

$$CO_2 + 4H_2 \rightarrow CH_4 + 2H_2O \qquad (6)$$

Methane has been reported from some meromictic saline lakes such as Ace Lake (Burton 1980; Franzmann et al. 1991) and the brackish meromictic Lake Fryxell in the McMurdo Dry Valleys (Smith et al. 1993). It has also been reported recently in subglacial Lake Whillans (Priscu et al. 2013b). Methane is more commonly found in the freshwater lakes, which have a smaller sulphate pool, the presence of which usually inhibits methanogenesis. For example, a number of lakes in the Maritime Antarctic display methanogenesis, such as on Signy Island (Ellis-Evans 1984), and the Bunger Hills (Galchenko 1994). Lake Untersee, a very large freshwater lake in Dronning Maud Land, also displays evidence for methanogenesis in one of its anoxic basins, showing extremely high methane concentrations in lake bottom waters (Wand et al. 2006). A combination of hydrogen- and acetate-utilizing methanogens have been isolated in enrichment cultures and support both pathways for methanogens in Antarctic lake sediments (Ellis-Evans 1984; Stibal et al. 2012). In most lakes where methanogenesis is reported it is accompanied by sulphate reduction and the oxic and anaerobic oxidation of methane occurs in less anoxic, upper parts of the lake water column (Wand et al. 2006).

In general, rates of activity of sulphate-reducing bacteria and methanogens are low in Antarctic lakes, and can be several orders of magnitude lower than those reported in coastal sediments of similar temperature and organic carbon content (Franzmann et al. 1988, 1991; Stibal et al. 2012). These low rates may reflect the fact that many of these lakes act as closed systems, where the depletion of electron acceptors and accumulation of toxic reaction products such as sulphides occurs over time (Franzmann et al. 1988).

1.6.3 Nutrient and organic carbon supply

Essential ingredients for the existence of life include the presence of liquid water and the availability of specific nutrients, without which most life processes cannot function. Primary nutrients required for life include, nitrogen, phosphorus, and both organic and inorganic carbon; the former being necessary to support heterotrophic and some mixotrophic organisms, and the latter essential for photosynthesis (Equation 3). Minor nutrients required for life include silica, iron, potassium, and a range of other cations. The vast majority of lakes in Antarctica can be classified as oligotrophic or ultra-oligotrophic, i.e. they contain very low levels of nutrients required to sustain life. The freshwater lakes or the fresher upper water column of saline lakes (e.g. Lake Vanda) (Vincent and Vincent 1982) are ultra-oligotrophic. For example, typical concentrations of soluble reactive phosphorus and nitrogen may be <1 µmol l^{-1} (Andreoli et al. 1992). New nutrient inputs to the lakes may only occur during very short time periods, a few weeks in the summer, as freshwater inputs supply the lakes with new nutrients (Spigel and Priscu 1996).

Higher nutrient levels may occur locally, often in association with penguin colonies or seal wallows, where there are substantial inputs of nutrients in faecal material to lake waters, e.g. Heywood Lake and Sombre Lake on Signy Island (Christie 1987; Hawes 1983). Unlike the continental lakes, where the catchments are effectively bare rock that provide little in the way of nutrient inputs, vegetated catchments in the Maritime Antarctic have the potential to provide inputs of allochthonous carbon, nitrogen, and phosphorus. In these lakes, which are undergoing eutrophication, continued input of nutrients from seal/penguin colonies enable phytoplankton blooms to be sustained through the summer (Hawes 1983).

Because of the very limited external input of organic carbon to many lakes, a large proportion of the dissolved organic carbon (DOC) pool is produced in situ (or autochthonous) (Matsumoto 1989; McKnight et al. 1991). This DOC is produced by phototrophic microorganisms and subsequently supports bacterial heterotrophic activity. This is particularly true of perennially ice-covered lakes such as those found in the McMurdo Dry Valleys and the Vestfold Hills (Hand and Burton 1981; McKnight et al. 1993), and results in a dissolved organic matter signature that is rich in microbially produced components such as fulvic acids and organic acids (McKnight et al. 1993). This contrasts with temperate lakes, where the bulk of the dissolved organic matter originates from 'allochthonous' organic matter (e.g. plant material) washed into the catchment by streams and rivers. Lakes which have a greater allochthonous source of dissolved organic matter (DOM) are generally those that receive significant glacial or snow melt input (e.g. Lake Hoare in the McMurdo Dry Valleys).

The DOC profiles of lakes vary depending on their stratification and the degree of vertical mixing of lake waters. For example, a very stable chemical stratification is observed in meromictic lakes with saline bottom waters, such as Lakes Vanda and Fryxell (McMurdo Dry Valleys). Here, high DOC is found in lake bottom waters and is thought to represent the microbial degradation of organic matter in lake sediments (McKnight et al. 1993). In more dilute lakes, where there is greater advective mixing, DOC concentrations are generally lower (0.1–0.3 mmol l^{-1}), reflecting DOC inputs in dilute meltwater entering the lake (McKnight et al. 1993). DOC concentrations in subglacial lakes are still unknown. The accretion ice of Lake Vostok has registered DOC concentrations of 7–43 µmol l^{-1}. Calculations suggest that DOC concentrations of up to 100 µmol l^{-1} occur in the lake water. This is believed to be capable of supporting heterotrophic life in the lake (Priscu et al. 1999a).

The cycling of nitrogen and phosphorus are important biogeochemical processes in all lakes and both have been found to limit the activity of phytoplankton (Priscu 1995). A schematic depicting the nitrogen cycle in Antarctic lakes is shown in Figure 1.16. Nitrogen inputs to lakes derive from the inputs of dissolved nitrogen (largely nitrate) in snow and ice-melt, where these lakes have stream inflows, e.g. Lake Vanda is fed by the glacially fed Onyx River. In the case of subglacial lakes, external nutrient inputs are from the melting of basal ice and upstream subglacial riverine inputs (Vaughan et al. 2007). In ice marginal lakes, additional nitrogen may be fixed from the atmosphere by Cyanobacteria in the upper water column, generating particulate organic nitrogen, which may subsequently be converted to nitrate by nitrification (Figure 1.16). Nitrifying bacteria have often been found to be particularly active at depth in some lakes just above the oxic/anoxic boundary, and it has been suggested that these microorganisms respond to a diffusional input of nitrogen from the anoxic zone below (Vincent and Vincent 1982; Canfield and Green 1985). As a result, nitrate concentrations in the oxic upper layers of some lakes can be quite high (Matsumoto 1993), reaching a maximum concentration just above the oxycline, but falling to low or zero values below this transition as denitrification takes place (Canfield and Green 1985). These patterns are observed for several lakes in the McMurdo Dry Valleys, such as Lake Vanda, Lake Fryxell, and Lake Joyce (Figure 1.16). An exception is the east lobe of Lake Bonney, where low oxygen and high redox potentials enable large pools of nitrite and nitrate to persist (Priscu et al. 1996). High concentrations of nitrous oxide (N_2O) gas close to the redox boundary are also reported from many of these lakes, associated with nitrification (Priscu 1997).

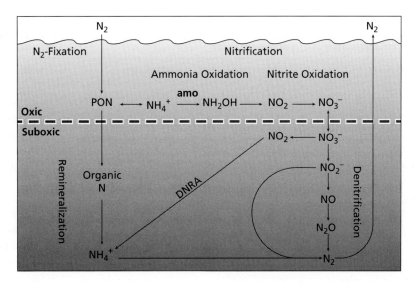

Figure 1.16 Nitrogen cycling in Dry Valley Lakes. Adapted from Francis et al. (2007), with the permission of the Nature Publishing Group.

Ammonium concentrations, a product of the decomposition of organic matter under anoxic conditions, are often low in the upper profiles of lakes but increase at the oxic/anoxic boundary and become further enhanced in the deeper anoxic waters of the lake. In some saline lakes, these deeper lake bottom waters are said to be in a steady state with respect to nitrogen cycling, such that rates of ammonium release from anoxic degradation of organic matter are balanced by the flux of ammonia from the anoxic zone (Canfield and Green 1985; Priscu 1997).

Like nitrogen, phosphorus is a major element required for microbial growth and has been found to be limiting to phototrophs in some Antarctic lakes (Vincent and Vincent 1982; Dore and Priscu 2001). Dissolved phosphate (or soluble reactive phosphorus, SRP as it is often termed, Table 1.1) is the most readily available P species for microbes, but because of the higher abundance of P in organic matter and the tendency of dissolved P to adsorb to sediments, SRP generally accounts for a small proportion of the total P pool in an Antarctic lake. A schematic depicting the phosphorus cycle in a typical lake is shown in Figure 1.17. Small inputs of phosphorus may occur via streamflow into lakes around the ice margins, particularly during high discharge events when soil and rock material are suspended in stream waters. This might be expected to be similar beneath the ice sheet, where sediments are mobilized by fast-flowing meltwaters, e.g. during lake drainage events (Wingham et al. 2006; Fricker et al. 2007). This may occur during short periods, for example most phosphorus supply to Lake Bonney has been shown to occur during a two-week period when high flow rates enable mobilization of phosphorus-containing sediments and organic matter by meltwaters (Weand et al. 1977). Additional phosphorus may be added to lake ice cover in aeolian debris, which gradually migrate downwards lake waters over time as successive ice layers accumulate (Priscu 1998). As for DOC and nitrogen species, elevated concentrations of phosphorus species are often found in association with the anoxic bottom waters of meromictic lakes, for example Lake Vanda, where the degradation of formerly surface-dwelling organisms occurs (Figure 1.18). Hence, phosphorus is then supplied by upward diffusion across the lake chemocline (Priscu 1995). This upward diffusion of nutrients is believed to be important in driving phytoplankton activity in the upper water column. For example, the response of phytoplankton carbon assimilation to nitrogen and phosphorus enrichments in the McMurdo Dry Valley lakes matched the flux ratios of dissolved inorganic nitrogen (DIN):soluble reactive phosphorus (SRP) (Dore and Priscu 2001).

22 ANTARCTIC LAKES

Table 1.1 Nomenclature of carbon, nitrogen, and phosphorus species, including abbreviations and their distributions in Antarctic lakes.

Group of nutrients and abbreviation	Component species	Distribution in lake
Dissolved organic carbon (DOC)	All organic compounds in the dissolved phase	Dissolved in lake waters
Dissolved inorganic carbon (DIC)	CO_2, HCO_3^-, H_2CO_3	Dissolved in lake waters
Particulate inorganic carbon (PIC)	Inorganic carbon present in the solid phase, e.g. $CaCO_3$	In lake sediments, lake ice, some sinking within the water columns
Particulate organic carbon (POC)	Organic carbon present in the solid phase, e.g. microbial cells and other detritus	In lake sediments, lake ice, some sinking within the water columns
Dissolved inorganic nitrogen (DIN)	Nitrite (NO_2^-) Nitrate (NO_3^-) Ammonium (NH_4^+)	Dissolved in lake waters
Dissolved organic nitrogen (DON)	Organic molecules containing N, e.g. amino acids	Dissolved in lake waters
Particulate inorganic nitrogen (PIN)	N species included within rock minerals in various forms	In lake sediments, lake ice, some sinking within the water columns
Particulate organic nitrogen (PON)	Nitrogen present in solid phase organic matter, e.g. microbial cells	In lake sediments, lake ice, some sinking within the water columns
Dissolved inorganic phosphorus (DIP)	PO_4^{3-} or HPO_4^{2-}, (also termed soluble reactive phosphorus or orthophosphate)	Dissolved in lake waters
Dissolved organic phosphorus (DOP)	Small or large dissolved organic molecules containing P, e.g. P-esters	Dissolved in lake waters
Particulate phosphorus (PP)	P that is bound to organic particles or metal oxides	Associated with lake sediments

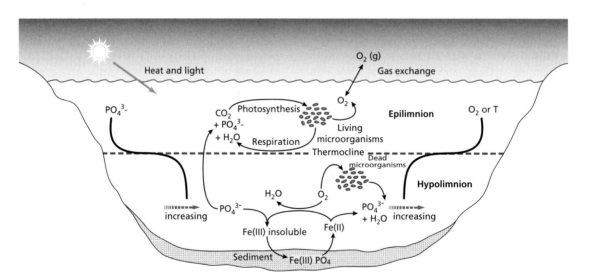

Figure 1.17 Schematic of the phosphorus cycle in freshwater lakes.

1.6.4 Geochemical indicators of lake history

The analysis of the geochemical composition of lake waters often provides vital clues to the history, age, and genesis of a lake. For example, seawater and fresh meteoric water sources have vastly different solute compositions, as discussed earlier (see Box 1.1). Isotope ratios for $\delta^{18}O$ and δD can also be used to infer lake water sources and their evolution. This analysis works on the premise that different

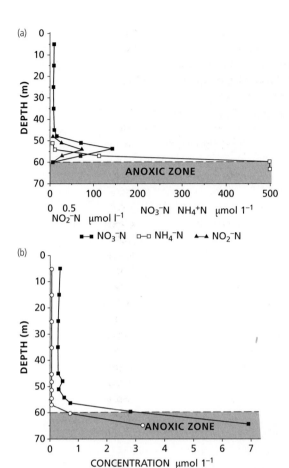

Figure 1.18 (a) Distribution of nitrite, nitrate, and ammonium in Lake Vanda. (b) Total phosphorus and dissolved phosphorus in Lake Vanda. Redrawn from Canfield and Green (1985).

water sources (e.g. snow, ice-melt, seawater) have contrasting isotopic signatures, which may also vary geographically. Any phase changes (e.g. freezing or evaporation) are also manifested in a change in the isotopic signature. A good example of where these techniques have been applied is the early study of Matsubaya et al. (1979), which is now discussed.

Figure 1.19 shows the $\delta^{18}O$ and δD compositions of lake waters for the Swoya Coast and the Vestfold Hills area (Matsubaya et al. 1979). The meteoric water line is also plotted. One group of lakes displaying isotope ratios for δD (–150 to 300‰) and $\delta^{18}O$ (–20 to –40‰) receive their meltwaters directly from the ice sheet and hence exhibit a depleted isotopic signature, but one which still aligns with the meteoric water line. The extreme range of values reflects mixing between meltwaters derived from coastal precipitation sources (i.e. snow) and those from the ice sheet. A second group of lakes plot in the region δD (–125 to 170‰) and $\delta^{18}O$ (–9 to –21‰), which indicates a similar δD to precipitation (δD = –150‰, $\delta^{18}O$ = –20‰), but enrichment in $\delta^{18}O$ in coastal areas. When plotted against chloride ion concentration, it appears that this enrichment in $\delta^{18}O$ correlates with increased Cl⁻ content, stabilizing at a Cl⁻ content of approximately 100 g l⁻¹ (Figure 1.20). The hypothesized explanation for this is that lakes with higher salinities remain ice-free for a greater part of the year, and undergo greater evaporation. Since the light isotopes tend to be lost during evaporation, lake waters gradually become enriched in the heavy isotopes, and waters become shifted to the right of the meteoric water line. Lakes with a Cl⁻ content of 100 g l⁻¹ remain ice-free all year round and hence, the isotope ratios stabilize at this point. A third group of just four lakes (Lake Oyayubi, Nurume, Lake 1, and Rookery Lake) have exceptionally high isotope ratios and are mostly located close to beaches. Lake Oyayubi, Lake 1, and Rookery Lake still have occasional seawater inflows, and hence maintain salinities close to that of seawater. Lake Nurume is a rather unusual case since its basal waters are of a similar composition to seawater, but surface waters are close to equilibrium with in situ conditions indicating a composition that has evolved via rapid flooding of a seawater basin. When these open system lake bodies become isolated from their seawater input by isostatic uplift, one might expect them to evolve towards the chemistry of the second group of closed system lakes discussed above.

This type of analysis of the isotopic composition and salinity of lake waters can be seen to have immense value in inferring lake water sources and evolution. A similar analysis was performed by Matsubaya et al. (1979) for the McMurdo Dry Valleys lakes and is discussed in Chapter 3.

1.7 Geomorphology of Antarctic lakes

Worldwide, lakes are formed by a variety of processes among which glaciation and tectonic activity are

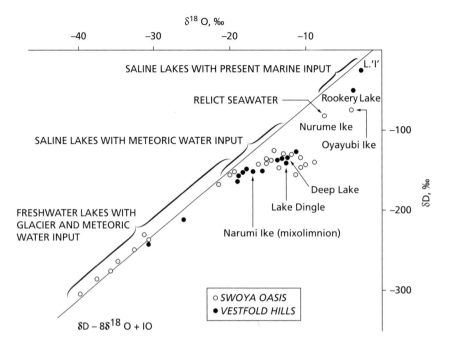

Figure 1.19 Relationship between δD and δ¹⁸O of lake waters in Vestfold Hills and Syowa Oasis, from Matsubaya et al. (1979), with the permission of Elsevier.

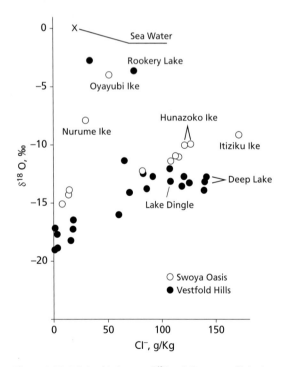

Figure 1.20 Relationship between δ¹⁸O and Cl⁻ content of lakes in the Vestfold Hills and Syowa Oasis, from Matsubaya et al. (1979), with the permission of Elsevier.

major forces. For example, the Great Lakes of North America and the lakes of the English Lake District resulted from the last major glaciation, when vast sheets of ice gouged depressions and steep-sided valleys in the landscape. Tectonic activity has also produced some of the world's largest and deepest lakes, for example Lake Baikal and Rift Valley lakes of Africa. Lakes often form in the craters or calderas of quiescent or extinct volcanoes, for example Crater Lake in the USA. Ox-bow lakes are formed when the meander of a river is cut off in its flood plain. In Antarctica, the major forces in lake formation are glacial processes, but tectonic processes have also been important. Lake Vostok and other subglacial lakes lie in valleys resulting from tectonic activity, and are likely to have been present beneath the ice sheet since the time of its formation. This makes these subglacial lakes potentially very old.

Glacial erosion caused by continental ice sheets during glacial periods has played a major role in the formation of Antarctic lakes. Depending on the topography and the nature of the underlying geology, ice flows generate quasi-linear troughs and irregular depressions in the landscape. Typical glaciated

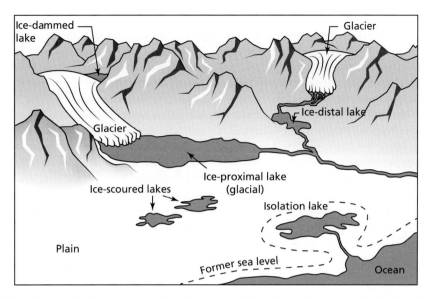

Figure 1.21 Different lake types found in Antarctica. Redrawn from Pienitz et al. (2008), with the permission of Oxford University Press.

valleys are steep-sided and U-shaped. The proglacial environment is associated with high rates of sediment transport and deposition. Sediment deposited at the mouth of a valley can block an existing drainage system and form a dammed lake (Figure 1.21) (Pienitz et al. 2008). Ice distal lakes are separate from the ice but are fed primarily by glacial meltwater via outwash streams (Figure 1.21). Sometimes they occupy an existing basin, in which case they may be long-lived, whereas if impounded by sediment dams they are likely to be short-lived (Ashley 2002). Epiglacial lakes are common in Antarctica and are usually fed by glacial meltwater. They can persist for years but may experience changes in water levels and morphology as a result of glacial movement and variations in glacial meltwater inputs (Pienitz et al. 2008).

Coastal lakes may form by a process of glacioisostatic uplift or rebound, which occurs when the land previously depressed by the weight of the ice sheet, rises as the ice retreats. This results in fjords being cut off from the sea, which if flushed by glacial melt eventually evolve into freshwater lakes. Portions of marine waters in hollows may be uplifted and separated from the sea. Where these are closed basins they become saline lakes of varying salinity depending on the degree of snow or glacial meltwater input and the level of evaporation. The evolution of these uplift lakes is complicated by oscillations in relative sea level. Effectively, in the absence of vertical tectonic movement, sea level changes at the Antarctic margin over thousands of years are the sum of a eustatic component that relates to global ice volume (a function of time) and a glacioisostatic component responding to the movements of ice and water masses on the surface of the Earth. This component varies with time and position, depending on a site's proximity to changing water and ice loads (Zwartz et al. 1998) (Figure 1.22). In some cases, low-lying freshwater lakes have suffered marine incursions and changed from freshwater to saline lakes, for example Ace Lake, one of the most

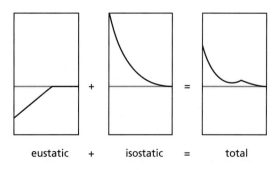

Figure 1.22 Isostatic and eustatic effects in lake formation, from Zwartz et al. (1998), with the permission of Elsevier.

studied lakes in the Vestfold Hills. During the Holocene (0–11,700 yrs B.P., which marks the end of the last glacial epoch), Ace Lake started its existence as a freshwater lake in a glacially formed hollow fed by snow or glacial melt. Approximately 9400 yrs B.P., relative sea level rise resulted in it being invaded by the sea and becoming a marine basin. Then at about 5000 yrs B.P., the sea level dropped below the sill isolating the modern saline Ace Lake. Today the lake is permanently meromictic and appears to have undergone two phases of meromixis in its history interspersed by a period of holomixis (Cromer et al. 2005a). In some instances saline lakes have been progressively flushed by water released from glacially ice-dammed lakes. A good example of this is Watts Lake in the Vestfold Hills which was a marine-derived lake flushed by waters from Glacial Crooked Lake (Gore et al. 1996).

The age of the coastal lakes has been the focus of considerable research in the Vestfold and Larsemann Hills. It is evident that some lakes predate the Last Glacial Maximum (LGM). The sediment cores of lakes are effectively a history book and their analysis and interpretation is termed palaeolimnology. Marine- and freshwater- derived sediments can be discriminated, the remains of some crustaceans and rotifers, as well as some testate protozoans and diatoms provide an idea of past biodiversity and environmental conditions. The transformation products of photosynthetic pigments can be identified by high pressure liquid chromatography and atmospheric chemical ionization liquid chromatography–mass spectrometry/mass spectrometry (APCI LC-MS/MS), providing a picture of the autotrophic community in the plankton and benthos during a lakes's history. Layers in sediment cores can be dated using ^{14}C dating techniques. Palaeolimnology has shown that part of the Larsemann Hills were ice-free during the LGM and that some of the lakes date back to the last interglacial period at around 125,000–115,000 yrs B.P. (Hodgson et al. 2001a; Squier et al. 2005; Hodgson et al. 2006a. At least some of the lakes in the Vestfold Hills may predate the LGM (Gibson et al. 2009).

The history of the McMurdo Dry Valley lakes is complex. The Dry Valleys are a large inland ice-free area lying between the Ross Sea and the interior ice sheet of East Antarctica, covering some 4000 km^2 (Figure 1.1). Polar lakes existed in this part of Antarctica throughout the late Quaternary, extending back at least 300,000 years (Hendy 2000). The Taylor Valley provides the most detailed history for the reconstruction of the Pleistocene and Holocene events that led to the formation of the Dry Valleys suites of lakes that we see today. Most are the remnants of much more extensive glacial lakes. The lakes of the Taylor Valley, which include Lakes Bonney, Hoare, and Fryxell, are the remains of a large glacial lake called Glacial Lake Washburn, that occupied all of the Taylor Valley during the last glacial period into the early Holocene. At that time its estimated area was 75 km^2 and its depth around 300 m (Chinn 1993; Hall and Denton 2000). A tongue of the Ross Sea Ice Shelf projected into the Taylor Valley blocking drainage and forming Lake Washburn, bounded by the tongue, the Taylor Glacier, and steep valley walls. The most likely source of water for Lake Washburn was from beneath the grounded Ross Ice Shelf in the lower Taylor Valley. When the lake was at its highest volume, it would have received additional flux from meltwater from local alpine glaciers and possibly also from the surface of the Ross Sea Ice Shelf (Stuiver et al. 1981). Following the retreat of the Ross Sea Ice Shelf and changes in climatic conditions influencing local hydrology, Lake Washburn lowered, eventually leaving behind a series of lakes. Based on oxygen and deuterium stable isotope analysis, these lakes and Lake Vanda in the Wright Valley appear to have lost their ice covers and dried down to hypersaline pools by around 1000–1200 yrs B.P. They have subsequently refilled during a warmer period (Lyons et al. 1998a). Lake Hoare either completely disappeared or did not exist prior to the desiccation. More recent analysis of Lake Hoare sediment cores suggests a complete desiccation in the middle Holocene, and by the late Holocene (<3300 yrs B.P.), establishment of the conditions seen today. The later desiccation event seen in other Taylor Valley lakes did not occur in Lake Hoare (Wagner et al. 2011). A large glacial lake (Glacial Lake Wright) existed in the Wright Valley during the LGM and the early Holocene (Hall et al. 2001) and Glacial Lake Trowbridge in the Miers Valley (Clayton-Green et al. 1988).

A more detailed picture of the formation of lakes and ponds will be given in subsequent chapters.

1.8 Antarctic lake biota

Antarctic lakes experience constant low temperatures, low annual levels of photosynthetically active radiation (PAR), and minimal nutrient inputs. As a general rule, as conditions become more severe food webs become truncated. Fish are not found in Antarctic lakes and invertebrate metazoans are sparse in both the plankton and benthos. Antarctic lake plankton is dominated by prokaryotic and eukaryotic microorganisms, including Archaea, Bacteria, Algae, and Protozoa, while the benthos is often dominated by cyanobacterial mats and in some cases mosses. Viruses have been recorded in the plankton; the majority are bacteriophage, i.e. parasites of bacteria.

The importance of bacteria and smaller protozoans has only been understood since the early 1980s. Limnological studies previously used nets with different sized meshes to collect phytoplankton and zooplankton. These nets did not retain small microorganisms. Then a seminal change occurred, and whole water samples started to be routinely analysed. In the 1980s, fluorochrome dyes became available, such as 4′,6-diamidino-2-phenylindole (DAPI), that were very effective at staining bacteria and flagellated protozoans, allowing them to be viewed and counted under epifluorescence microscopy. Previous studies of bacteria by dilution plating and growth on various types of agar plate media resulted in calculated numbers that were much lower than from direct epifluorescence counts. This suggested that many bacteria were not amenable to culture. The supposition is that only around 2–3% of bacteria are culturable using current culture technology.

As a result of this advance in technology a new theory was postulated; that of the microbial loop (Pomeroy 1974; Azam et al. 1983). At that time the data were limited. Now we have a huge database on the organisms and their functional ecology and it is clear that the term microbial loop is something of a misnomer. In fact aquatic food webs are a complex network of microorganisms and higher organisms, rather than a microbial loop operating in tandem with the so-called classical food chain (Figure 1.23). Microorganisms play a crucial role in the biogeochemical cycling of carbon and other elements. Table 1.2 shows the terminology and size ranges of microscopic aquatic organisms.

The microscopic organisms that form the phytoplankton are taxonomically diverse, including Algae (e.g. diatoms, desmids, chlorophytes), photosynthetic

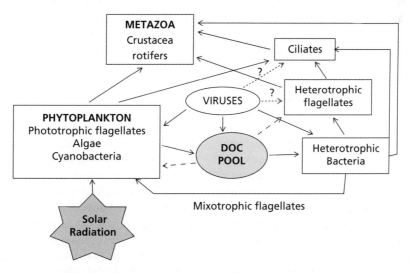

Figure 1.23 Typical planktonic food web in Antarctic lakes. The system is dominated by eukaryotic and prokaryotic microorganisms with few Metazoa. The dissolved organic carbon pool (DOC) is largely autochthonous, derived from exudation from the phytoplankton and recycling resulting from the viral lysis of bacteria and other infected cells. Some heterotrophic flagellates can take up DOC as well as feeding on bacteria. Some phototrophic flagellates are mixotrophic, feeding on bacteria and DOC.

Table 1.2 Size categories of microorganisms. Organisms 0.2–2.0 µm often have the prefix pico when found in the plankton (picoplankton), organisms 2.0–20 µm the prefix nano (nanoplankton), and larger organisms up to 200 µm have the prefix micro. Note that some groups, such as the ciliates and the dinoflagellates, exhibit a wide size range among their species.

Constituent organisms	Size range
Viruses	Less than 0.02 µm
Bacteria, coccoid Cyanobacteria	0.2–2.0 µm
Heterotrophic and autotrophic nanoflagellates, small ciliates, small naked amoebae, small dinoflagellates	2.0–20 µm
Ciliates, amoebae, dinoflagellates, diatoms, larger Cyanobacteria	20–200 µm

Protozoa called phytoflagellates (e.g. dinoflagellates, chrysophytes, cryptophytes, euglenoids), and Cyanobacteria. Thus the term phytoplankton is generic. The phytoplankton carries out photosynthesis, the first step in the carbon cycle. Photosynthesis is the process whereby autotrophic organisms fix inorganic carbon (CO_2), combining it with water to produce new living material or biomass. Various photosynthetic pigments harvest light energy within specific wavelengths of the visible light spectrum and this provides the energy to drive the reaction. Chlorophyll *a* is the most important and ubiquitous of the photosynthetic pigments, and as a consequence its concentration in lake water samples is a good analogue for phytoplankton biomass. Chlorophyll *a* concentration is one indicator of the trophic status of a water body. In eukaryotic organisms, photosynthetic pigments are contained in cellular organelles called chloroplasts. The process of photosynthesis is summarized in Equation 3, Section 1.6.2.

Autotrophs also require sources of major inorganic nutrients such as phosphorus and nitrogen, and trace elements such as iron and manganese to create new biomass. In many aquatic environments these essential nutrients become limiting at stages in the annual cycle and constrain photosynthesis. Aquatic organisms effectively live in a shaded environment, because light is attenuated as it enters a water column following an exponential decrease until only 1% of total PAR is present. This is called the compensation point, where the carbon fixed during photosynthesis equals the carbon released during respiration, so there is no net gain of carbon to the organism. Antarctic lakes have ice cover for all or most of the year and this further reduces the levels of PAR to the underlying water column (see Section 1.8).

As indicated in Figure 1.23, during the process of photosynthesis the phytoplankton releases dissolved organic carbon (DOC) into the environment. This is a portion of the photosynthate and forms a crucial energy source for the bacteria. The released photosynthate forms part of a pool of DOC, some of which is derived from other sources, for example sloppy feeding by zooplankton, allochthonous inputs from the catchment, and from decay of faecal pellets. However, in most Antarctic lakes zooplankton are sparse or absent and allochthonous inputs are minimal, so the majority of the DOC pool is likely to be derived from the phytoplankton. However, the cycling of carbon may be enhanced by viral lysis of bacteria (see Section 1.8.2).

The level of bacterial growth that can be achieved depends upon the nature of the DOC pool and the availability of nitrogen and phosphorus. Bacteria and phytoplankton compete for these nutrients. The conventional view is that the DOC pool accessible to bacteria is composed of two distinct fractions. The first of these can be termed 'light fuel' and consists of a wide array of organic compounds including dissolved free amino acids, dissolved free carbohydrates, and low molecular weight (<200 Da) organic acids, lipids, vitamins, hydrocarbons, polyphenols, and enzymes. This pool is rapidly utilized and turned over by heterotrophic bacteria communities (Fuhrman 1987). However, this represents only a small fraction of the total DOC pool (Amon and Benner 1996). The bulk of the DOC pool is composed of higher molecular weight material in the form of proteins, polysaccharides, and humic substances. These forms of DOC are less nutritionally favourable to bacteria because they require hydrolysis by bacterial enzymes into monomers and oligomers before they can be assimilated (Chróst et al. 1989; Coffin 1989). Bacteria excrete these enzymes into the medium and then assimilate the products of hydrolysis. Nevertheless there is clear evidence that bacteria can exploit this high molecular weight faction to sustain growth (Amon and Benner 1996; Meyer et al. 1997).

1.8.1 Archaea and Bacteria

These organisms are prokaryotes, so called because they lack a nucleus. In eukaryotic organisms the nucleic acid is contained in chromosomes within a nucleus that is bounded by a nuclear membrane, whereas in prokaryotes nucleic acids appear as a central mass lacking a nuclear membrane. In photosynthetic bacteria, like the Cyanobacteria, the photosynthetic pigments are also free within the cell and not contained in chloroplasts as they are in eukaryotes. There are two distinct groups of prokaryotes, the Archaea and the Bacteria. Archaea are characteristic of extreme environments, being very tolerant of extreme salinities and high temperatures. Many are methanogens (methane producing), and both aerobic and anaerobic forms occur. They can be divided broadly into three groups: the methane producing Methanoarchaea, the Haloarchaea that are highly salt tolerant, and a group of extreme thermophiles. The Bacteria are a very diverse group physiologically and include photosynthetic, chemoautotrophic, and heterotrophic species. Some are facultative anaerobes capable of living in the absence of oxygen, others are obligate anaerobes, and others are entirely aerobic.

The Cyanobacteria are an extremely important photosynthetic group that is very conspicuous in Antarctic lakes, especially in benthic mats. They include filamentous forms (Figure 1.24a) and unicells, some of which are very small, termed the picocyanobacteria (see Table 1.2). Some species of Cyanobacteria can fix atmospheric nitrogen, thereby gaining an advantage in environments where inorganic nitrogen is limited. These Cyanobacteria possess cells called heterocysts and include species like *Nostoc, Anabaena,* and *Nodularia* (Figure 1.24b). Nitrogen fixation is mediated by the enzyme nitrogenase. Dinitrogen is inert because of the strength of its N–N triple bond, thus breaking one of the nitrogen atoms from the other requires all three of the chemical bonds to be broken. Consequently nitrogenase has a high energy requirement. Nitrogen fixation occurs under anoxic conditions and must therefore be separated from oxygenic photosynthesis within the organism. Heterocystous cells do not undertake photosynthesis. They possess a thickened cell wall that resists the diffusion of gases.

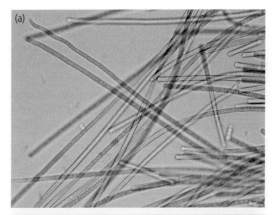

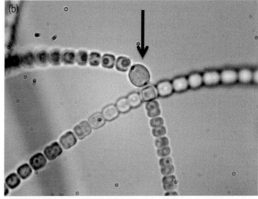

Figure 1.24 (a) Filamentous cyanobacterium, (b) a heterocystous cyanobacterium showing heterocyst cell indicated by arrow. Photos courtesy of A. Wilmotte, University of Liége.

Inside the heterocyst, respiratory activity renders the cell virtually anoxic. Effectively, nitrogen fixing heterocystous Cyanobacteria have functionally separated oxygenic photosynthesis from nitrogen fixation, and the two processes can occur in tandem.

The majority of Cyanobacteria from the Arctic and Antarctic are not psychrophiles (adapted to low temperatures), but psychrotrophs that have optimum growth temperatures between 15 °C and 35 °C (average 19.9 °C). Their growth at low temperatures is slow, but they are nonetheless a very common component of the polar biota. They have attributes that confer a competitive advantage in low temperature environments, including their tolerance to desiccation, freeze–thaw cycles, and adaptations to high levels of solar and UV radiation (Tang et al. 1997).

Our understanding of the taxonomy of prokaryotes is still limited, although molecular technology is making major strides towards building our knowledge. The exciting advent of environmental metagenomics, involving the genomic analysis of microorganisms by direct extraction and cloning of DNA from whole assemblages of organisms in environmental samples, has provided a much better picture of biodiversity (Handelsman 2004). Since only around 2% of bacteria appear culturable, being able to develop a picture of natural community composition allows all sorts of interesting questions to be posed about the physiological potential and function of bacterial assemblages.

At the time of writing, there are over a thousand published complete genome sequences for prokaryotes, leading to the development of comparative genomics, allowing generalizations on genomic organization and evolution to be developed. An important phenomenon has become apparent in prokaryote evolution, that of horizontal gene transfer (HGT). One of the major tenets of modern evolutionary biology, the Darwinian–Mendelian model of parent to offspring gene flow (vertical gene flow), has been challenged for microbes (Charlebois et al. 2003). The transfer of genes between organisms other than by vertical gene transfer was recognized before the genomics revolution; however, it was not until the sequencing of many prokaryote genomes that the importance of horizontal gene transfer became fully apparent. It is a dominant force in prokaryote evolution and may also occur in eukaryote microorganisms. HGT is the transfer of genes between related and totally unrelated organisms. This process is mediated by mobilomes, which are viruses, plasmids, and other elements (Koonin and Wolf 2008). There are many examples, but perhaps one of the most striking is the photosynthetic gene clusters of Cyanobacteria and other photosynthetic bacteria. There are five phyla that contain photosynthetic species; these include the Cyanobacteria, the Proteobacteria (purple bacteria), green sulphur bacteria, green filamentous bacteria, and Gram-positive Heliobacteria. Comparisons of genomes of these phototrophs from different phyla have shown that the genetic components of the photosynthetic apparatus have crossed species lines in a non-vertical manner (Raymond et al. 2002). As we will see, viruses are common in Antarctic lakes (Säwström et al. 2008) where they can act as a mobilomes in HGT. It is possible that HGT is important in these extreme ecosystems.

1.8.2 Viruses

Viruses are not strictly speaking living organisms, because they cannot reproduce. They can be defined as a bit of nucleic acid, RNA or DNA, surrounded by a protective protein coat (the capsid) that can regenerate only by commandeering the metabolic machinery of an appropriate prokaryotic or eukaryotic host. At the core of a virus is the viral genome, which can be composed of either RNA or DNA. Viruses differ from prokaryotic and eukaryotic cells that contain both RNA and DNA, in that they rarely contain both (Perry et al. 2002). Figure 1.25a shows a typical virus from an Antarctic lake and Figure 1.25b shows viruses inside a bacterial cell prior to the cell bursting or lysing. When a host cell is lysed the viral particles are released into the water where they find further hosts to infect. The lysis of bacterial cells short circuits the microbial food chain, returning carbon to the organic carbon pool before it can be ingested by a consumer, such as a heterotrophic flagellate. In some environments the destruction of bacterial cells by viruses can equal the predation impact of flagellated nanoflagellates (Weinbauer 2004). The majority of viruses so far described from aquatic environments appear to be parasites of bacteria.

As well as undergoing the lytic cycle that results in the lysis and death of the host cell, viruses can live within a host without causing its destruction. Viruses are passed by bacteria to their progeny during cell division. This is termed the lysogenic cycle, during which viral nucleic acid recombines with the host genome. Once the virus has inserted itself into the host chromosome it is termed a prophage and the cell harbouring the prophage is called a lysogen. The lysogenic cycle can continue until one or more factors trigger the lytic cycle (Wommack and Colwell 2000).

The study of viruses in aquatic environments started in the late 1970s when Torella and Morita (1979) reported concentrations of 10^3 to 10^4 ml^{-1} in the sea. Subsequent improvements in analytical

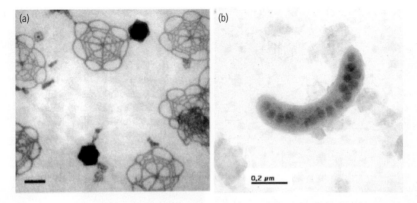

Figure 1.25 (a) Viruses from a Vestfold Hills lake together with scales from the cell wall of phyoflagellate. (b) A bacterial cell containing viruses. Photos J. Laybourn-Parry and C. Säwström.

methods showed that viruses were much more numerous in the marine environment, with densities ranging from 10^3 to 10^7 ml^{-1} (Bergh et al. 1989; Børsheim et al. 1990; Suttle et al. 1990). Studies on lacustrine environments followed, but the study of viruses in Antarctic lakes is more recent, dating from the late 1990s. Generally, virus concentrations are higher in freshwaters worldwide ranging between 4×10^7 and 2.5×10^8 ml^{-1} (Maranger and Bird 1995). Virus concentrations in Antarctic lakes range between 16×10^4 and 9.9×10^7 ml^{-1}, varying with the trophic status of the lake (Säwström et al. 2008).

Antarctic lake viral dynamics appear to differ from those seen in lower latitude lakes. Firstly, a much higher proportion of the bacterial community appears to be infected, up to 34% compared with an average of 2.2% for lower latitude lakes. However, burst sizes or the number of viruses released from a bacterial cell when it lyses are typically lower, with a mean of 4 compared with an average of 26 for lakes elsewhere (Säwström et al. 2007a). Both the lysogenic and lytic cycles occur in Antarctic lakes, but in ultra-oligotrophic freshwater lakes lysogeny is less frequent.

1.8.3 Protozoa

The Protozoa are eukaryote single-celled microorganisms that are extremely common in the plankton and benthos of lakes worldwide, as well as in marine environments (Laybourn-Parry 1992). The Protozoa used to be classified in the Animal Kingdom (Protozoa means first animals), while at the same time photosynthetic flagellates were classified as Algae in the Plant Kingdom. This confusing state of affairs was resolved when a new classification of the living world was proposed (Whittaker 1969; Margulis 1974). In the new system, the living world is divided into five kingdoms: the Monera (prokaryotes), the Protists which includes the sub-kingdom Protozoa, the Plantae, Fungi, and Animalia. The kingdom Protista is an artificial grouping that contains many organisms with no evolutionary affinities. The sub-kingdom Protozoa contains numerous phyla, some of which are entirely parasitic. Some protozoan parasites are of huge medical importance, for example the malarial parasites *Plasmodium*. Among the protozoan phyla there are two that are relevant to aquatic environments and soils. These are the phylum Sarcomastigophora and the phylum Ciliophora. The majority of their members are free-living, but they do contain some parasitic species among which the flagellated trypanosomes that cause sleeping sickness are particularly important.

The phylum Sarcomastigophora contains the amoebae (the sub-phylum Sarcodina), a very complex group both physiologically and morphologically, and the flagellated protozoa (the sub-phylum Mastigophora). Amoebae are not common in plankton, they require a surface for attachment. However, the heliozoan actinopods are often conspicuous in the plankton (Figure 1.26a). Amoebae are much more common in benthic environments where the

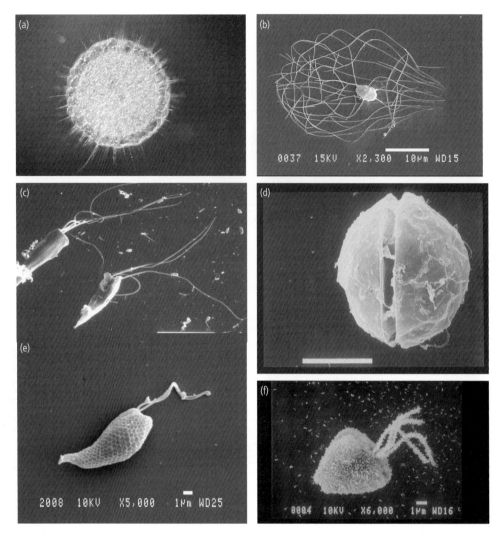

Figure 1.26 (a) A heliozoan; (b) choanoflagellate *Diaphanoeca grandis*; (c) amoeba-flagellate *Tetramitus* sp; (d) a dinoflagellate; (e) phytoflagellates *Cryptomonas* sp. (f) phytoflagellate *Pyramimonas gelidicola*. Photos J. Laybourn-Parry and G. Nash, Australian Antarctic Division.

necessary surfaces for attachment, allowing movement, and feeding are available. The sub-phylum Mastigophora is divided into two classes: the Zoomastigophorea that contains the heterotrophic flagellates (Figure 1.26b, c) and the Phytomastigophorea or the phytoflagellates (Figure 1.26 d, e, f).

Heterotrophic flagellates are usually the major consumers of bacteria in aquatic environments. Some of the heterotrophic flagellates are described as collared (Figure 1.26b). The complex silica basket seen in *Diaphanoeca grandis* is typical of marine species, where it acts as a flotation mechanism. The specimen shown in the figure comes from one of the marine-derived brackish lakes in the Vestfold Hills (Highway Lake), where it is extremely abundant. The central flagellum draws water through the collar of pseudopodia where bacterial food particles become trapped. The collar then rolls down and the collected particles are ingested at the collar base. In other heterotrophic flagellates, food particles can be ingested at any point on the cell, as the flagellates do not possess a cell mouth or cytostome. The flagellum or flagella are also used to effect locomotion.

Autotrophic flagellates, particularly cryptophytes, are the major planktonic component in Antarctic lakes (Figure 1.26 d–f). Algal species (diatoms, desmids) are relatively sparse. However, diatoms are common in the benthos. The water columns of ice-covered lakes are not subject to wind-driven turbulence, which is a major phenomenon in lower latitude lakes. Turbulent mixing is important in maintaining phytoplankton cells in the euphotic zone, particularly cells that lack motility. In the unmixed water columns of Antarctic lakes, organisms must be motile to maintain their position in a suitable light climate for photosynthesis.

Many of the phytoflagellate species found in Antarctic lakes are capable of mixotrophy, or mixed nutrition. Thus primarily phototrophic protozoans also feed on bacteria or take up DOC. Cryptophytes and *Pyramimonas* are mixotrophic in Antarctic lakes (Bell and Laybourn-Parry 2003; Laybourn-Parry et al. 2005). Phytoflagellates practise mixotrophy for a number of reasons, firstly when light is limited, as it often is in Antarctic systems, it provides an additional source of carbon or energy, and secondly when nutrients for photosynthesis are limited, eating bacteria provides a source of phosphorus and nitrogen. Such nutritional versatility confers an important survival advantage in extreme environments. Dinoflagellates are often a conspicuous component of Antarctic lake plankton, especially in brackish and saline lakes. The dinoflagellates are an interesting group because they are classified as Phytomastigophorea, but only around half of the extant species possess photosynthetic pigments. In the case of dinoflagellates, the green colour of the chlorophyll is masked by the brown colour of carotenoid pigments. The colourless dinoflagellates are heterotrophic, feeding on bacteria, other protozoans, and DOC.

The ciliates or Phylum Ciliophora are common in the plankton and benthos of Antarctic water bodies. They are one of the most taxonomically uniform groups of protozoans. Some are large enough to be discerned by the naked eye, particularly if they contain coloured pigments, while others are extremely small, only 5–10 μm. The ciliate cell is typified by the possession of cilia; complex rows of hair-like structures that produce coordinated beating that effects movement and feeding (Figure 1.27a). Most ciliates possess a cell mouth or cytostome, in some species surrounded by complex ciliary structures whose function is to create currents for filter feeding. Other species are raptorial feeders consuming larger particles. Some are predators of other ciliates. Among these are the suctorian ciliates that are sedentary ambush predators, trapping their prey with tentacles that extract the prey cell contents transporting it into the cell for digestion in food vacuoles (Figure 1.27b). Suctorians can be regarded as polystomate or mouthed, as each tentacle represents a mouth. To date, suctorians have only been recorded in the McMurdo Dry Valley lakes.

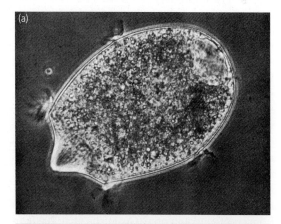

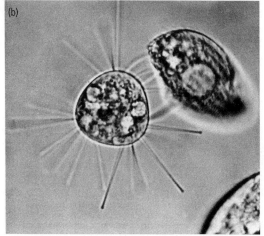

Figure 1.27 (a) The ciliate *Didinium*; (b) the suctorian ciliate *Podophyra*. Photos J. Laybourn-Parry.

Mixotrophy is also seen in the ciliates, but in this case the organisms are primarily heterotrophic, undertaking photosynthesis as a secondary mode of nutrition. Mixotrophy can take one of two forms, either the ciliate contains symbiotic zoochlorellae (algae) (Figure 1.28a), or the cell sequesters the plastids of ingested phytoflagellates and uses them for photosynthesis (Figure 1.28b). In the latter case the chloroplasts have to be replaced regularly, as they do not replicate in the ciliate but progressively senesce. The oligotrichs, a very common planktonic ciliate group, exemplify this type of mixotrophy. Where ciliates harbour zoochlorellae, the algae gain from living in a highly motile organism that can position itself in an optimum light climate in the water column, and have access to the ammonium and orthophosphate excretory products of the ciliates that are required for photosynthesis.

One ciliate dominates the plankton of brackish to hypersaline lakes in the Vestfold Hills and contains symbiotic cryptophytes (Perriss et al. 1995; Bell and Laybourn-Parry 1999a). This ciliate is *Mesodinium rubrum* (Figure 1.28c), a ubiquitous marine species found worldwide in the open sea and estuaries, where it is one of the causes of red tides. *Mesodinium rubrum* is a species complex. It lacks a cell mouth and apparently derives all of its nutritional needs from its endosymbiont (Lindholm 1985). However, some investigators have shown that at least some members of this species complex sequester plastids from cryptophycean prey (Gustafson et al. 2000). In the Vestfold Hills marine-derived lakes, *Mesodinium rubrum* is found in lakes with salinities as low as 4‰ and survives in lakes up to 62‰ (27% higher than seawater).

1.8.4 Algae

Among the Algae, the diatoms (Class Bacillariophyceae), the desmids (Order Desmidiales), and some members of the green algae (Classes Chlorophyceae and Euglenophyceae) are found in Antarctic lakes (Figure 1.29). Like the Protozoa they are also members of the Kingdom Protista. The diatoms and desmids lack locomotory appendages, so they tend not to be found in any significant numbers in the plankton, but benthic diatoms are common in benthic algal mats. Many of the green algae do have flagella and are consequently motile.

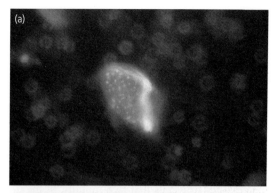
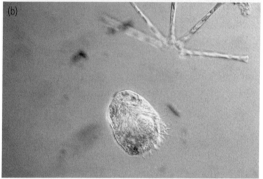
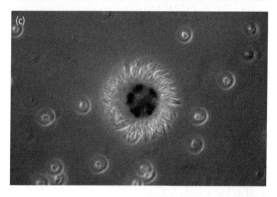

Figure 1.28 (a) *Euplotes* sp. containing symbiotic zoochlorellae from Lake Hoare. Stained with DAPI and viewed under epifluorescence microscopy; the long blue area is the ciliate macronucleus, the red areas are autofluorescing zoochlorellae with their individual blue DAPI stained nuclei; (b) *Strombidium* sp. containing sequestered (green) plastids from a temperate lake; (c) *Mesodinium rubrum* fixed in Lugol's iodine from Ace Lake, note deeply stained cryptophycaen endosymbiont within the cell. Photos J. Laybourn-Parry. (See Plate 9)

The diatoms possess a pectinaceous cell wall that is impregnated with silica, constructed in two halves or valves (Figure 1.29a). They can be solitary or colonial forming chain-like colonies. Because silica is an important component of their cell walls the availability of silica in their habitat can be a limiting

AN INTRODUCTION TO ANTARCTIC LAKES 35

1998; Unrein and Vinocur 1999). Euglenoids (Figure 1.29c) are flagellated cells capable of maintaining their position in the water column.

1.8.5 Rotifers

Rotifers are microscopic multicellular organisms common in the plankton and benthos of lakes worldwide (Figure 1.30). The body is cylinder shaped with a corona of cilia (also known as the wheel organ) at the front or head end, which is used

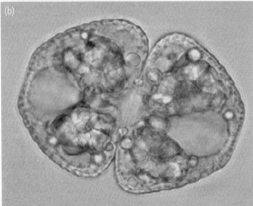
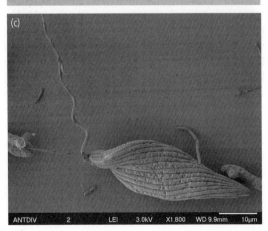

Figure 1.29 (a) A diatom; (b) a desmid *Cosmarium*; (c) a euglenoid flagellate. Photos J. Laybourn-Parry.

factor to their growth. Desmids are not common in Antarctic lakes, but members of the genus *Cosmarium* (Figure 1.29b) have been reported from the plankton of lakes in the Larsemann Hills and from the South Shetland Islands (Ellis-Evans et al.

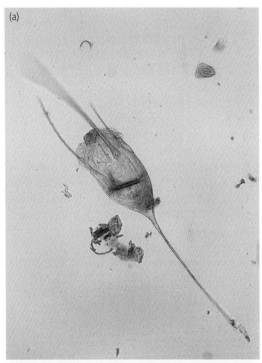
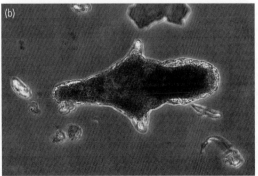

Figure 1.30 (a) The rotifer *Kellicottia* sp. from Lake Hoare; (b) the rotifer *Philodina alata* from Lake Hoare. Photos courtesy of J-E, Svensson and L-A, Hansson.

to create feeding currents and for locomotion. They feed on phytoplankton, bacteria, and heterotrophic protozoa and some species are predatory on other rotifers. Rotifers mostly reproduce asexually by a process of parthenogenesis, where females produce young without requiring fertilization. Some species are entirely parthenogenic while others are capable of sexual reproduction. Sexual reproduction is usually only resorted to when conditions become adverse and results in the production of resting eggs that are highly resistant and can remain dormant (in diapause) until favourable conditions return. When sexual reproduction is initiated, a special type of female called a mictic female is produced. These females produce haploid eggs which if unfertilized develop into males. The males lack a digestive system and are short-lived. Their role is solely to fertilize females. If mictic or haploid eggs are fertilized, restoring the diploid condition, they develop into resting eggs that eventually hatch out into amictic or parthenogenic females. Among the Rotifera there are two broad groups, the Bdelloidea that reproduce exclusively by parthenogenesis, and the Monogononta that are capable of sexual reproduction. Both groups are found in Antarctic lakes (Dartnall 1980, 1995, 2000; Nedzarek and Pociecha 2010). Information on the life cycles of Antarctic rotifers is very limited, so we do not know if the Monogononta species resort to sexual reproduction in Antarctic lakes and ponds.

1.8.6 Crustaceans

The Crustacea is a class in the Phylum Arthropoda, which contains organisms characterized by jointed limbs and an exoskeleton. Various groups of microcrustaceans are found in the zooplankton of lakes and the sea worldwide. As we shall see in subsequent chapters, these invertebrates are not very conspicuous in continental Antarctic lakes, being more abundant in maritime locations. However, species diversity is low across Antarctica. A note of caution should be made, the Crustacea of Antarctic lakes are not well studied when compared with the wealth of information on benthic microbial mats and microbial plankton. Where individuals of a species are not abundant they may be overlooked. Large volume samples are required to gain a proper picture.

The subclass Copepoda is the most common group found in Antarctica (Figure 1.31a). They reproduce sexually, the females bearing posterior egg sacs. The eggs hatch out into individuals known as nauplii (singular nauplius) that do not resemble the adult (Figure 1.31b). There are usually five naupliar stages each terminated by a moult (ecdysis) or casting off of the exoskeleton. Since the exoskeleton is rigid it does not grow, so development requires successive moults and a reforming of the external exoskeleton. The final naupl[i]ar stage metamorphoses into the first copepodid stage that resembles the adult form. There are a further five moults, the final stage being the adult. Both calanoid and cyclopoid copepods occur in Antarctic lakes. At lower latitudes, cyclopoid copepods can enter a diapause in one of the copepodid stages, while calanoid copepods are able to produce resting eggs.

Harpacticoid copepods are mainly benthic in habit, but both benthic and planktonic species have been reported from continental and maritime lakes and ponds. Most of the saline lakes of the Vestfold Hills possess one planktonic calanoid copepod (*Paralabidocera antarctica*); however, there are exceptions. Marine-derived Lake Abraxas has been isolated from the sea since before the LGM and contains both *Paralabidocera antarctica* and the planktonic harpacticoid *Amphiascoides* sp. *P. antarctica* was the most abundant of the two species and they showed different distributions in the water column (Bayly and Burton 1981; Bayly and Eslake 1987. Ace Lake has *P. antarctica* in the plankton and a benthic harpacticoid, *Idomene scotti*, in its benthic algal mats (Rankin et al. 1999).

The order Cladocera differs from the Copepoda morphologically and in having a life cycle that is largely parthenogenic or asexual (Figure 1.31c). The females have a dorsal brood pouch in which the eggs develop and hatch. The released young cladocerans resemble the adult and undergo a series of moults or ecdyses before attaining maturity. Under adverse conditions, males may appear in the population and fertilization results in the production of resting or ephippial eggs. These are very distinctive dark structures that are enclosed in the ephippium, a chitinous structure formed of two valves. These resting eggs can remain dormant in a diapause for long periods eventually hatching

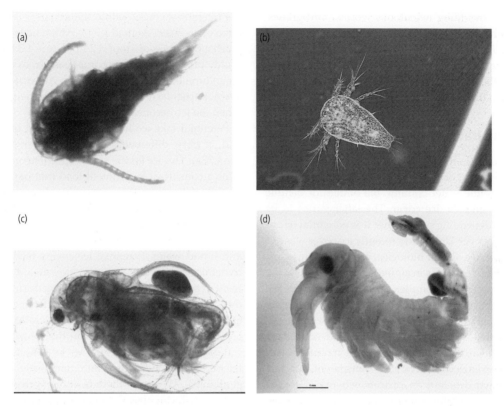

Figure 1.31 (a) Calanoid copepod *Boeckella poppei* from Beaver Lake; (b) a nauplius: (c) the cladoderan *Daphniopsis studeri*; (d) the fairy shrimp *Branchinecta*. Photos (a–c) J. Laybourn-Parry and (d) courtesy of the Biodiversity Institute of Ontario.

into parthenogenic females. Cladocera are not common in Antarctic lakes. *Daphniopsis studeri* (Figure 1.31c) is found in the freshwater and slightly brackish lakes of the Vestfold and Larsemann Hills (Laybourn-Parry and Marchant 1992; Ellis-Evans et al. 1998). Males and ephippial eggs have been reported in the populations from some of these lakes (Gibson et al. 1998). However, the males may be non-functional and the ephippial eggs produced parthenogenically, as happens in some Arctic cladocerans Hebert (1981). The density of *Daphniopsis* in some of the freshwater lakes is very sparse, so the chances of encountering a mate are very low and in such resource-limited environments one would expect a predominance of asexuality.

In the Maritime Antarctic the cladoceran *Alona* sp. is present in the benthos of numerous lakes on Signy Island, together with ostracod crustaceans (Heywood et al. 1979). The planktonic calanoid copepod *Boeckella poppei* is common in the South Shetland Islands, Signy Island, and the Antarctic Peninsula (Heywood 1977; Heywood et al. 1979; Izaguirre et al. 2003; Toro et al. 2007) and is largely confined to western Antarctica.

The fairy shrimp *Branchinecta gaini*, an anostracan crustacean, is commonly found in small lakes and ponds in the Maritime Antarctic (Figure 1.31d). It is a benthic detritivore/herbivore and often dominates the crustacean biomass of lacustrine environments in summer (Paggi 1996). The anostracans reproduce sexually, the eggs hatching as into nauplii, which transform into the next development stage the metanauplii before maturing into adults. They are able to produce resting eggs that overwinter in the sediment (Pociecha and Dumont 2008).

1.8.7 Other invertebrates

The benthic habitat of Maritime Antarctic water bodies supports a range of other invertebrate species including nematodes worms, tardigrades, oligochaetes, and chironomid larvae (Heywood 1977; Toro et al. 2007). At least two species of tubellarian (flat worms) and tardigrades are reported from the benthos of both freshwater and saline lakes in the Vestfold Hills (Dartnall 2000).

1.9 Habitats in Antarctic lakes

There are three habitats in a polar lake or pond. The plankton or pelagic environment, which is effectively the water column, the bottom or benthic habitat, and the ice cover. The latter varies in its nature depending on whether it is perennial or annual ice cover. Until fairly recently ice cover was not considered a site for biological activity, but research on annual ice covers in alpine lakes has revealed the development of complex microbial communities (Felip et al. 1995, 1999). There are no published data on annual ice covers in Antarctica; however, the nature of this ice closely resembles the ice described for Lake Baikal, the world's largest lake in terms of volume and its deepest. The ice is transparent, but during winter develops an intricate network of cracks, some of which form into crevasses. Diatoms have been described from the ice–water interface and in the interstitial cracks within the ice cover (Bondarenko et al. 2006). Annual ice on Antarctic lakes can reach a thickness of around 2 m (Figure 1.32b). It is highly probably that microorganisms from the water column become associated with Antarctic annual ice, particularly during the summer melt phase, when the ice becomes opaque.

Perennial lake ice has been the subject of study and is very different to annual ice cover (Figure 1.32a). This ice is very thick (up to 6 m) and contains a considerable sediment load that has blown onto the lakes from the surrounding bare rock surfaces of valley floors and mountains. This dark material absorbs the heat of the sun and melts into the ice matrix to create layers and patches of aggregates and liquid water. Perennial ice covers represent a dynamic equilibrium between downward movement of sediment as a result of melting during summer, and the upward movement of ice from ablation at the surface and freezing at the bottom of the ice layer. Liquid water is present in the ice for about 150 days during summer and up to 40% of the total ice cover volume may be liquid water (Priscu et al. 1998). Wherever there is liquid water microbial life will colonize. Cyanobacteria dominate the biomass of perennial ice covers (Priscu et al. 2005a).

Figure 1.32 (a) the perennial ice cover of Lake Hoare; note the rough debris strewn surface. (b) Annual ice cover of Crooked Lake with drill auger; note the smooth surface and the transparency. Photos J. Laybourn-Parry.

Aggregates of particles found in the ice contain complex bacterial communities. Both solitary and colonial cyanobacterial taxa are found, including *Phormidium, Oscillatoria, Nostoc, Scytonema*, and *Synechococcus*. Very few eukaryotics colonize the ice, unlike the underlying water where cryptophytes, chrysophytes, and chlorophytes predominate. The ice-associated aggregate microbiota form web-like matrices in which a variety of microorganisms and small amorphous minerals and organic particles become trapped (Paerl and Priscu 1998).

The water columns of lakes are effectively shaded environments as photosynthetically active radiation (PAR) attenuates with depth, following an exponential decay (Figure 1.33). In polar lakes the presence of ice cover further attenuates light penetration. Transparent annual ice allows reasonably good light transmission, for example in Lake Druzhby (Vestfold Hills), between 18% and 25% of surface irradiation was transmitted to the top 2–4 m of the water column over the course of a year (Bayliss et al. 1997) (Figure 1.33a). In contrast, the perennial debris laden ice of many Dry Valley lakes severely attenuates light, such that in Lakes Bonney, Hoare, and Fryxell, on average, only between 1.34% and 2.73% of incident surface photosynthetically active radiation penetrates to immediately below the ice (Figure 1.33b). In contrast with this, the perennial, less debris laden ice of Lake Vanda transmits an average of 13% of surface PAR to the water column (Howard-Williams et al. 1998). In effect, this means that the plankton of Antarctic lakes has to be very shade adapted. Moreover, the lack of wind driven turbulence requires that planktonic organisms must have some means of locomotion, so that they can maintain their position in the water column.

Snow cover is not a common feature of continental lakes, where any snow tends to be dispersed by the katabatic winds that blow off the continental ice sheet. In the Dry Valleys, which is a polar desert, snowfall is minimal. Maritime Antarctic lakes, for example on Signy Island, do have seasonal snow cover that strongly attenuates light, which in turn has a very marked impact on levels of primary production (Hawes 1985a).

As one moves southwards the plankton communities become progressively more microbial. In the Maritime Antarctic, crustacean zooplankton are common and more than one species commonly occurs, for example in lakes on Livingston Island and other Maritime Antarctic locations, both *Boeckella poppei* (calanoid copepod) and the fairy shrimp, *Branchinecta gainii*, as well as a number of rotifer species are found (Camacho 2006; Toro et al. 2007).

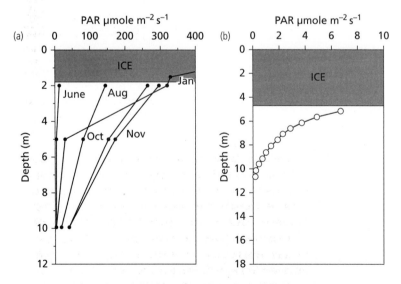

Figure 1.33 Typical photosynthetically active radiation (PAR) curves for (a) Crooked Lake, Vestfold Hills during summer and winter and (b) Lake Fryxell during summer. Note much higher summer PAR penetration in Crooked Lake.

In the extreme lakes of the McMurdo Dry Valleys, rotifers belonging to the genus *Philodina* have been reported since the 1990s (Laybourn-Parry et al. 1997), but recently a study that took very large 20 l water samples from each depth in the water column of Lake Hoare found nine species of rotifer (Hansson et al. 2011). They also found one intact juvenile copepod belonging to the genus *Boeckella*, suggesting that the species is present in this lake but extremely sparse. There is only one other report of a crustacean in the Dry Valley lakes; a copepod nauplius from Lake Joyce (Roberts et al. 2004a). Thus crustaceans are few and far between in the Dry Valley lakes. In continental lakes further north, for example those of the Vestfold Hills, the saline lakes usually support populations of a single planktonic copepod species (*Paralabidocera antarctica*), while the freshwater and slightly brackish lakes have planktonic populations of the cladoceran *Daphniopsis studeri*. Both the saline and freshwater lakes support a few species of rotifer. In the Maritime Antarctic where there may be significant zooplankton predation pressure on the phytoplankton and microbial plankton, significant top-down effects are evident.

The benthic communities of Antarctic lakes can be spectacular, with extensive algal or moss mats that in some cases have a complex morphology of pillars (Hawes and Schwarz 1999; Kudoh et al. 2009). Algal mats are composed of filamentous cyanobacteria (*Oscillatoria* and *Leptolyngbya*), and pennate diatoms (Hawes and Schwarz 1999). Among these algal mats and in moss pillars heterotrophic bacteria are common (Nakai et al. 2012; Peeters et al. 2012) as are protozoans (ciliates, testate amoebae, and heterotrophic and phototrophic flagellates), as well as rotifers (Cathey et al. 1981). In the Maritime Antarctic, where biological diversity tends to be higher, benthic mats also harbour metazoans such as oligochaetes, cladocerans, chironomids, nematodes, and tardigrades (Toro et al. 2007). As we shall see in later chapters, mats vary with depth in the lake and their latitude on the Antarctic continent.

CHAPTER 2

Freshwater lakes

2.1 Introduction

Globally, freshwater lakes are the most common lake type. Antarctica has many such lakes and freshwater ponds. Some of the latter may be ephemeral water bodies. Antarctica also has a large number of brackish through saline to hypersaline lakes (see Chapter 3). As we have seen in Section 1.6, some freshwater lakes have evolved from saline lakes and some saline lakes were freshwater at some time in their history. Globally, freshwater lakes and running waters usually have salt concentrations between 0.01 and 1.0 g l^{-1} with the majority lying between 0.1 and 0.5 g l^{-1} (Wetzel 2001). Worldwide, freshwater lakes range in their trophic status on a continuum from ultra-oligotrophic (extremely unproductive) through oligotrophic, mesotrophic, eutrophic, to hyper-eutrophic (extremely productive). Trophic status relates to the productivity of the lake, which in turn relates to the concentration of major nutrients such as phosphorus and nitrogen, needed for plant and bacterial growth. Over the years a range of factors have been used to classify lakes, including nutrient concentrations, chlorophyll *a*, and Secchi disk depth. Chlorophyll *a* is a proxy for phytoplankton biomass and hence growth, and clearly relates to nutrient concentration, but also to levels of photosynthetically active radiation (PAR). Secchi disk depth is a now defunct method for measuring the euphotic depth (Z_{eu}) in the water column, where less than 1% of PAR remains. It is a 25 cm disk painted with black and white segments that is lowered until it disappears from view–at the Secchi depth. These days we use light meters that provide an accurate measure of PAR in µmol m^{-2} s^{-1}, the current SI unit for visible light energy in the water column. Clearly, where there is an actively growing dense phytoplankton population and high concentrations of organic matter in a productive lake, light penetration in the water column will be impeded by self-shading; so Z_{eu} will be shallow. Conversely, in an unproductive or oligotrophic lake, the depth to which light penetrates will be much greater and Z_{eu} will be much deeper in the water column. Most of the freshwater lakes of Antarctica are ultra-oligotrophic to oligotrophic. Exceptions are those lakes that are adjacent to penguin rookeries or seal wallows which receive allochthonous nutrient and organic carbon inputs, and some rare examples of lakes used by research stations as their water supply. Ultra-oligotrophic lakes typically have chlorophyll *a* concentrations of less than 2.5 µg l^{-1} and oligotrophic lakes with between 2.5 and 8 µg Chl *a* l^{-1} with low concentrations of major nutrients; particularly soluble reactive phosphorus (PO$_4$–P). Typically, as Figure 1.33 shows, PAR penetrates to great depth in clear unproductive water where there is a clear ice cover.

The ice cover of freshwater lakes is usually transparent allowing good light transmission, then becoming opaque in summer as it begins to break up. There are exceptions to this, such as Lake Hoare in the Taylor Valley which has a perennial debris-laden ice cover. Here only 2.6%–5.5% of surface irradiation reaches the water column immediately under the ice, and during the summer light transmission can vary by a factor of 3. This is because as summer progresses the ice melts and destroys fine cracks and other scattering structures, allowing increased light transmission. As winter encroaches, the ice temperature decreases allowing cracks and fissures to reappear, leading to a decrease in transmission (Howard-Williams et al. 1998). Lake Miers in the McMurdo Miers Valley is also freshwater with a

Antarctic Lakes. Johanna Laybourn-Parry and Jemma L. Wadham.
© Johanna Laybourn-Parry and Jemma L. Wadham 2014. Published 2014 by Oxford University Press.

perennial thick, heavily debris-laden ice cover. This ice cover has numerous melt pools with dirt cones up to a height of 5 m that severely attenuate light transmission to the water column (Bell 1967). There are limited data for Lake Miers but a scalar irradiance value (see Glossary) of 0.110 was measured in January 1995. This compares with scalar irradiance values in the range of 0.127–0.216 between November and January for Lake Hoare (Howard-Williams et al. 1998). These are, however, atypical freshwater Antarctic lakes in terms of light transmission.

The coastal oases and the Maritime Antarctic carry a large number of freshwater lakes that range in surface area from 20 km^2 to small water bodies. Those that have been researched are listed in Table 2.1 together with some of their major morphometric characteristics. It should be noted that there are many more freshwater lakes that are unnamed and have not

Table 2.1 Antarctic freshwater lakes (including epishelf lakes) on which there is published limnological data. Note that there are many more named and unnamed freshwater lakes that have not been sampled in detail. In some cases information on depth and area are not available and such lakes have been excluded from the table. Lakes are listed by decreasing surface area.

Lake and notes	Location	Maximum depth (m)	Surface area (km^2)
1. Beaver Lake[1,2] (*Epishelf, freshwater interface at 220–260 m^2*)	Amery Oasis (70°48′S, 68°15′E)	435	800
2. Radok Lake[3]	Amery Oasis (70°52′S, 67°00′E)	~346	20
3. Transkripsii Gulf[4] (*Epishelf- freshwater interface at 86–88 m*)	Bunger Hills (66°15′S, 100°35′E)	122	14.4
4. Lake Untersee[5]	Queen Maud Land (71°21′S, 13°28′E)	169	11.4
5. Lake Figurnoye[6] (*also known as Algae Lake*)	Bunger Hills	145	Unknown 25 km long
6. Crooked Lake[7]	Vestfold Hills (68°37′S, 78°22′E)	150	9.5
7. Ablation Lake[8] (*Epishelf, freshwater interface at 55–60 m*)	Alexander Island (70°49′S, 68°27′W)	70	Approx. 9
8. Lake Druzhby[9]	Vestfold Hills (68°36′S, 78°16′E)	40	7
9. Lake Zvezda	Vestfold Hills (68°32′S, 78°23′E)	Unknown	Approx. 4
10. Lake Obersee[10]	Queen Maud Land (71°17′S, 13°39′E)	>40	3.43
11. Lake Hoare[11]	Taylor Valley (McMurdo)	34	2.9
12. Lake Bisernoye[12]	Vestfold Hills (68°32′S, 78°30′E)	>35	Approx. 2
13. Lake Terrasovoje[13]	Amery Oasis (70°33′S, 68°01′E)	31	Approx. 2
14. Lake Pol'anskogo[4] (*Epishelf*)	Bunger Hills (66°19′S, 100°30′E)	69	2.0
15. Chelnok Lake (*Epiglacial*)	Vestfold Hills (68°38′S, 78°19′E)	15.5	Approx. 2
16. Lake Miers[11]	Miers Valley, McMurdo (78°07′S, 163°54′E)	21	1.3

continued

Table 2.1 Continued

Lake and notes	Location	Maximum depth (m)	Surface area (km²)
17. Pauk Lake[12]	Vestfold Hills (68°35′S, 78°30′E)	35	Approx. 1
18. Moutonée Lake[8] (*Epishelf–freshwater interface at 36 m*)	Alexander Island (70°51′S, 68°21′W)	70	Approx. 1
19. Northern Lake[4] (*Epishelf*)	Bunger Hills (66°13′S, 100°39′E)	22	0.55
20. Watts Lake*[14]	Vestfold Hills (68°36′S, 78°13′E)	35	0.38
21. Lake Nottingham (*Unofficial name—abuts ice sheet*)	Vestfold Hills (68°28′S, 78°26′E)	12	0.22
22. Lichen Lake	Vestfold Hills (68°28′S, 78°25′E)	11	0.21
23. Lake Prival'noye[10] (*Epishelf*)	Schirmacher Oasis (71°S, 12°E)	47.5	0.156
24. Higashi-yukidori Ike[15]	Syowa Oasis (69°S, 39°E)	19.5	0.148
25. Lake Glubokoye[10]	Schirmacher Oasis (71°S, 12°E)	34.5	0.145
26. Lake Nella[16]	Larsemann Hills (69°24′S, 76°22′S)	18	0.13
27. Kirisjes Pond[16]	Larsemann Hills (69°22′S, 76°09′E)	9	0.12
28. Lake Prival'noye[10] (*Epishelf*)	Schirmacher Oasis (71°S, 12°E)	34	0.115
29. Progress Lake[16]	Larsemann Hills (69°24′S, 76°24′E)	34	0.10
30. Namazu Ike[15]	Syowa Oasis (69°29′S, 39°41′E)	20.8	0.091
31. Lake Spate[16]	Larsemann Hills (69°25′S, 76°07′E)	11	0.09
32. Lake Heidi[16]	Larsemann Hills (69°24′S, 76°06′E)	5	0.075
33. Tres Hermanos Lake[17]	King George Island (62°02′S, 58°21′W)	6.5	0.073
34. Boeckella Lake[18] (*Water supply for Esperanza St.*)	Antarctic Peninsula (63°24′S, 57°00′W)	4	0.067
35. Midge Lake[19]	Livingston Island (62°38′S, 61°06′W)	9	0.065
36. Hyotan Ike[20]	Syowa Oasis (69°S, 39°E)	12.4	0.053
37. Smirnov Lake[10] (*Epiglacial*)	Schirmacher Oasis (71°S, 12°E)	25	0.056
38. Long Lake[16]	Larsemann Hills (69°24′S, 76°07′E)	11	0.05

continued

Table 2.1 *Continued*

Lake and notes	Location	Maximum depth (m)	Surface area (km²)
39. Oyake Ike[20]	Syowa Oasis (69°28'S, 39°33'E)	8	0.048
40. Naga Ike[20]	Syowa Oasis (69°S, 39°E)	10.8	0.048
41. Oyako Ike[20]	Syowa Oasis (69°S, 39°E)	8	0.048
42. Heywood Lake[21] (*Receives allochthonous inputs from seal wallow*)	Signy Island (60°43'S, 45°38'W)	6.4	0.045
43. Lake Pomornik[10]	Schirmacher Oasis (71°S, 12°E)	11.8	0.044
44. Kami-kama Ike[20]	Syowa Oasis (69°S, 39°E)	4.5	0.044
45. Lake 21[22]	Terra Nova Bay (74°50'S, 162°30'E)	2–3	0.042
46. Yukidori Ike[20]	Syowa Oasis (69°S, 39°E)	8.6	0.041
47. Lake Burgess[16]	Larsemann Hills (69°25'S, 76°07'E)	16	0.04
48. Ten-no-kama Ike[20]	Syowa Oasis (69°S, 39°E)	4.7	0.035
49. Knob Lake[21]	Signy Island (60°43'S, 45°38'W)	4	0.035
50. Twisted Lake[21]	Signy Island (60°43'S, 45°38'W)	4	0.031
51. Sombre Lake[21]	Signy Island (60°43'S, 45°38'W)	11.2	0.027
52. No Worries Lake[16] (*Water supply for Chinese Station*)	Larsemann Hills (69°22'S, 76°23'E)	3.8	0.025
53. Lake Limnopolar[19]	Livingston Island (62°38'S, 61°06'W)	5	0.022
54. Emerald Lake[21]	Signy Island (60°43'S, 45°38'W)	15	0.022
55. Lake Verkhneye[10]	Schirmacher Oasis (71°S, 12°E)	9	0.021
56. Tranquil Lake[21]	Signy Island (60°43'S, 45°38'W)	8	0.021
57. Discussion Lake[16]	Larsemann Hills (69°23'S, 76°22'E)	4	0.02
58. Mago Ike[20]	Syowa Oasis (69°S, 39°E)	5.8	0.017
59. Moss Lake[21]	Signy Island (60°43'S, 45°38'W)	10.4	0.015
60. Lake 10[22]	Terra Nova Bay (74°50'S, 162°30'E)	2–3	0.015

continued

Table 2.1 Continued

Lake and notes	Location	Maximum depth (m)	Surface area (km²)
61. Himebati Ike[20]	Syowa Oasis (69°S, 39°E)	2.5	0.013
62. Pumphouse Lake[21]	Signy Island (60°43′S, 45°38′W)	4	0.012
63. Lake 9[22]	Terra Nova Bay (74°50′S, 162°30′E)	2–3	0.010
64. Nyorai Ike[20]	Syowa Oasis (69°S, 39°E)	3	0.010
65. Changing Lake[21]	Signy Island (60°43′S, 45°38′W)	5.4	0.009
66. Bosatsu Ike[20]	Syowa Oasis (69°S, 39°E)	4.8	0.009
67. Lake Wujka[23]	King George Island (62°02′S, 58°21′W)	1.4	0.008
68. Hotoke Ike[20]	Syowa Oasis (69°28′S, 39°33′E)	3	0.006
69. Amos Lake[21]	Signy Island (60°43′S, 45°38′W)	4.3	0.005
70. Lake 1[23]	King George Island (62°02′S, 58°21′W)	3	0.004
71. Chico Lake18	Antarctic Peninsula (63°24′S, 57°00′W)	5.5	0.003
72. Unnamed[24] (*Water supply to Gabriel de Castilla Station*)	Deception Island (62°57′S, 60°38′W)	3.6	0.0026

*Watts Lake is slightly brackish 0.3–4.1 mS cm⁻¹, but contains a freshwater community and is therefore included in this chapter.
1: Galton-Fenzi et al. 2012; 2: Bardin et al. 1990; 3: Adamson et al. 1997; 4: Gibson and Andersen 2002; 5: Wand and Perlt 1999; 6: Verkulich et al. 2002; 7: Laybourn-Parry et al. 1992; 8: Heywood 1977; 9: Laybourn-Parry and Bayliss 1996; 10: Richter and Bormann 1995; 11: Spigel and Priscu 1998; 12: Dartnall 2000; 13: Wagner et al. 2004; 14: Heath 1988; 15: Kudoh et al. 2009; 16: Sabbe et al. 2004; 17: Unrein and Vinocur 1999; 18: Izaguirre et al. 2003; 19: Hodgson et al. 1998; 20: Kimura et al. 2010; 21: Heywood et al. 1979; 22: Andreoli et al. 1992; 23: Nedzarek and Pociecha 2010; 24: Llames and Vinocur 2007; Lakes 15, 21, and 22 unpublished data Laybourn-Parry.

been investigated or have only been subject to superficial study. For example in the Vestfold Hills (Figure 2.1) there are over 300 lakes and ponds of which only nine are listed in Table 2.1 and 27 in Table 3.1 (saline lakes, see Chapter 3). Similarly there are hundreds of large and small lakes in the Bunger Hills (Klokov et al. 1990) and 150 freshwater lakes and ponds in the Larsemann Hills (Gillieson et al. 1990), only a small number of which have been researched (Figure 2.3 and 2.2). Epishelf lakes are also included in Table 2.1 for comparative purposes as they are essentially freshwater. They are dealt with in detail in Chapter 4, but it is evident from Table 2.1 that they are among the largest surface lakes in Antarctica.

Table 2.2 shows biological data for lakes where it is available. With very few exceptions all of the lakes, both large and small, have chlorophyll *a* concentrations in the ultra-oligotrophic range (<2.5 μg l⁻¹). A number of small lakes (lakes 33, 51, 60, 61, 62 in Table 2.1) have chlorophyll *a* values that range into the oligotrophic category (2.5–8.0 μg l⁻¹). Heywood Lake on Signy Island has a fur seal wallow on its shores that provides allochthonous inputs of organic carbon and nutrients which significantly enhance productivity. In the 1980s, chlorophyll *a* ranged between 3 and 30 μg l⁻¹ over an annual cycle (Hawes 1985a), but by the late 1990s it had risen to a maximum of 297 μg l⁻¹ in summer,

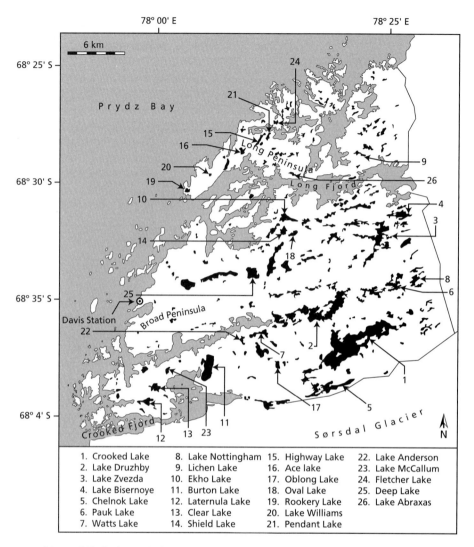

Figure 2.1 A map of the Vestfold Hills showing the lakes and the position of Davis Station (Australia). The lakes numbered 1–9 are freshwater, those numbered 10–26 are hyposaline, saline, or hypersaline. For more details about the freshwater lakes see Table 2.1 and for saline lakes see Table 3.1.

reflecting an increase in the seal population and enhanced nutrient inputs to the lake (Butler 1999). The oligotrophic nature of the majority of Antarctic freshwaters reflects low nutrient inputs from their catchments and the fact that inflows, and potential nutrient inputs, occur for only a short period each summer, when there is melt from glaciers, the ice sheet, and snow accumulations. While low temperatures and low annual PAR undoubtedly constrain production, productivity is enhanced where there are significant nutrient inputs. This is particularly the case in the Maritime Antarctic, where climatic conditions are less extreme than on the continent.

A number of lakes are used as water supplies to research stations. Among these are Boeckella Lake on the Antarctic Peninsula, No Worries Lake in the Larsemann Hills, and an unnamed lake on Deception Island (lakes 34, 52, and 72 respectively in Table 2.1). In all three cases chlorophyll a concentrations are elevated compared with other lakes in their respective regions. Prior to the establishment of Zhongshan Station in the Larsemann Hills in 1989,

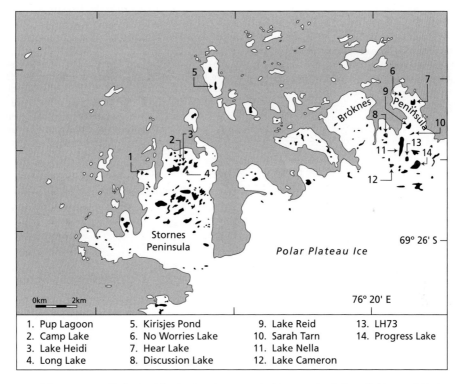

Figure 2.2 A map of the Larsemann Hills showing lakes that have been researched. For more detail of specific lakes see Table 2.1.

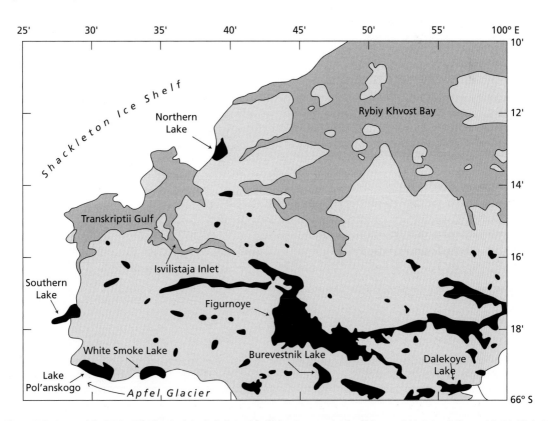

Figure 2.3 A map of the Bunger Hills showing lakes including epishelf lakes. For more details of lakes see Table 2.1 and Chapter 4 (epishelf lakes).

48 ANTARCTIC LAKES

Table 2.2 Chlorophyll a (µg l⁻¹), primary production (µg C l⁻¹ day⁻¹), and bacterial production (µg l⁻¹ day⁻¹) in freshwater lakes. Unless otherwise indicated data relate to a single summer. *: over an annual cycle; §: McMurdo Long Term Ecosystem Research (LTER) program summer limno runs from 1993 to 2004; ∂: over summers 1993/94 to 1996/97; Φ: unofficial name. Note the lake number corresponds with the numbers in Table 2.1.

Lake and location	Chl a µg l^{-1}	Primary production µg C l^{-1} day^{-1}	Bacterial production µg C l^{-1} day^{-1}
4. Lake Untersee (Queen Maud Land 71°21′S, 13°28′E)	0.08–0.27[1]		
5. Lake Figurnoye (also known as Algae Lake) (Bunger Hills 66°10′S, 101°00′E)	0.1–0.22[2]		
6. Crooked Lake (Vestfold Hills 68°37′S, 78°22′E)	0.6–0.05*[3]	0–38.5*[3]	0–11.5*[4]
8. Lake Druzhby (Vestfold Hills 68°36′S, 78°16′E)	0.15–1.1*[5]	0.24–37.7*[3]	0–8.5*[4]
9. Lake Zvezda (Vestfold Hills 68°32′S, 78°23′E)	0.42[6]		
10. Lake Obersee (Queen Maud Land 71°17′S, 13°39′E)	0.71–0.76[1]		
11. Lake Hoare (Taylor Valley, McMurdo Dry Valleys 77°38′S, 162°55′E)	0.2–6.0[7]	0.0066–2.30§ 0.06–0.81[8]	0.05–0.7∂[9]
12. Lake Bisernoye (Vestfold Hills 68°32′S, 78°30′E)	0.05[6]		
15. Chelnok Lake (epiglacial lake) (Vestfold Hills 68°38′S, 78°19′E)	0.86–1.56[10]		
16. Lake Miers (Miers Valley, McMurdo Dry Valleys 78°07′S, 163°54′E)	0.3–2.9[7]		
17. Pauk Lake (Vestfold Hills 68°35′S, 78°30′E)	0.12[6]		
20. Watts Lake (Vestfold Hills 68°36′S, 78°13′E)	0.9–2.8*[11]	2.4–154.6*[11]	
21. Lake Nottingham^Φ (epiglacial) (Vestfold Hills 68°28′S, 78°26′E)	0.23–0.33[10]	0–5.2[10]	0.08–0.15[10]
22. Lichen Lake (Vestfold Hills 68°28′S, 78°25′E)	0.1–1.8[10]	0–7.02[10]	0.18–0.51[10]
25. Lake Glubokoye (Schirmacher Oasis 71°S, 12°E)	0.21–2.08*[1]	0–17.7*[1]	
26. Lake Nella (Larsemann Hills 69°24′S, 76°22′E)	0.10[12]	0.14–0.41[12]	
27. Kirisjes Pond (Larsemann Hills 69°22′S, 76°09′E)	0.26[12]		
29. Progress Lake (Larsemann Hills 69°24′S, 76°24′E)	0.17[12]		
30. Namazu Lake (Syowa Oasis 69°29′S, 39°41′E)	0.02[13]		
32. Lake Heidi (Larsemann Hills 69°24′S, 76°06′E)	0.30[12]		
33. Tres Hermanos Lake (King George Island 62°02′S, 58°21′W)	1.6–8.3[14]		
34. Boeckella Lake (water supply for Esperanza Station) (Antarctic Peninsula 63°24′S, 57°00′W)	0.0–8.71[15]		
35. Midge Lake (Livingston Island 62°38′S, 61°06′W)	0.07–0.15[16]		
42. Heywood Lake (receives allochthonous inputs from a seal wallow) (Signy Island 60°43′S, 45°38′W)	3–30[17] 297.6[18]	0–250[19] 0–840[18]	4.8–26.4[18]
43. Lake Pomornik (Schirmacher Oasis 71°S, 12°E)	0.12–0.16[1]		
45. Lake 21 (Terra Nova Bay 74°50′S, 162°30′E)	0.62[20]		
51. Sombre Lake (Signy Island 60°43′S, 45°38′W)	0–7*[17]	0–50[19]	
52. No Worries Lake (Larsemann Hills 69°22′S, 76°23′E) (water supply to Chinese Station)	0.59–3.96[21]	4.8–55.2[21]	
53. Lake Limnopolar (Livingston Island 62°34′S, 60°54′W)	0.15[22]		
54. Emerald Lake (Signy Island 60°43′S, 45°38′W)	2.0[23]		
55. Lake Verkhneye (Schirmacher Oasis 71°S, 12°E)	0.05–0.46*[1]	0–3.7*[1]	
56. Tranquil Lake (Signy Island 60°43′S, 45°38′W)	0–4.2*[24]	<12–107.23*[24]	<2.4–8.64*[24]

continued

Table 2.2 Continued

Lake and location	Chl a µg l^{-1}	Primary production µg C l^{-1} day^{-1}	Bacterial production µg C l^{-1} day^{-1}
57. Discussion Lake (Larsemann Hills 69°23'S, 76°22'E)	0.40[12]	0.72[12]	
59. Moss Lake (Signy Island 60°43'S, 45°38'W)	0.4–8.0*[25]	0–28*[25]	0–12.5*[25]
60. Lake 10 (Terra Nova Bay 74°50'S, 162°30'E)	8.58[20]		
62. Pumphouse Lake (Signy Island 60°43'S, 45°38'W)	12.0[23]		
63. Lake 9 (Terra Nova Bay 74°50'S, 162°30'E)	4.82[20]		
68. Hotoke Ike (Syowa Oasis 69°28'S, 39°33'E)	0.18[13]		
71. Chico Lake (Antarctic Peninsula 63°24'S, 57°00'W)	0–2.45[15]		
72. Unnamed (water supply for Gabriel de Castilla Station) (Deception Island 62°57'S, 60°38'W)	0.44–4.61[26]		

1: Kaup 1995; 2: Klokov et al. 1990; 3: Henshaw and Laybourn-Parry 2002; 4: Laybourn-Parry et al. 2004; 5: Laybourn-Parry and Bayliss 1996; 6: Laybourn-Parry and Marchant 1992; 7: Roberts et al. 2004a, b; 8: Priscu 1995; 9: Takacs and Priscu 1998; 10: Laybourn-Parry unpublished data; 11: Heath 1988; 12: Ellis-Evans et al. 1998; 13: Kudoh et al. 2009; 14: Unrein and Vinocur 1999; 15: Izaguirre et al. 2003; 16: Toro et al. 2007; 17: Hawes 1985a; 18: Butler 1999; 19: Hawes 1983; 20: Andreoli et al. 1992; 21: Ellis-Evans et al. 1997; 22: Rochera et al. 2010; 23: Pearce et al. 2007; 24: Butler et al. 2000; 25: Ellis-Evans 1981; 26: Llames and Vinocur 2007.

No Worries Lake was a clear shallow lake similar to others in the area. By 1993 the lake was surrounded by buildings and there was significant vehicle activity. In addition, water from the lake was used to cool the station generators and then recycled to the water body, causing heating of the lower water layers. These activities resulted in marked changes in the plankton and elevated production within a timeframe of 4–5 years (Ellis-Evans et al. 1997; Riddle and Muir 2008). The use of natural water bodies as water supplies in Antarctica and its impact are not well researched. A number of nations (e.g. USA, UK, and Australia) use reverse osmosis treatment of seawater as a source of potable water. In many cases there is no locally convenient lake or pond to serve as a water supply to a station. For example the British base on Signy Island and Davis Station in the Vestfold Hills. In the latter case, prior to the commissioning of a reverse osmosis plant, the station relied on melted snow for its water supply.

A glance at Table 2.2 reveals a relative paucity of process-related data for primary production and bacterial production in Antarctic freshwaters. However, there are numerous publications on community composition, particularly for the phytoplankton and benthic mats. The measurement of primary and bacterial production requires relatively sophisticated laboratory facilities in Antarctica, which are outside the resources of many nations. Nevertheless, the available data span a latitudinal gradient from the McMurdo Dry Valleys to the Maritime Antarctic and across a wide spectrum of lake size.

2.2 Formation of freshwater lakes

Lakes are features of the landscape that undergo constant change and evolution, much of which is related to sedimentation. Usually lakes become progressively shallower as sediment brought in by rivers and streams from the catchment is laid down, eventual filling the lake basin and incorporating it into the terrestrial landscape (Wetzel 2001). As indicated in Section 1.7, the usual pattern of ontogeny is from oligotrophic to eutrophic, although there are exceptions. Antarctic lakes experience low rates of sedimentation for two reasons, firstly they receive inputs for only a short time each year and secondly they are not fed from extensive catchments as is the case elsewhere in the world and in the Arctic. Melt streams from glaciers tend to be relatively short. The longest river in Antarctica is the Onyx River in the Wright Valley which feeds Lake Vanda. It is 40 km in length. In the Dry Valleys the discharge of streams is low, summer flows are typically less than 1 m^3 s^{-1}. Thus stream-flow is only sufficient to erode and transport appreciable quantities of the surficial material (McKnight et al. 2008). This is the case elsewhere in Antarctica. In the Vestfold Hills the short outflow from Lake Druzhby through the Ellis Rapids to the Ellis Fjord has a maximum discharge rate of 2 m^3 s^{-1}, while the Algae River in the Bunger Hills (25 km long) has a maximum discharge of 2.31 m^3 s^{-1} (Gibson et al. 2002a). There are considerable yearly, seasonal, and daily variations in flow reported in Antarctic streams and rivers (Conovitz et al. 1998; Bronge 1996). For example, the Ellis Rapids show marked inter-annual variation and in some years flow is virtually zero (Bronge 1996). It is evident that Antarctic lakes have a different pattern of evolution to those of lakes elsewhere. While many postdate the LGM, some have survived from the previous interglacial period.

The reconstruction of ice advances and retreats during the Quaternary in the coastal areas of Antarctica, and their impacts on lake formation are based on the interpretation of palaeontology, geomorphology, and marine and lacustrine sediments. As indicated above, lacustrine sedimentation in Antarctica is low, even when compared with Arctic and Alpine systems. Sedimentation rates that characterize the deglaciation of lake basins in Antarctica show facies that are a matter of a few decimetres, while in Arctic and alpine lakes there may be several metres of glaciolacustrine and glaciofluvial sediments that are derived from deglaciation of the catchment (Wagner et al. 2004). Palaeolimnological studies based on carbon dating fossil pigments and algal remains provide a picture of lake formation and episodes of enhanced and decreased productivity that relate to periods when the climate was warming or cooling.

There is debate about the extent of the East Antarctic Ice Sheet during the LGM. CLIMAP (1981) reconstruction shows the East Antarctic Ice Sheet extending to the edge of the continental ice shelf at the LGM (21,000 yrs B.P.), with deglaciation of the coastline occurring around the start of the Holocene, 10,000 yrs B.P. (Hughes 1998). However, there is now clear evidence that some areas of Antarctica were ice-free during the last glacial maximum and that deglaciation was diachronous. Lake sediments provide unique records of environmental and ecological change through the last glacial interglacial cycle. A number of the lakes in the Larsemann Hills possess sediments that predate the last glaciation. Sediment cores show well-preserved structured layers. The basal dates of these cores provide minimum ages for the onset of sedimentation and confirm that on the Broknes Peninsula (Figure 2.2) some lakes possess sediments that predate the LGM, and at higher elevations some lakes have probably existed since at least the late Pleistocene. The Stornes Peninsula (Figure 2.2) only became ice-free in the mid to late Holocene and carries young lakes that have ages ranging from 3800 to 1800 yrs B.P. (Hodgson et al. 2001a). Progress Lake is one of the older lakes and was covered by a layer of in situ firnified snow and ice during the last glaciation, resulting in a marked decline in sedimentation. The sediment core of Progress Lake shows two main stratigraphic zones separated by a transition zone. The oldest zone (57–24 cm) corresponds to 125,000–115,000 yrs B.P. and has a fossil photosynthetic pigment record that suggests a community of Cyanobacteria and/or chlorophytes together with purple photosynthetic bacteria. The presence of chlorin steryl esters (CSEs) suggests that there was grazing activity by animals. CSEs are the secondary transformation products of chlorophyll and occur in the faecal pellets of grazers. If this is the case, it indicates very different conditions in the previous interglacial period, as metazoan grazers are rare in continental lakes today. This sediment layer is covered by a transition zone (24–22 cm) dated to 115,000–20,920 yrs B.P. which equates to the glacial phase when the lake was covered by firnified snow and ice. During this period of reduced sedimentation there was low primary production evidenced by a weak pigment signal. Above this layer there are two distinct subzones. The lower one (22–16 cm) equates to between 24,644–25,571 and 3470–3643 yrs B.P. and has low pigment content, suggesting that the lake was still ice-covered, while the upper sediment layer (16–0 cm, age range between 3470–3653 B.P. and the present) is linked to the late Holocene retreat of the firnified snow and ice and the return of higher biological activity. Here the pigment content is less complex than the earliest part of the core. The indications are that benthic mats formed of Cyanobacteria and possibly some chlorophytes predominated (Squier et al. 2005).

As indicated above, the lakes on the Stornes Peninsula of the Larsemann Hills are young and were formed after the LGM. The freshwater lakes have evolved from oligotrophic proglacial lagoons. Kirisjes Pond has a sedimentary record covering the last 9000 years. The record indicates disturbance within the catchment with periodic inwashing of coarse sand and mud of proglacial origin and finer sand derived from aeolian processes. Prior to 6000 yrs B.P. the fossil diatom community was freshwater, indicative of a proglacial freshwater environment. Between 6000 and 4000 yrs B.P. the community changed to one indicative of brackish to marine water, suggesting a postglacial rise in sea level and a marine incursion into the lake. Following the stabilization of sea level, the lake returned to being a meltwater system with associated low conductivity and freshwater diatoms. Recent inwashing of seawater and salt spray have prompted the re-establishment of polyhalobous diatom species tolerant of higher salinities (Gillieson 1991).

During the late Holocene deglaciation the Vestfold Hills contained a large ice-dammed glacial lake, Glacial Crooked Lake. The contemporary Crooked Lake remains and ranks as one of the largest surface lakes in Antarctica (Table 2.1, lake 6; Figure 2.1). Glacial Crooked Lake had a volume of $250 \pm 45 \times 10^6$ m^3. A series of previous shorelines indicate the extent of the lake, which at its maximum was 46 m above sea level. The present-day Crooked Lake is 22 m above sea level. Water from Glacial Crooked Lake was periodically released through the Ellis Rapids into Ellis Fjord, which occasionally could not accommodate the flow. Backponded water would have flowed over a low flat sill from the Druzhby River into Watts Lake, which at that time was a saline lake formed as a result of isostatic uplift from a marine embayment. The evidence suggests that Watts Lake was flushed of its salt water between 2700 and 1700 yrs B.P. Even today water periodically flows into Watts Lake from the Druzhby River, during times when the Ellis Rapids are temporarily blocked by a snow dam (Gore et al. 1996).

Lake Druzhby lies at the head of the Ellis Fjord (Figure 2.1) and was at one time part of that marine inlet, as indicated by the presence of marine sediments. The upper 0–30 cm of a core taken from the lake was lacustrine, while below 30 cm the sediments were of marine origin. The transition from marine to freshwater occurred between 5500 and 5940 yrs B.P. The lake became freshwater as a result of isostatic uplift and changes in sea level (see Section 1.7 and Figure 1.22). Lake Druzhby is separated from the Ellis Fjord by a sill. During the transition period the saline waters were progressively flushed by glacially derived freshwater, creating the multi-basin large lake we see today (Zwartz et al. 1998).

The evidence suggests that the Bunger Hills (Figure 2.3) may have been largely ice-free at the last LGM (Gore 1997; Gore et al. 2001). Gore et al. (2001) suggest that the East Antarctic Ice Sheet had begun retreating from the Bunger Hills, exposing the hilltops, no later than 30,000 yrs B.P. and possibly as early as 40,000 yrs B.P. By Marine Isotope Stage (MIS) 2 and through the LGM most of the present topography was exposed. High-level glacial lakes filled many of the valleys. By the latest Pleistocene, further recession had resulted in the high-level glacial lakes draining into more extensive lower glacial lakes. By the start of the Holocene many of these lakes had also drained to reveal a picture similar to that seen today.

The largest and deepest lake in the Bunger Hills is Lake Figurnoye (also known as Algae Lake) (lake 5, Table 2.1; Figure 2.3). It is up to 145 m deep and has a surface area of 14.3 km^2. The lake receives inflow from ice sheets and snowfields through streams and from other lakes. It outflows into a marine inlet. Lake Figurnoye clearly existed throughout the Holocene, because 100 cm sediment cores taken from the lake provide a picture of the climatic conditions through the early Holocene to the present day (Verkulich et al. 2002). Three clear units were evident in the cores. The oldest (unit 3) was laid down in the early Holocene and is characterized by low organic carbon and some freshwater diatom species. A prevalence of terragenic sediment suggests input from floating glacier ice masses or blanketing snow. Unit 2, covering the period c. 9000–5500 yrs B.P., had a decreased grain size and a predominance of organic matter with benthic mosses and low diatom numbers and diversity. This suggests reduced nutrient concentrations and limited light climate, possibly due to perennial ice cover. Some marine diatoms suggest a marine influence in parts of the lake. The upper section of the cores (unit 1) covering the period 5500–500 yrs B.P. is highly biogenic, with algal mats and abundant diatoms particularly in the upper strata, indicating conditions resembling the present-day environment. The lower section of this unit had decreased diatoms and a greater sand fraction with some isolated thin clastic layers reflecting mixing, cooling, and enrichment of the lake water by terragenous material, which might indicate an increase in meltwater inflow indicative of climate warming. The indications are of a relatively warm climate between 5500 and 2000 yrs B.P. with a short phase of cooling at around 2000 yrs B.P., followed by a subsequent warming (Verkulich et al. 2002). The climatic conditions indicated in the core sequence are shown in Figure 2.4.

The Amery Oasis in eastern Antarctica is located 200 km inland of the seaward edge of the Amery Ice Shelf (see Figure 1.1). The oasis is bordered by the Charybdis Glacier to the north, the ice sheet to the west and southwest, and the drainage system of the Amery Ice Shelf/Lambert Glacier to the south (Figure 2.5). Lake Terrasovoje lies in a glacial basin and is fed by surrounding snow banks (lake 13, Table 2.1). Its outlet in the northern corner feeds into an embayment onto which the lateral tongue of the Charybdis Glacier flows (Figure 2.5). A 552 cm sediment core taken from the deepest part of the lake covers a period extending back to 12,374 ± 426 yrs B.P., which equates to the late Pleistocene and the Holocene. Prior to 12,400 yrs B.P. the basin

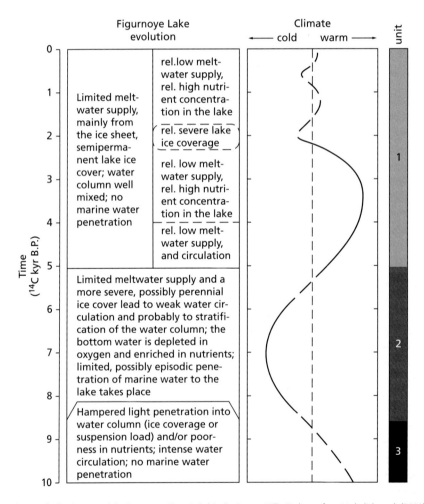

Figure 2.4 The evolution of Lake Figurnoye (also known as Algae Lake) in the Bunger Hills. Redrawn from Verkulich et al. (2002).

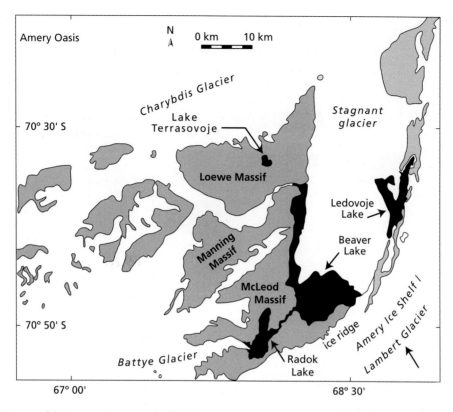

Figure 2.5 A map of the Amery Oasis. For more detail of Lake Radock, Lake Terrasovoje, and epishelf Beaver Lake; see Table 2.1 and Chapter 4.

is documented by clastic sediment after which biogenic soft matter was laid down following the development of the lake (Wagner et al. 2004). The early stages of lake development during the late Pleistocene are characterized by low amounts of organic matter and no diatoms (Figure 2.6). This record implies low productivity, probably related to perennial ice cover and cold climatic conditions. In the early Holocene diatoms make an appearance and there are increases in total carbon and nitrogen. Diatoms increase as do total carbon and nitrogen (Figure 2.6), indicating warmer temperatures between 8600 and 8400 yrs B.P. with periods of less ice cover on the lake. In the middle Holocene (6700–3600 yrs B.P.) there is a period of few or no diatoms and reduced total carbon and nitrogen deposition, suggesting a period of cooling and harsher environmental conditions. In the late Holocene (3600–0 yrs B.P.) the sediment record shows changing contents of organic matter and shifts in the carbon/nitrogen and carbon/sulphur ratios, as well as large variations in diatom abundance. Increased abundance of sulphur in sediments is indicative of reduced conditions in lake bottom waters. The data and diatom record suggest two periods of warming in the late Holocene. Warming periods in the late Holocene are evident from other studies in eastern Antarctica (Wagner et al. 2004).

Lake Radok in the Amery Oasis (Figure 2.5, lake 2; Table 2.1) lies 7 m above tidal, epishelf Beaver Lake in an overdeepened basin with cliffs rising 400 m above the lake. It is the deepest surface freshwater lake in Antarctica. The lake is fed by the Battye Glacier, which projects over 2 km into the lake. The outflow through the Pagadroma Gorge extends 4.5 km onto the floor of epishelf Beaver Lake. The overdeepening of the Radok basin is likely to have occurred along a tectonically weak zone (the Amery Fault), which follows the axis of the lake (Wagner and Cremer 2006). Lake Radok was excavated by

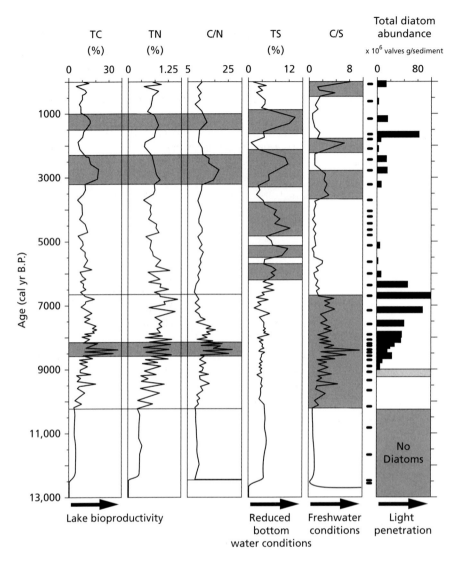

Figure 2.6 The evolution of Lake Terrasovoje, Amery Oasis. TC : total carbon, TN : total nitrogen, C/N : carbon to nitrogen ratio, TS : total sulphur, C/S : carbon to sulphur ratio. Redrawn from Wagner et al. (2004).

the palaeo-Nemesis Glacier and its ancestral equivalents flowing southwards during the late Cenozoic (Adamson et al. 1997). While the sediments provide a record of changes in allochthonous sediment input and productivity, it is not possible to date them because of finely disseminated coal fragments derived from coal-bearing Permo-Triassic beds that outcrop to the east of the lake. This material would have given an inaccurate age. Other dating methods were also discounted. The sedimentary record shows glaciation and deglaciation of the lake basin and transition to the present limnetic setting (Wagner and Cremer 2006).

The McMurdo Dry Valleys (Figure 2.7) lie on the western side of the TransAntarctic Mountains and contain a long lacustrine sediment record going back at least 300,000 years. The laying down of lacustrine sediment is cyclical and occurred over 100,000-year periods. During the last cycle, relatively small lakes existed adjacent to the East Antarctic

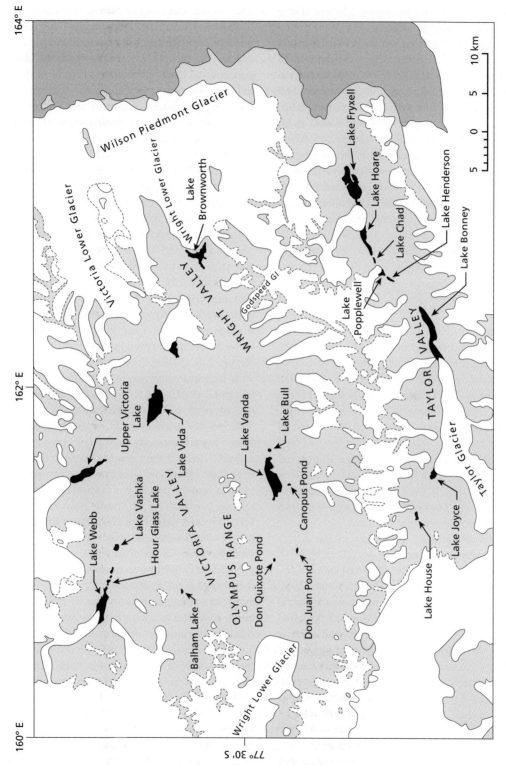

Figure 2.7 A map of the McMurdo Dry Valleys showing major lakes. For more details of lakes see Table 2.1 (freshwater lakes) and Table 3.1 (saline lakes).

Ice Sheet and were fed by alpine glaciers. These lakes then gave way to large proglacial lakes that were dammed by the Ross Sea Ice Sheet. The same pattern probably occurred during previous 100,000-year cycles (Hendy 2000). The expansion of the West Antarctic Ice Sheet near the continental shelf margin during the last glaciation between 20,000 and 10,000 yrs B.P. caused McMurdo Sound to fill with grounded ice, forming the Ross Sea Ice Sheet. As the Ross Ice Sheet expanded it forced tongues of ice into the Dry Valleys. On the landward side of these tongues meltwater was impounded forming proglacial lakes, including Glacial Lake Trowbridge in the Miers Valley and Glacial Lake Washburn in the Taylor Valley.

Lake Miers (lake 16, Table 2.1) in the Miers Valley is the remains of Glacial Lake Trowbridge that occupied both basins in the valley from at least 23,000–10,000 yrs B.P. The lake was 80 m deep and extended to more than 156 m above present sea level at its maximum. The U-shaped valley we see today is covered by extensive lacustrine deposits and glacial drift, that are the remnants of Glacial Lake Trowbridge. At around 18,000 yrs B.P. Lake Trowbridge started to recede. It is thought that the western basin that contains the present-day Lake Miers was isolated from the eastern basin at around 15,000 yrs B.P. The current ice cover of the lake contains glacial debris from this time. Lake Miers is fed by meltwater from the Miers and Adams Glaciers. The lake's outflow is the Miers River, which flows most years. The Miers River feeds into the Alph River which terminates in the sea.

Lake Hoare (lake 11, Table 2.1) in the Taylor Valley, is separated from Lake Fryxell (a brackish lake—see Chapter 3) by the Canada Glacier. Lake Hoare is mainly fed by meltwater from the Canada Glacier, but also receives meltwater from the Suess Glacier which overflows through Lake Chad further up the valley. Lake Hoare has no outflow and loses water through sublimation of its ice cover and by evaporation from the open water moats that form around the perennial ice cover each summer. The lake has two lobes, the deeper of which is dammed by the Canada Glacier. It is suggested that the lake would drain into Lake Fryxell if the Canada Glacier were to retreat. Equally a rise in the level of the lake of only a few metres would allow it to flow around the Canada Glacier into Lake Fryxell. Abandoned stream channels suggest that this may have occurred in the past (Hendy 2000).

Analysis of a sediment core from the deepest part of Lake Hoare allowed the history of the lake to be determined. Inconclusive radiocarbon dating prevented the establishment of reliable age depth models, however cross correlations with sediment records from Lake Fryxell indicated that the lowest two units in the Lake Hoare sedimentary record were probably deposited during the final stages in the history of proglacial Lake Washburn. Five distinct units were evident in the sedimentary record. The basal unit is younger than 23,000 ^{14}C yrs B.P., when Lake Washburn occupied the Taylor Valley. The lower largely silty sediments (unit I) were probably deposited in the late Holocene at around 9500 yrs B.P. when the Ross Sea Ices Sheet had largely retreated (Figure 2.8). Unit II is dominated by silt, implying relatively calm sedimentary conditions, during which time Lake Washburn receded and Lakes Hoare and Fryxell became individual basins (Figure 2.8). During the deposition of unit III there were significant changes in the deposition environment of Lake Hoare. A lowering of the lake levels due to evaporation probably occurred at around 4000 yrs B.P. leaving small remnant lakes in the Lake Hoare basins. The initial lowering relates to a slight retreat of the Canada Glacier and drainage from Lake Hoare into Lake Frxyell. A complete desiccation and the presence of subaerial conditions in the lake basin during the middle Holocene (5000 yrs B.P.) are evidenced by a dominance of sand and gravel and low amounts of silt (Figure 2.8). At this time the Ross Ice Sheet had retreated, there were open water conditions in the Ross Sea and warmer temperatures (6000–4000 yrs B.P.). Unit IV has a grain size comparable with that of the sediments in the present-day ice cover of the lake, suggesting conditions similar to the present. At this time the lake was probably refilled with freshwater following the period of low lake levels. Radiocarbon dating of the microbial mats that alternate with coarse grained sand layers indicates continuous sedimentation since *c.* 2500 yrs B.P. (Wagner et al. 2011). There is very weak evidence for a desiccation event at around 1200 yrs B.P. which has been suggested by stable isotope (δ^{18}O and δD) analysis of waters

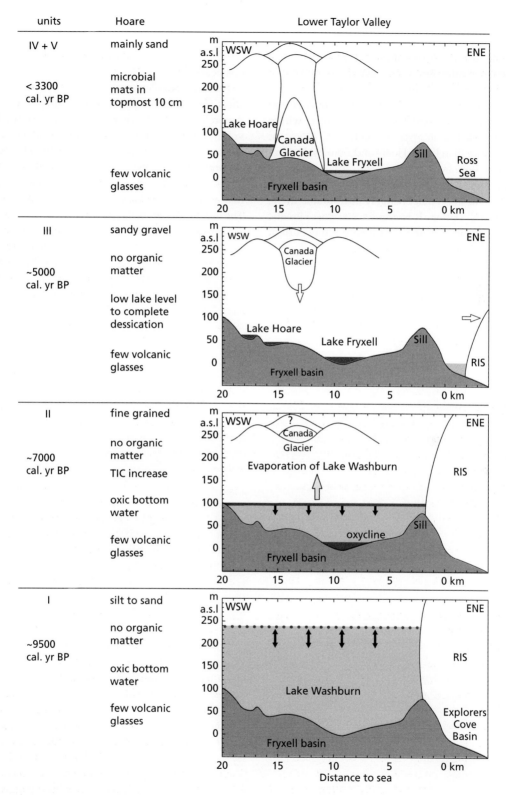

Figure 2.8 A pictorial evolution of Lake Hoare in the Taylor Valley, McMurdo Dry Valleys based on sediment cores. RIS – Ross Ice Shelf. From Wagner et al. (2011) with the permission of Cambridge University Press.

from the lake water column (Lyons et al. 1998a), at a time when other Dry Valley lakes had dried down to hypersaline pools.

The lakes of the Taylor Valley experience different climatic conditions. For example during winter Lake Frxyell has an approximately 5 °C lower air temperature than Lakes Hoare and Bonney. In summer Lake Bonney is approximately 1°C warmer on average than Lakes Fryxell and Hoare. Thus there is a spatial variation that is largely attributable to the topographic setting. It is argued that past climatic conditions have strongly influenced the current ecological conditions of the Taylor Valley lakes (Lyons et al. 2000).

The Antarctic Peninsula (Maritime Antarctic) ice sheet (Figure 2.9) is part of the marine-based West Antarctic Ice Sheet, where the major control on ice volume is sea level. There is geomorphological evidence of a more extensive ice cover in the Maritime Antarctic than is seen today from ice-abraded ridge crests at high altitudes, striated bedrock on islands that are currently ice-free, erratics, and glacial tills. The exact onset of deglaciation is not known but the supposition is that it was triggered by sea level rise

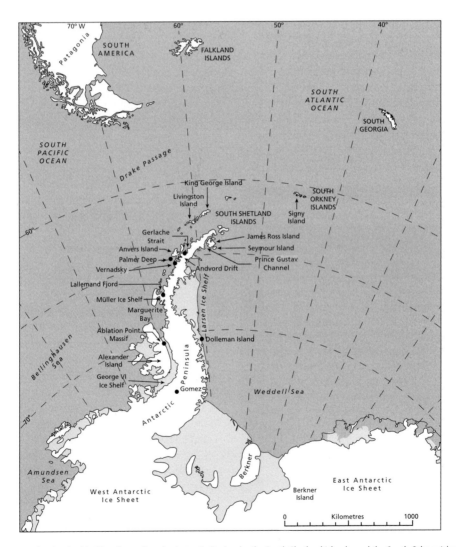

Figure 2.9 A map showing the Maritime Antarctic—the Antarctic Peninsula, the South Shetland Islands, and the South Orkney Islands.

some time after 14,000 yrs B.P. Deglaciation of Alexander Island occurred sometime before 6000 yrs B.P. and there is evidence of an advance and expansion of the George VI ice shelf east of the island (Ingólfsson et al. 1998 and references therein). Two warm events are clearly evident in the palaeoclimate record of the Antarctic Peninsula, one during the early Holocene (c. 11,000– 9500 yrs B.P.) and a second during the mid Holocene (c. 4500–2800 yrs B.P.). The speculation is that these warm periods may have occurred because of relatively abrupt shifts in the position of south westerlies superimposed on slower solar insolation changes (Bentley et al. 2009).

Signy Island in the South Orkneys has been a site of considerable lake research by the British Antarctic Survey over many years (Figure 2.10). The lakes on this island are relatively young. At the end of the LGM most of Signy Island was overridden by the South Orkney ice cap which disappeared sometime before 7000 yrs B.P. Lakes then formed in glacial scoured hollows under cool conditions at around 7000–5900 yrs B.P. Palaeolimnological analysis

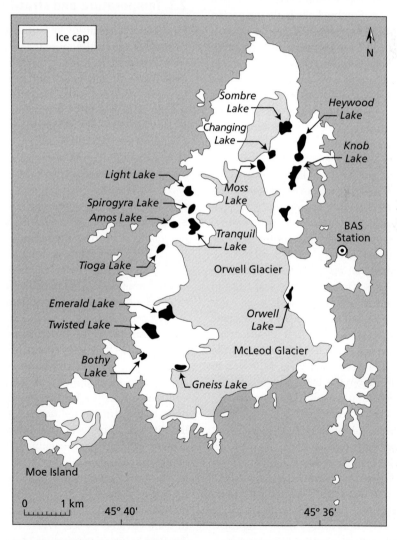

Figure 2.10 A map of Signy Island showing the lakes (see Table 2.1 for more detail), the ice cap, and the position of the British Antarctic Survey (BAS) Station.

of cores from Heywood Lake and Sombre Lake indicate a warm period, a late Holocene climate optimum at around 3300–1200 yrs B.P., followed by cooling, instability, and recent environmental change from c. 1300 yrs B.P. to the present (Jones et al. 2000). Evidence of a decoupling between the climate of the West Antarctic Peninsula and the South Orkney Islands appears to have occurred in the last 1900 years, attributed to change in the regional atmospheric–oceanographic circulation patterns (Noon et al. 2003).

As indicated in Section 2.1, Heywood Lake has undergone recent eutrophication related to fur seal wallows on its shore. Sombre Lake is also impacted by fur seals, but to a lesser extent. Commercial sealing in the Southern Ocean between the nineteenth and early twentieth centuries caused a decrease in the populations of various seal species, including the fur seal (*Arctocephalus gazella*). This species has recovered very considerably in the sub-Antarctic (e.g. South Georgia) and Maritime Antarctic. On Signy Island only a few dozen fur seals were recorded each year during the 1950s and 1960s, but by the 1980s the population had increased by 728% and in the 1990s had risen to over 20,000 (Smith 1988; Hodgson et al. 1998). The presence of such large populations has impacted on both terrestrial and lacustrine ecosystems. Analysis of fur seal hair abundance in sediment cores from Sombre Lake (Figure 2.10) shows that fur seals have visited the island for 6000 years. The number of seal hairs in younger sediments indicates that seals were present prior to the 1820s and then appeared to decline. There was a brief increase in the first decade of the twentieth century, but from around 1911 until the late 1970s virtually no sea hairs were deposited in the lake sediments. From the 1980s onwards the increase in hair deposition mirrored documented observations on the seal population.

The South Shetland Islands were deglaciated at around 6000–5000 yrs B.P. Thus the lakes of the region postdate this event. Analysis of sediments from Lake Åso on the Byers Peninsula, Livingston Island reveals that this lake was formed around 5000 years ago. Today it is a shallow water body with a depth of 1 m or less. Throughout its history there is repeated evidence of tephra fallouts. Tephra is material produced by volcanic eruptions. As is seen elsewhere in the Maritime Antarctic there were clear climatic changes throughout the history of the lake; a warm period starting at around 4000 yrs B.P. with an optimum at c. 3000 yrs B.P., followed by drier colder conditions at 2500 yrs B.P. (Björck et al. 1993). Midge Lake on the Byers Peninsula was formed around 4000 years ago. The sediment record suggests milder climatic conditions between 3200 and 2700 yrs B.P. with colder drier conditions between 1500 and 500 yrs B.P., comparable with the palaeoclimatic record seen in Lake Åso (Björck et al. 1991).

2.3 Temperature and stratification

Freshwater lakes at lower latitudes that are of sufficient depth (usually more than 6 m) undergo seasonal thermal stratification. Water has its greatest density at 4 °C and above, and below this temperature its density decreases. This fact is responsible for a wide range of different patterns of stratification worldwide. By way of illustration let us imagine a temperate latitude water body where there is complete water circulation or mixis. As solar radiation increases in the spring the upper waters warm and become less dense than the underlying lower waters. As this heating of the upper waters progresses a marked transition between the upper (epilimnion) and lower waters (hypolimnion) develops and the lake becomes thermally stratified with warmer waters above and colder waters below. The transitional boundary zone between the two layers is called the thermocline. In temperate regions lakes are typically dimictic, meaning that they are thermally stratified twice a year, separated by periods of mixis. Thus they have complete circulation in the spring when the water rises above 4 °C. They are stratified in the summer with temperatures in the epilimnion being higher than those in the hypolimnion. In the autumn the upper waters cool so that the differential between the epilimnion and hypolimnion decreases and eventually the stratification breaks down and circulation or the overturn occurs. In the winter inverse stratification develops when the lower waters in the hypolimnion are warmer (around 4 °C) than the upper epilimnion waters that are covered by ice. Where ice cover does not develop, winter stratification does not occur and the water column remains mixed throughout the winter months. In this case the lakes are described as monomictic.

Continental Antarctic lakes experience continuous low temperatures. For example Lake Hoare (Dry Valleys) has a maximum summer temperature of around 3.5 °C at 10.5 m and Lake Miers displays temperatures of 5.5 °C at the bottom of the water column (Spigel and Priscu 1998). Permanently ice-covered freshwater lakes are usually amictic, that is, having no thermal stratification (Wetzel 2001). Crooked Lake water column temperatures ranged between 0.2 °C and 2.2 °C and in Lake Druzhby, between 0.4 °C and 1.6 °C over a year (Laybourn-Parry et al. 1992, 2004). Under ice cover, Crooked Lake showed some indication of stratification with cooler water underlying the ice and slightly warmer water at depth. During the ice-free period the water column was mixed (Figure 2.11). In the example shown in Figure 2.11 the lake became completely ice-free in the summer, but in many years it retains varying degrees of ice cover, limiting wind-driven turbulence, which will affect the water column temperatures. Thus there is probably significant inter-annual variation in the temperature pattern seen in the water column. Lake Verkhneye (Schirmacher Oasis) becomes ice-free in the summer and is a relatively shallow lake compared with Crooked Lake, being only 9–10 m deep. Inverse thermal stratification develops under ice with temperatures up to 5 °C at the bottom of the water column. During the summer ice-free period, between January and February, the water column is mixed, with temperatures reaching 7 °C–8 °C (Kaup 1994). These lakes are cold monomictic lakes typical of high latitude or altitude, having one period of circulation in the summer (Wetzel 2001). This pattern of winter inverse stratification and summer mixis is also seen in Maritime Antarctic lakes that have been subject to year-long investigations, for example Sombre Lake and Tranquil Lake (Signy Island) (Hawes 1990a; Butler et al. 2000). Shallow lakes in the Maritime Antarctic can reach summer temperatures of between 4 °C and 7 °C when ice cover is absent (Izaguirre et al. 1993; Mataloni et al. 1998; Unrein and Vinocur 1999). Such elevated temperatures can have a marked effect on productivity. These shallow, more northerly water bodies tend to have higher chlorophyll *a* concentrations than both shallow and deep continental lakes (see Table 2.2).

Lake Untersee, which is permanently ice covered, has a complex stratification (lake 4, Table 2.1). In its shallower reaches the water column is homogeneous with respect to temperature. It has two deep basins, one of which is stratified with anoxic lower waters, while the other is oxygenated to 160 m (Andersen et al. 2011). The anoxic basin, lying in a trough below 94 m, exhibits both thermal and chemical stratification. Below 80 m the water column is anoxic, with increased conductivity and high concentrations of ammonium and orthophosphate and the presence of hydrogen sulphide. The upper 80 m of the water column was inversely thermally stratified at the time of the investigation in summer 1991/1992. The upper 40 m of the water column had a temperature of 0.4 °C, with temperatures of up to 4.5 °C below a sharp thermocline at 40–50 m (Figure 2.12) (Wand et al. 1997). This is an unusual type of stratification, but it is seen in large lakes elsewhere, for example in tropical Lake Tanganyika, the world's second largest freshwater lake. Here there is a permanently anoxic hydrogen sulphide-containing monomolimnion below a depth of 100–200 m. The upper oxygenated waters develop seasonal thermal stratification with a thermocline at a depth of 25–75 m and mixis during the windy season between May and September. In Lake

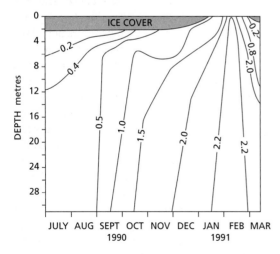

Figure 2.11 Annual temperature profile (°C) in Crooked Lake, Vestfold Hills during 1990 and 1991. From Laybourn-Parry et al. (1992) with the permission of Springer.

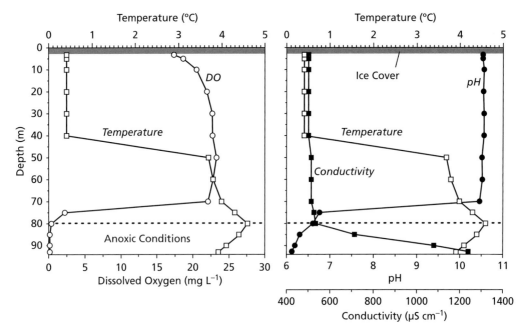

Figure 2.12 Stratification structure of a deep trough in Lake Unterzee, from Wand et al. (1997). Left-hand panel temperature and dissolved oxygen (DO), right-hand panel temperature, conductivity, and pH. With the permission of Cambridge University Press.

Unterzee the thermal stratification is probably continuous because of perennial ice cover.

2.4 Water chemistry

As outlined in Sections 2.1 and 2.2, Antarctic freshwater lakes are typically ultra-oligotrophic to oligotrophic. They have low concentrations of major inorganic nutrients (nitrogen and phosphorus) needed to support primary production and bacterial production (Table 2.3). It is important to remember that concentrations of ammonium (NH_4–N), nitrate (NO_3–N), and soluble reactive phosphorus (PO_4–P) in the water at any one time give no indication of fluxes or turnover rates. However, they can tell us if a particular nutrient might be limiting to production. Phosphorus is often regarded as the major limiting nutrient in freshwater lakes and as Table 2.3 shows, PO_4–P concentrations are low, some of the values, for example for Lake Nella and Progress Lake, are at the limit of detection. Phosphorus is generally the scarcest element in the Earth's crust among those needed to support plant growth. While present, soluble reactive phosphorus (PO_4–P) is readily precipitated as complexes of iron, aluminium, and other metals, or is scavenged to sediment by adsorption. There is no reservoir of gaseous phosphorus compounds available in the atmosphere, as there is for nitrogen (Moss 1980). For more detail of nitrogen and phosphorus dynamics see Chapter 1, Section 1.6.3.

Among the lakes listed in Table 2.3 Boeckella Lake and No Worries Lake have been subject to human impact, as they are used as a water supply to research stations (see Section 2.1). Boeckella Lake has elevated concentrations of dissolved inorganic nitrogen and phosphorus species compared with other maritime lakes and No Worries Lake has elevated ammonium concentrations compared with other Larsemann Hills lakes. Heywood Lake is undergoing eutrophication as a result of nutrient inputs from a fur seal wallow on its shores and is suffering increased nutrient loading, which has resulted in enhanced productivity (see Tables 2.2 and 2.3).

One way of determining whether a particular nutrient is limiting to primary production is to undertake incubations with enhanced levels of that

Table 2.3 Nutrients and organic carbon in euphotic zone. ND not detectable. * Over an annual cycle, other data are summer only and in some cases from a single determination. Lake numbers correspond with the numbers in Table 2.1.

Lake	NH_4–N $\mu M\ l^{-1}$	NO_3–N $\mu M\ l^{-1}$	PO_4–P $\mu M\ l^{-1}$	DOC $mg\ l^{-1}$	POC $mg\ l^{-1}$
11. Hoare (Dry Valleys)[1]	0.07 ± 0.1	2.63 ± 3.4	0.02 ± 0.02	1.9 ± 0.5	0.13 ± 0.05
16. Miers (Dry Valleys)[1]	0.14 ± 0.2	ND	0.05 ± 0.01	1.2 ± 0.5	0.19 ± 0.08
6. *Crooked (Vestfold Hills)[2,3]	0.44 ± 0.16	0.42 ± 0.30	0.07 ± 0.02	1.6 ± 0.5	–
8. *Druzhby (Vestfold Hills)[2,3]	0.34 ± 0.14	0.45 ± 0.31	0.13 ± 0.05	1.5 ± 0.6	–
12. Bisernoye (Vestfold Hills)[4]	1.24	–	0.05	0.4	–
9. Zvezda (Vestfold Hills)[4]	0.34	–	0.03	0.5	–
22. Lichen (Vestfold Hills)[5]	0.71 ± 0.21	0.49 ± 0.09	0.52 ± 0.15	0.9 ± 0.17	–
29. Progress Lake (Larsemanns)[6]	0.47	0.17	0.003	0.51	0.94
26. Lake Nella (Larsemanns)[6]	0.09	0.15	0.004	1.28	0.51
52. No Worries (Larsemanns)[7,8]	1.9	0.49	0.03	ND	–
4. Lake Untersee (Donning Maud Land)[9]	ND	1.77	0.10	–	–
45. Lake 21 (Terra Nova Bay)[10]	0.43	0.12	0.20	–	–
34. Boeckella (Antarctic Pen.)[11]	42.4	15.1	3.55	–	–
71. Chico (Antarctic Pen.)[11]	2.49	4.49	0.82	–	–
51. Sombre (Signy Island)[12]	0.71	2.93	0.02	–	–
59. Moss Lake (Signy Island)[12]	0.31	1.79	0.02	–	–
42. Heywood (Signy Island)[12]	3.11	5.27	0.33	–	–
35. Midge (Livingston Island)[12]	0.36	0.03	0.02	–	–

1: Roberts et al. 2004a, 2: Laybourn-Parry et al. 2004, Henshaw and Laybourn-Parry 2002, 4: Säwström et al. 2007b, 5: Laybourn-Parry unpublished data, 6: Ellis-Evans et al. 1998, 7: Ellis-Evans et al. 1997, 8: Sabbe et al. 2004, 9: Wand et al. 1997, 10: Andreoli et al. 1992, 11: Unrein et al. 2005, 12: Jones et al. 1993.

nutrient and to compare the resulting rates of photosynthesis with unamended controls. This approach has been adopted for Lake Hoare where both ammonium (20 µM) and soluble reactive phosphorus (2 µM) were added together and singly to incubations at 5 m and 14 m in the water column (Dore and Priscu 2001). Both nitrogen and phosphorus additions together had nearly the same effect as phosphorus alone at 14 m, but not at 5 m. At 5 m there appeared to be a synergistic stimulatory effect of nitrogen plus phosphorus, but no effect when they were added singly. This possibly suggests a co-deficiency of both nutrients. A complex statistical analysis of a 15-year LTER data set for Lake Hoare revealed that rates of primary production are strongly correlated with PAR at 12–15 m and to soluble reactive phosphorus at 8–22 m (Herbei et al. 2010).

Concentrations of dissolved organic carbon (DOC) are generally low in Antarctic freshwater lakes (Table 2.3), they all lie below 2 mg l^{-1}; see Section 1.6.3. Saline Antarctic lakes have higher DOC concentrations (see Chapter 3). The DOC pool is exploited by the bacterioplankton as an energy source, and it is the lability of the DOC fractions that is important for bacterial utilization of DOC (see Section 1.8). Most investigations simply measure total DOC without determining its constituents. Analysis of the DOC pool from Crooked Lake and Lake Druzhby showed that bulk dissolved amino acids and bulk dissolved carbohydrates constituted 5%–25% and 5%–64% of the DOC pool respectively. Concentrations of inorganic nitrogen were not limiting, so the bacterioplankton were not dependent on amino acids for a nitrogen source. Low concentrations of dissolved carbohydrates relative to dissolved amino acids, suggest that the bacterioplankton were mainly exploiting the former (Laybourn-Parry et al. 2004).

In lakes worldwide, DOC is derived from autochthonous sources (i.e. biological activity within the water body) and allochthonous sources (input from the catchment). The latter are considered to be the major component of DOC in lakes. The range of DOC from a survey of 7514 lakes worldwide ranged from 0.1 mg l^{-1} to 332 mg l^{-1} (mean 7.58 ± 0.19 mg l^{-1}); 87% of the lakes had DOC concentrations between 1 and 20 mg l^{-1}, while 8.3% were less than 1 mg l^{-1} (Sobek et al. 2007). In contrast, Antarctic lakes typically receive little or no allochthonous input. They differ from lakes at lower latitudes and many Arctic lakes in having a DOC pool derived largely from autochthonous sources. Antarctic freshwater lakes in particular lie at the lower end of the global spectrum (Table 2.3). Alpine lakes above the tree line have some features in common with Antarctic lakes, in that they lie in catchments with poorly developed soils and sparse vegetation. Thus there is little allochthonous carbon input to the lakes. They have DOC concentrations that range from 0.24 to 2.48 mg l^{-1} (derived from 57 lakes). Interestingly, glacier-fed alpine lakes had lower DOC concentrations than lakes not fed by glaciers (Sommaruga et al. 1999). In a summer survey of freshwater lakes in the Larsemann Hills the majority had undetectable DOC (Sabbe et al. 2004). Clearly, single samplings only take a snapshot and DOC may have been at measurable concentrations at other times. Indeed the two lakes (Progress Lake and Lake Nella) had measurable DOC during the summer of 1993 (see Table 2.3). DOC has been shown to vary seasonally in Antarctic freshwater lakes (see Figure 2.21, Section 2.6.2).

2.5 The planktonic biota of freshwater lakes

2.5.1 Heterotrophic bacteria

Bacteria are a very important component of the heterotrophic community in lakes (see Section 1.8.1). There are a significant number of studies that have focused on the abundance and physiology of bacteria in Antarctic lakes across the continent and Maritime Antarctic (See Tables 2.2 and 2.4). However, we have little information on the taxonomic make-up of the planktonic communities in freshwater systems. At the time of writing there is considerably more information on saline lakes and on benthic microbial mats in both freshwater and saline lakes. The data we have indicate that the communities are made up of genera similar to those seen in freshwater lakes at lower latitudes (Pearce 2003; Pearce et al. 2003; Michaud et al. 2012). Based on a range of culture-dependent and culture-independent molecular techniques, communities have been shown to possess a predominance of β-Proteobacteria, the Flavobacteria–Cytophaga group (Phylum Bacteriodetes) Actinobacteria, with lesser taxa from the α-Proteobacteria and γ-Proteobacteria. The community composition changes over the summer in maritime lakes on the Peninsula, King George Island, and Signy Island, and this has been attributed to changes in physico/chemical parameters (Pearce 2005; Schiaffino et al. 2009). Many of the bacteria identified have their closest matches with sequences and classes found in polar regions and there is some indication of endemism. The trophic status of a water body affects its community structure. For example, a lake on Inexpressible Island in Northern Victoria Land near the Mario Zucchelli Station (previously known as the Italian Station at Terra Nova Bay) that suffers occasional marine incursions, differed from two other freshwater lakes in the region (Michaud et al. 2012). A range of lakes on the Peninsula and King George Island (South Shetlands) showed differences in diversity in relation to trophic status in lakes with chlorophyll *a* values that ranged from 0.2 to 55 μg l^{-1}, and on Signy Island lower diversity was seen in eutrophic Heywood Lake compared with Sombre Lake and Moss Lake (Pearce 2005; Schiaffino et al. 2009).

Interestingly some of the bacteria encountered in Antarctic lakes are not psychrophiles, that is, adapted to growing at low temperatures with an optimum growth rate below 15 °C, but are psychrotrophs. For example, *Flavobacterium hibernum*, a new species isolated from Crooked Lake (Vestfold Hills), had an optimum growth temperature of 26 °C and is consequently a psychrotroph (McCammon et al. 1998). Such bacteria are psychrotolerant, capable of growing and reproducing at low temperatures but with optimum growth temperatures well above those they encounter in polar waters. There is more information on the temperature

Table 2.4 Bacteria, heterotrophic flagellated protozoan (HNAN) and phototrophic flagellated protozoan (PNAN) concentrations in Antarctic freshwater lakes. * indicates a year-long study; all other data relate to summer only as indicated. Lake numbers correspond with Table 2.1.

Lake and notes	Bacteria $\times 10^8$ l^{-1}	HNAN $\times 10^4$ l^{-1}	PNAN $\times 10^4$ l^{-1}
a) 11. Lake Hoare (Dry Valleys)[1, 2] Summers 1996/1997; 1997/1998 Summer Nov 2007–Apr 2008	mean 11.8 2.7–8.4	mean 33.8 7.0–22.0	mean 56.5 75.0–225.5
16. Lake Miers (Dry Valleys)[1] Summers 1996/1997; 1997/1998	mean 2.57	mean 7.29	mean 25.9
6. Crooked Lake (Vestfold Hills)*[3] Dec 1992–Nov 1993	1.19–4.46	0–50.9	
8. Lake Druzhby (Vestfold Hills)*[4] Dec 1992–Dec 1993	0.75–2.5	0.9–14.0	1.0–9.5
Corner Lakes (Vestfold Hills)[5] Feb 1991 (one sampling)	3.6	1.15	1.33
12. Bizernoye (Vestfold Hills)[5] Feb 1991 (one sampling)	2.2	0.7	0.5
9. Zvezda (Vestfold Hills)[5] Feb 1991 (one sampling)	1.7	0.9	0.4
Nicholson (Vestfold Hills)[5] Feb 1991 (one sampling)	1.4	0.5	4.7
17. Pauk (Vestfold Hills)[5] Feb 1991 (one sampling)	1.4	0.5	2.6
22. Lichen Lake (Vestfold Hills)[6] Nov 1999–Feb 2000	0.85–4.56	13.9–20.7	1.5–36.7
21. Lake Nottingham (Vestfold Hills)[6] Nov 1999–Feb 2000	1.5–10.8	23.0–51.5	4.6–93.5
26. Lake Nella (Larsemann Hills)[7] Dec 1992–Feb 1993	mean 1.8	mean 57.0	mean 15.8
29. Progress Lake (Larsemann Hills)[7] Dec 1992–Feb 1993	mean 2.49	mean 57.0	mean 2.8
57. Discussion Lake (Larsemann Hills)[7] Dec 1992–Feb 1993	mean 3.7	mean 55.1	mean 1.0
32. Lake Heidi (Larsemann Hills)[7] Dec 1992–Feb 1993	mean 2.6	mean 183	not detectable
27. Kirisjes Pond (Larsemann Hills)[7] Dec 1992–Feb 1993	mean 2.9	mean 155	not detectable
52. No Worries Lake (Larsemann Hills)[8] Nov 1992–Jan 1993	10.5–23.0	19.4–150	12.6–216.0
53. Lake Limnopolar (Livingston Island)[9] summer 2001–2002	12.2–21.6		
35. Midge Lake (Livingston Island)[9] Summer 2001–2002	8.1–14.3		
56. Tranquil Lake (Signy Island)*[10] Dec 1994–Feb 1996	3.6–190	0–1600	0–1200
42. Heywood Lake (Signy Island)*[11] Dec 1994–Feb 1996	100–700	1.1–2350	0–4900
51. Sombre Lake (Signy Island)[12] Dec 1994–Jan 1995	2.0–31.8	1.0–174	40–950

1: Roberts et al. 2004a; 2: Roberts et al. 2004b; 3: Laybourn-Parry et al. 1995; 4: Laybourn-Parry and Bayliss 1996; 5: Laybourn-Parry and Marchant 1992; 6: Laybourn-Parry unpublished data; 7: Ellis-Evans et al. 1998; 8: Ellis-Evans et al. 1997; 9: Toro et al. 2007; 10: Butler et al. 2000; 11: Butler 1999; 12: Laybourn-Parry et al. 1996.

related growth physiology of bacteria from saline lakes. Of 13 strains isolated from Burton Lake in the Vestfold Hills, 12 were psychrophiles and one a psychrotroph (Nichols et al. 1993) while *Halomonas subglaciescola*, *Halomonas meridiana*, *Flavobacterium gondwanense*, and *Flavobacterium slaegens* isolated from Organic, Ekho, Fletcher, and Ace Lakes (Vestfold Hills) all had optimum growth temperatures above 18 °C, with *H. meridiana* able to grow between 0 °C and 45 °C (James et al. 1994).

The concentrations of heterotrophic bacteria reported in Antarctic lakes mostly pertain to the austral summer (Table 2.4), and where lakes are not suffering allochthonous inputs or being used as a station water supply they support cell densities that fit within the lower range reported for lower latitude oligotrophic lakes. Across five temperate latitude oligotrophic lakes, bacterial abundance ranged between 1.0 and 80×10^8 l^{-1} (Pick and Caron 1987; Vaqué and Pace 1992; Laybourn-Parry et al. 1994; Hofer and Sommaruga 2001; Lymer et al. 2008). In High Arctic lakes in Svalbard the range is 0.5–2.9×10^8 l^{-1}, which is comparable with the Antarctic (Säwström et al. 2007a). Numbers of bacteria remain high during the winter months. In two separate year-long studies of Crooked Lake (lake 6,

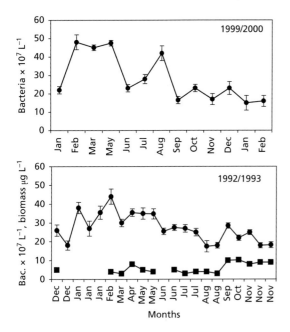

Figure 2.13 Bacterial abundances in Crooked Lake over two separate year-long studies: 1999/2000 and 1992/1993 (●), the latter also showing bacterial biomass (■). Data from Laybourn-Parry et al. (1995, 2004).

Table 2.1) highest concentrations of bacteria were recorded during autumn and winter (Figure 2.13), while in Lake Druzhby (lake 8, Table 2.1) highest abundance occurred in winter (Laybourn-Parry et al. 1995, 2004).

One factor that is rarely recorded is the mean cell volume (MCV) of the bacterial community. This is an important variable that can change in relation to the taxonomic make-up of the community and its physiological state. Where MCV is high, biomass may be high even if cell numbers are relatively low. Between December 1992 and November 1993 biomass in Crooked Lake varied between 0.2 and 10 μg C l^{-1} (Figure 2.13). While bacterial numbers were lower from September to November, MCV volume reached its maximum; consequently, biomass reached its highest level during spring and summer. In common with all lakes the water column of Antarctic lakes is characterized by microscopic aggregates often known as lake snow. In marine environments they are known as marine snow. These are small particles 20 μm or larger. Often aggregates are formed from mineral material to which bacteria attach. These bacteria exude a kind of biological glue that creates an extensive network of extracellular carbon around a particle. Other bacteria attach and they in turn attract the predators of bacteria, heterotrophic flagellates, and amoebae. In this way a microbial consortium is created. Bacteria associated with aggregates or microbial consortia often have larger mean cell volume and higher growth or production than freely floating bacteria. This was the case in both Lake Druzhby and Crooked Lake, and has been observed in both marine and freshwater systems at lower latitudes (Laybourn-Parry et al. 2004; Allredge and Silver 1988; Rogerson and Laybourn-Parry 1992; Grossart and Simon 1993).

2.5.2 Viruses

While viruses are now widely investigated in lower latitude lakes, there are considerably less data on Antarctic lakes (see Figure 1.25). There are few studies that cover an entire year or more than one sampling of a water body, so our picture is rather fragmentary. Exceptions are year-long studies of Crooked Lake and Lake Druzhby in the Vestfold Hills (Table 2.5). In these large ultra-oligotrophic systems highest virus abundance occurs in summer (Säwström et al. 2007a). As Table 2.5 shows virus numbers in Antarctic freshwaters are low. In lower latitude lakes virus concentrations are higher; for example in a Swedish oligotrophic lake concentrations were between 50 and 60 × 10^9 l^{-1} (Lymer et al. 2008) and in oligotrophic/mesotrophic Lake Pavin (France), viruses ranged between 800 and 5100 × 10^9 l^{-1} (Colombet et al. 2009). They are also higher in high Arctic lakes on Svalbard (78°56′N), where across three lakes virus abundances ranged between 4.3 and 28.9 × 10^9 viruses l^{-1} in summer (Säwström et al. 2007c). The highest reported values in Antarctica occur on Signy Island in two lakes (Heywood and Sombre) that are undergoing eutrophication as a result of nutrient inputs from fur seal wallows. As a general rule, the more productive lakes, as indicated by chlorophyll a concentration, support higher concentrations of viruses (Table 2.5). Most of the viruses appear to be parasites of bacteria and there is a good correlation between virus and bacterial concentrations in polar lakes, as is the case at lower

Table 2.5 Virus abundance and virus to bacterium ratio (VBR) in freshwater lakes. Chlorophyll *a* concentration is given as an indication of trophic status. *: 4.5–19.25 m in water column, +: February 1979–1981, §: February 1995/1996. Lake numbers correspond with Table 2.1.

Lake and location	Month(s) sampled	Viruses ×10^9 L^{-1}	VBR	Chloro *a* µg l^{-1}
17. Pauk (Vestfold Hills)[1]	Sept 2004	0.8	6.0	0.30
12. Bisernoye (Vestfold Hills)[1]	Sept 2004	1.7	1.2	0.43
22. Lichen (Vestfold Hills)[1,2]	Sept 2004	2.5	15.0	0.17
	Dec 1999	2.2	3.53	0.93
9. Zvezda (Vestfold Hills)[1]	Sept 2004	1.6	6.5	0.37
20. Watts (Vestfold Hills)[1]	Sept 2004	12.3	32.5	4.54
Nicholson (Vestfold Hills)[1]	Sept 2004		13.5	0.55
21. Nottingham (Vestfold Hills)[2]	Dec 1999	1.5	0.15	0.39
6. Crooked (Vestfold Hills)[3]	Dec 2003 to Nov 2004	0.16–0.92	1.2–7.0	0–0.33
8. Druzhby (Vestfold Hills)[3]	Dec 2003 to Nov 2004	0.30–1.56	1.4–8.4	0.01–0.46
11. Hoare (Taylor Valley)[4,6]	Dec/Jan 96/97.	10.9	8.5	0.2–6.0[5]
	Nov 1999	0.22–0.52*	0.6–2.3*	0.4–5.1*
51. Sombre (Signy Island)[7]	Feb 1999	12.0	2.4	3.0–4.0[8+]
42. Heywood (Signy Island)[7]	Feb 1999	27.4	3.6	11–22[9§]
56. Tranquil (Signy Island)[7]	Feb 1999	7.6	2.8	1.8–2.5[10§]

1: Säwström et al. 2007b; 2: Laybourn-Parry et al. 2001a; 3: Säwström et al. 2007d; 4: Kepner et al. 1998; 5: Roberts et al. 2004a; 6: Lisle and Priscu 2004; 7: Wilson et al. 2000; 8: Hawes 1985b; 9: Butler 1999; 10: Butler et al. 2000.

latitudes (Wommack and Colwell 2000; Säwström et al. 2008) (Figure 2.14).

Where detailed electron microscopy has been undertaken, a high morphological diversity is apparent suggesting a wide range of hosts in maritime lakes (Wilson et al. 2000). In the Vestfold Hills lakes, including both freshwater and brackish water bodies, a range of morphologies was observed including phages with short tails (Podoviridae), long tails, icosahedric viruses up to 300 nm, and star-like particles of around 80 nm (Laybourn-Parry et al. 2001a). In Lake Hoare large icosahedral viruses (50–175 nm) were found that are morphologically similar to double-stranded DNA viruses isolated from temperate environments that infect flagellates (Kepner et al. 1998). There are little genomic sequence data for viruses in Antarctic lakes; however, a detailed study of Lake Limnopolar (Livingston Island; lake 53, Table 2.1) revealed a high diversity with viral genotypes distributed across 12 virus families (López-Bueno et al. 2009). This exceeds viromes isolated from other aquatic environments, most of which contain only 3–6 viral families. There was a clear seasonal succession in the composition

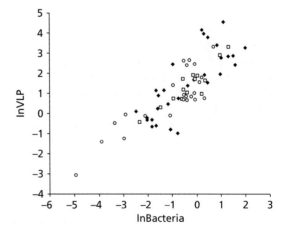

Figure 2.14 Correlation of virus abundance with bacterial abundance in 68 Antarctic, Arctic, and sub-Arctic water bodies. Antarctic lakes: filled diamonds, Arctic waters: open circles, sub-Arctic lakes: open squares. From Säwström et al. (2008) with the permission of Springer.

of the virus community. In the spring, while ice cover was intact, the virome was dominated mainly by ssDNA viruses and secondarily by dsDNA Caudovirales. When the ice cover thawed and broke out in summer, the viral assemblage was dominated by dsDNA viruses mainly of the family Phycodnaviridae with some Caudovirales and Mimiviridae. What is particularly interesting is that, in contrast with most other aquatic environments, Lake Limnopolar is dominated by viruses known to infect eukaryotes, rather than bacteria.

Virus to bacterium ratios (VBR) in Antarctic lakes range between 0.15 and 32.5 (Table 2.5). The range reported for freshwaters at lower latitudes is between 4.9 and 77.5 (mean 20), and in polar and sub-polar waters, including cryoconite holes on glaciers, it lies between 1 and 34 (Maranger and Bird 1995; Säwström et al. 2008). In marine planktonic environments the VBR range is 0.38–53.8 (mean 10). The differences seen in VBR are attributed to the increased dependence of freshwater bacteria on allochthonous inputs of carbon and nutrients and higher relative contributions of carbon substrates from cyanobacteria in freshwaters (Weinbauer 2004). Continental Antarctic lakes do not normally receive allochthonous inputs and are systems that function on autochthonous carbon production, which may in part explain the low VBR seen in these systems.

The lysogenic cycle (see Section 1.8.2) was only detected in Crooked Lake and Lake Druzhby in August, September, and December when lysogenic bacteria contributed between 18.2% and 73.0% of the community (Säwström et al. 2007d). For most of the year the lytic cycle appeared to predominate, suggesting that suitable actively growing hosts were available, even in winter. Bacterial production continues during winter in these lakes, despite low temperatures (Laybourn-Parry et al. 2004). The presence of lysogenic bacteria was determined in Lake Hoare in November 1999 and ranged between 2% and 18.6% of the community, varying with depth in the water column, with the highest incidence at 14 and 19.5 m (Lisle and Priscu 2004).

When a bacterial cell is lysed as a result of viral infection its viral load is dispersed into the water where the viruses will encounter and infect new hosts. The number of viruses released from a bacterial cell is called the burst size and is usually determined by counting viruses within bacterial cells that are close to lysing from electron micrographs (see Figure 1.25b). Burst sizes in Antarctic freshwater viruses are low compared with those seen in lower latitude lakes. The mean burst size across 10 temperate latitude lakes is 26, whereas in Crooked Lake and Lake Druzhby it is 4 ± 0.1. A similar low value is seen in cryoconite holes on Arctic glaciers (3 ± 0.2) (Säwström et al. 2007a). These low burst sizes reflect the low productivity of bacteria in cold Antarctic waters. On infection the bacterial cell becomes a virus producing factory and its ability to produce viruses is clearly controlled by temperature and the availability of carbon substrate and nutrients. The factors that control bacterial growth control the production of viruses.

While burst size is low, the number of bacterial cells infected in Antarctic freshwaters is high compared with lower latitude lakes. Across 10 temperate lakes the frequency of visibly infected bacteria (FVIB%) gave an average of 2.2%. In Crooked Lake and Lake Druzhby it was 34.2% ± 3.2% and 22.7% ± 2.0% respectively. In Arctic cryoconite holes the figure was 11.2% ± 3.1%. The data are limited, but it appears that in extreme polar aquatic habitats viral dynamics are characterized by high infection rates and low burst sizes (Säwström et al. 2007a). As we shall see in Section 2.6.2, high infection rates result in viruses playing an important role in carbon cycling in Antarctic lakes.

2.5.3 Protozoa

Ciliated protozoans

Planktonic ciliated protozoa have been reasonably well studied across the Antarctic continent and the Maritime Antarctic (Table 2.6; Figures 1.27 and 1.28). The diversity of Antarctic ciliated protozoan communities is much lower than is seen in lower latitude lakes and the Arctic (Petz et al. 2007). Given the extreme conditions of the Dry Valleys lakes it is perhaps surprising that Lake Hoare has a high diversity compared with other continental lakes. Within the Dry Valleys Lake Hoare has a greater diversity and abundance of ciliates compared with Lake Miers. During the summers of 1996–1997 and 1997–1998 Lake Hoare supported between 330 and 790 cells l^{-1}, while Lake Miers had a mean ciliate concentration of 38 ± 27. This difference reflects the fact that Lake Hoare is the more productive of the

Table 2.6 Planktonic ciliated protozoa in Antarctic freshwater lakes. Vestfold Hills lakes include Crooked Lake, Lake Druzhby, Lichen Lake and two unnamed lakes close to the ice sheet; Larsemann Hills lakes include lakes 26, 27, 29, 32, 38, and 57 listed in Table 2.1; Signy Island lakes are Heywood Lake and Tranquil Lake; Terra Nova Bay includes a range of lakes and ponds; South Shetland Islands includes 10 lakes on Livingston Island and Deception Island.

GENUS/SPECIES	Lake Hoare	Lake Miers	Vestfold Hills	Larsemann Hills	Terra Nova Bay	South Shetlands Is.	Signy Island
Askenasia	1, 2	1	4	5			
Aspidisca cicada						6	
Blepharisma	1, 2	1				6	
Chilodonella algivora	2						
Chilodonella sp.						6	
Cinetochilum margaritaceum	1				6	6	7
Cyclidium glaucoma					6	6	7
Cyclidium	2						7
Didnium sp.	1						7
Enchelys mutans							
Euplotes spp.	1, 2	1					7
Frontonia acuminata						6	
Frontonia angusta					6		
Halteria	1	1			6	6	7
Holophyra sp.				5			7
Mesodinium	1						
Monodinium (large sp.)	1, 2	1	4	5			7
Monodinium (small sp.)	1, 2	1					
Plagiocampa	1, 2						
Spathidium	1						7
Sphaerophyra	1, 2					6	7
Strombidium	1, 2		3, 4	5	6		
Urotricha	1, 2					6	
Vorticella mayeri	1, 2						
Vorticella sp.	1, 2	1			6	6	7
Unidentified scuticociliates	1		4	5			7
Unidentified suctorian							7

1: Roberts et al. 2004a; 2: Kepner et al. 1999; 3: Laybourn-Parry et al. 1992, 4: Laybourn-Parry unpublished data; 5: Ellis-Evans et al. 1998; 6: Petz et al. 2007; 7: Butler 1999 and Butler et al. 2000.

two lakes, as indicated by chlorophyll *a* concentrations (see Table 2.2, lakes 11 and 16) (Roberts et al. 2004a). However, there are significant inter-annual variations in the abundance of bacteria and flagellated protozoa in the Dry Valley lakes and lakes elsewhere in Antarctica, which are apparent from the long-term databases. This is also likely to be the case for ciliates which have been less well studied in Lakes Hoare and Miers. During an extended season from mid-November until the end of March 2007–2008, ciliate abundances in Lake Hoare ranged between 44 and 265 cells l^{-1} with the highest number recorded in March (Thurman et al. 2012). These values are lower than occurred in the summers of

1996–1997 and 1997–1998, although the species diversity was the same.

The freshwater lakes of the Vestfold Hills are ultra-oligotrophic and the two lakes that have been studied in detail, Crooked Lake and Lake Druzhby, are among the largest surface lakes on the continent (Table 2.1). Ciliate species diversity is low in these systems as are abundances. Over an annual cycle ciliate abundances ranged from 5 to 143 l^{-1} in Lake Druzhby (Laybourn-Parry and Bayliss 1996) and from 60 to 490 l^{-1} in Crooked Lake (Laybourn-Parry et al. 1992). The Maritime Antarctic has reasonably high ciliate diversity and much higher abundance than continental lakes. In Heywood Lake (Signy Island) ciliate abundances reached a maximum of >40,000 l^{-1} in December, declining to low numbers during the winter, while in Tranquil Lake numbers peaked at 11,000 l^{-1} in the summer to undetectable in the winter (Butler 1999; Butler et al. 2000).

A number of ciliate species found in the plankton are mixotrophic either through a relationship with endosymbiotic zoochlorellae or by the sequestration of the plastids of phytoflagellate prey (see Section 1.8.3 and Figure 1.28). In Lake Hoare some of the *Euplotes* specimens contained zoochlorellae. Some *Strombidium* species that are common in the lakes of the Vestfold and Larsemann Hills, sequester the plastids of their prey. *Plagiocampa* is a common species in Lake Hoare, where it feeds on cryptophytes that clearly autofluorescence within the ciliate cells for a number of days without being digested. It is possible that *Plagiocampa* retains the ingested cryptophytes for some time without digesting them, exploiting their photosynthetic capacity in a form of temporary mixotrophy (Roberts and Laybourn-Parry 1999).

The differences in ciliate diversity across Antarctica can be attributed to a number of factors. Firstly, the trophic status of the lake. Lake Hoare and the Maritime lakes are more productive than ultra-oligotrophic continental lakes. Secondly, the Antarctic continent is isolated and consequently colonization potential is limited. Propagules can be brought to the continent by meteorological events on air masses (Marshall 1996) and by birds. Time for colonization is an important factor. The lakes of the Dry Valley are relatively old compared with those of the Vestfold Hills and the Maritime Antarctic.

Lake Hoare was formed from Glacial Lake Washburn at least 23,000 years ago. During its history it underwent a desiccation period at about 4000 year ago and then refilled. One can speculate that Glacial Lake Washburn provided the early ciliate community that would have been augmented over time. During the dry-down period resting cysts may have been cryopreserved in the exposed lacustrine sediments to excyst and recolonize hundreds of years later. We have no idea how long protozoan cysts are viable. Thirdly, suitable food resources are crucial in determining which species of the microbial community can survive within a water body. It used to be assumed that protozoa do not practise selectivity in feeding, but that they only discriminate on the basis of size. Using live bacteria containing different coloured fluorescence protein tags, it is clear that ciliates do select different bacterial species. They even select their preferred food species when its concentration in the medium is low compared with other species (Thurman et al. 2012). Thus the concentration of a food source is not in itself an indication of suitability.

The sparse ciliated protozoan communities are unlikely to impose any significant predation pressure on flagellated protozoans and bacteria in continental lakes. For example, brackish Lake Fryxell has a population of *Plagiocampa* feeding on cryptophytes, as also occurs in neighbouring Lake Hoare. *Plagiocampa* removed less than 3% of cryptophyte biomass per day in Lake Fryxell (Roberts and Laybourn-Parry 1999). *Askenasia* grazed only 0.25% of flagellate biomass per day and *Vorticella* 0.22% of bacterial biomass per day in Lake Hoare during the summer (Roberts et al. 2004a). However, in Maritime Antarctic lakes bacterivore ciliates together with heterotrophic flagellated protozoans may impose a significant impact in the summer. In Sombre Lake protozoans may remove up to 100% of bacterial production per day, while in Heywood Lake the impact is lower at around 10% per day (Laybourn-Parry et al. 1996).

Flagellated protozoa

Most flagellated protozoans are small (see Table 1.2) and often difficult to identify with light microscopy (Figure 1.26). Detailed taxonomic analysis involves electron microscopy and now the use of molecular analysis. Consequently the database on the species diversity of phototrophic and heterotrophic flagellates is less detailed than it is for ciliated protozoans. There are, however, many data on their abundances and biomass as well as their grazing and photosynthetic rates. Among the heterotrophic flagellate community bodonids and choanoflagellates are common components of the plankton communities in Signy Island lakes (Butler 1999; Butler et al. 2000). In the Vestfold Hills *Heteromita* aff. *globosa*, *Monosiga*, and *Paraphysomonas vestita* have been identified from Crooked Lake and Lake Druzhby (Tong et al. 1997). Among the phytoflagellates three to four species of cryptophyte, *Ochromonas*, *Pyramimonas*, and *Chlamydomonas* are found in Lake Hoare plankton. Here cryptophytes dominate and are highly successful because of their mixotrophic capability (Roberts and Laybourn-Parry 1999; Marshall and Laybourn-Parry 2002; Thurman et al. 2012). Cryptophytes and *Pyramimonas* also dominate many saline lakes as we shall see in Chapter 3. In the Maritime Antarctic, Signy Islands lakes contain the cryptophytes *Chroomonas acuta*, *Cryptomonas* sp., and *Cyanomonas* sp. with lesser numbers of chryosphytes, euglenoids, and prasinophytes (Butler 1999; Butler et al. 2000). Chryosphytes, a number of *Chlamydomonas* species, and *Ochromonas* have been recorded from Boeckella Lake and Chico Lake (see Table 2.1, lakes 34 and 71), and a pond at Hope Bay, Antarctic Peninsula (Izaguirre et al. 1993; Izaguirre et al. 2003). On Livingston Island, Chryosphytes (*Chrysococcus* spp., *Pseudokephyrion* sp., and *Ochromonas* sp.) and *Chlamydomonas* sp. were identified from a range of lakes, including Midge Lake and Lake Limnopolar (lakes 35 and 53, Table 2.1) (Toro et al. 2007).

Abundances of heterotrophic and autotrophic nanoflagellates are shown in Table 2.4. What is immediately apparent is that both autotrophic and heterotrophic nanoflagellates are most abundant in the less climatically extreme, smaller lakes of the Maritime Antarctic. Lake Hoare has higher autotrophic nanoflagellate abundance than other continental lakes, although lower heterotrophic flagellate abundance than other lakes. Lake Hoare consistently supports higher autotrophic nanoflagellate numbers, as data for four years between 1997 and 2008 show; although overall numbers of both flagellate groups vary (Thurman et al. 2012). As previously indicated, Lake Hoare has a nanoplankton with a dominance of cryptophytes that are mixotrophic and at times out-graze the heterotrophic community (Roberts and Laybourn-Parry 1999). This is not the case in other freshwater Antarctic lakes. In some shallow lakes in the Larsemann Hills (Lake Heidi and Kirisjes Pond) autotrophic nanoflagellates were so sparse that they were not detected. In these lakes the extensive benthic algal mats probably dominate primary production. Their carbon exudation supports the bacterioplankton and their bacterivore predators, the heterotrophic nanoflagellates (Ellis-Evans et al. 1998).

Autotrophic dinoflagellate species occur in freshwater lakes, sometimes becoming abundant (Figure 1.26d). However, in continental lakes this flagellate group is much more common in saline marine-derived lakes. *Gymnodinium* has been recorded in very low numbers in Lake Hoare during March (Thurman et al. 2012) and *Peridinium* in Crooked Lake (Laybourn-Parry et al. 1992). Dinoflagellates are much more common in the freshwater bodies of the Maritime Antarctic. They are also a conspicuous component of the plankton in lower latitude lakes. Signy Island lakes appear to support 'blooms' of a phototrophic *Gymnodinium* sp. in spring and summer. In Tranquil Lake it reached densities of $1.8 \times 10^4 \, l^{-1}$ in the surface waters in summer, in eutrophic Heywood Lake highest numbers were recorded in spring $7.8 \times 10^4 \, l^{-1}$ immediately under the ice (Butler 1999; Butler et al. 2000), and in Sombre Lake peak numbers occurred at around mid-summer, $0.72 \times 10^4 \, l^{-1}$ (Laybourn-Parry et al. 1996). Surprisingly dinoflagellates have not been recorded from many other Maritime Antarctic lakes that have focused on the species composition of the phytoplankton. The exception is Izaguirre et al. (1993), who noted *Peridinium* in Boeckella Lake.

Sarcodina

Naked amoebae are virtually impossible to identify and enumerate from lake water samples. The only way to determine their presence and potential abundance is by culture and the most probable number technique. Consequently they are a group that is overlooked in plankton and benthic studies. Where they have been investigated in lakes at lower latitude, abundance tends to be low. In eutrophic lakes numbers are typically less than 50 l^{-1} (Laybourn-Parry 1992). Amoebae have only been investigated in Heywood and Tranquil Lakes (Signy Island). Twenty-six morphotypes were identified from plankton samples in Heywood Lake, where estimated maximum numbers were 4800 l^{-1} during March. There was a close correlation between their numbers and their bacterial food supply (Butler 1999). In Tranquil Lake naked amoebae appeared in early winter where their numbers were usually low (<650 l^{-1}), although on one occasion relatively high numbers were found in January 1995 (1500 l^{-1}) (Butler et al. 2000).

Heliozoan sarcodines that resemble stylized suns are easily seen in plankton samples where they feed on flagellates and small ciliates (Figure 1.26a). These protozoans are commonly seen in lower latitude lakes where their occurrence tends to be sporadic. They have been recorded in in Signy Island and Vestfold Hills lakes (Butler 1999; Butler et al. 2000; Laybourn-Parry et al. 1992; Laybourn-Parry unpublished data). In Crooked Lake, heliozoan abundance ranged between 10 and 150 l^{-1} over a year, with peak numbers in the winter, and in Lichen Lake and Lake Nottingham, cell concentrations ranged from 0–140 l^{-1} to 0–20 l^{-1} respectively. Heliozoans represent a small fraction of the protozoan community and are unlikely to impose any significant predatory impact on flagellate communities.

2.5.4 The phytoplankton

Phytoplankton is a generic term for the photosynthetic community of the plankton. It includes the autotrophic flagellated protozoans outlined above (Table 2.4), photosynthetic bacteria, and algae. As indicated in Chapter 1, ice cover removes wind-driven turbulence in the water column, so phytoplankton species need to be motile in order to maintain their position in the euphotic zone when lakes are ice covered. While lakes in the Maritime Antarctic are ice-free for a significant part of the summer, continental lakes lose their ice for short periods only and sometimes do not become completely ice-free in particularly cold summers. Thus wind-driven turbulence is more commonly present in maritime latitudes allowing non-motile species to form actively growing populations in the summer.

Maritime Antarctic lakes appear to support a higher phytoplankton diversity than continental lakes. However, one needs to exercise some caution as many of the studies on Maritime freshwater bodies report analyses of littoral water samples or data from what are very shallow water bodies, so that the assemblages described contain benthic, epilithic, as well as planktonic species. Lakes on Signy Island have a phytoplankton with a significant phytoflagellate component including cryptophytes that are abundant in summer, while the chlorophyte *Ankistrodesmus falcatus* and other *Ankistrodesmus* spp. are common and dominate in the spring in Heywood and Sombre Lakes (Light et al. 1981; Hawes 1983; Butler 1999). *Ankistrodesmus* has bright green needle-shaped cells. The genus has been recorded in Corner Lakes in the Vestfold Hills. These small lakes lie close to the Sørsdal Glacier and possess turbid waters due to the glacial rock floor (Laybourn-Parry and Marchant 1992). Diatoms are widely reported in the Maritime Antarctic in samples from shallow lakes/ponds and from littoral samples in larger lakes, such as Boeckella and Midge Lakes, during the summer (see Figure 1.29a). Among the genera reported are *Fragilaria*, *Cyclotella*, *Navicula*, *Nitzschia*, *Achnanthes*, and *Gomphonema* (Izaguirre et al. 1993, 2003; Toro et al. 2007; Unrein and Vinocur 1999; Unrein et al. 2005). Continental freshwater lakes tend to be dominated by phytoflagellates. Diatoms are rare, although a small species of non-colony-forming *Fragilaria* occurs in very low numbers in the plankton of the Vestfold and Larsemann Hills lakes (Laybourn-Parry et al. 1992; Ellis-Evans et al. 1998). Desmids are a common component of lower latitude lakes, but are rare in Antarctic lakes. Several species of the genus *Cosmarium* occur in Tres Hermanos Lake on King George Island (Unrein and Vinocur 1999). The genus has also been reported from lakes in the Vestfold Hills and Larsemann Hills, but usually as a rare component (Laybourn-Parry and Marchant 1992a; Ellis-Evans et al. 1998) and from lakes in the Schirmacher Oasis (Richter 1995).

The Cyanobacteria are a diverse group of phototrophic bacteria with some species forming filamentous colonies (e.g. *Oscillatoria*, *Lyngbya*, *Phormidium*) and others, small rounded colonies of a small number of cells in a gelatinous matrix (e.g. *Gloeocapsa*) (see Figure 1.24). Cyanobacteria are common in samples from shallow Maritime water bodies and in samples from some larger lakes, including species belonging to the genera *Oscillatoria*, *Microcystis*, *Lyngbya*, and *Phormidium* (Izaguirre et al. 1993, 2003; Toro et al. 2007; Unrein and Vinocur 1999; Unrein et al. 2005; Butler 1999). They are rarely reported from continental lake plankton. Low numbers of *Phormidium* occur in the plankton of Lake Hoare, probably having colonized the plankton from the very well-developed benthic algal mats (Marshall and Laybourn-Parry 2002). Picocyanobacteria (see Table 1.2) are very small, solitary cyanobacteria that can be very abundant in lakes and the sea. They have not been well studied in Antarctic lakes, although it is likely that they do occur quite widely. The picocyanobacterium *Synechococcus* has been recorded in samples from Boeckella Lake (Antarctic Peninsula, lake 34, Table 2.1), an unnamed lake on Deception Island (Lake 72, Table 2.1), Heywood Lake (Signy Island), and lakes on Livingston Island (lakes 35, 53, Table 2.1) (Izaguirre et al. 1993, 2003; Llames and Vinocur 2007; Butler 1999; Toro et al. 2007). Picocyanobacteria reached abundances of 2×10^7 cells l^{-1} in Lake Limnopolar (lake 53, Table 2.1) and 36.4×10^7 l^{-1} in Boeckella Lake (lake 34, Table 2.1), where they formed a dominant component of the phytoplankton (Toro et al. 2007; Allende and Izaguirre 2003). The fact that they have not been recorded from continental freshwater lakes is probably because they have not been studied. Combined eukaryotic and prokaryotic picoplankton can constitute up to 50% of the phytoplankton in freshwater lakes in the Terra Nova Bay region (lakes 45, 60, and 63, Table 2.1). In most instances the < 2 µm component represented the major proportion of total chlorophyll *a* (Andreoli et al. 1992). A molecular study of the plankton from three shallow lakes in the region revealed that picocyanobacteria constituted 10.2%–16.3% of the bacterial community; they were detected by fluorescence in situ hybridization (FISH) (Michaud et al. 2012).

2.5.5 The zooplankton

The major component of the heterotrophic plankton in Antarctic lakes, especially continental lakes, is the protozooplankton discussed in Section 2.5.2. Metazoan plankton is sparse and represented by a few crustaceans and a number of rotifer species (see Chapter 1, Sections 1.8.5 and 1.8.6). In freshwater Eastern Antarctic continental lakes the cladoceran *Daphniopsis studeri* occurs in the plankton (Figure 1.31c). While it is essentially a freshwater species it has invaded some of the slightly brackish lakes of the Vestfold and Larsemann Hills. It has low fecundity, with a mean of 1.7 ± 1.0 to 2.8 ± 1.2 eggs/embyros in its brood pouch across six freshwater lakes in the Vestfold Hills (Laybourn-Parry and Marchant 1992). In slightly brackish Highway Lake (lake 11, Table 3.1) it achieves a higher fecundity of 7.7 ± 3.0 eggs/embryos per individual. *Daphniopsis* stays active during the winter months feeding in the plankton and using its endogenous energy reserves. The latter are conspicuous fat globules within the organism. During the summer, *Daphniopsis* relies on the phytoplankton in Lake Druzhby (Vestfold Hills), but during winter it is able to feed on the bacterioplankton when it can remove up to 34% of biomass, although its impact is usually low at only 0.6%–4.6% of bacterial biomass removed per day (Säwström et al. 2009). The life history appears to vary between lakes. In Lake Druzhby it reproduces parthenogenically in the summer, giving rise to an adult overwintering population, whereas in neighbouring Crooked Lake there is a gradual development of juveniles over the winter (Bayliss and Laybourn-Parry 1995). It appears that only one generation is produced per year, but in an ecosystem where there are no predators the population is sustained.

Cladocerans typically reproduce parthenogenically, resorting to a sexual phase when conditions become extreme with the production of ephippial resting eggs. Males and ephippial eggs have been noted in some lakes (Watts Lake and hyposaline Highway Lake), and in some cases ephippial eggs were found but no males (Lake Druzhby) (Gibson et al. 1998). However, the presence of ephippial eggs is not proof of a sexual episode. In both the Antarctic and Arctic, cladocerans overwinter as adults while ephippial eggs are also produced. This has been noted on Signy Island (Heywood 1972) and in the Arctic (Edmondson 1955). The presence of ephippial eggs does not necessarily require the presence of males and fertilization. *Daphnia middendorffiana*, a common species in the Arctic and alpine zones in the northern hemisphere, is an obligate parthenogen. Both asexual or subitaneous eggs and ephippial eggs are produced ameiotically and males do not occur (Hebert 1981). The density of *Daphniopsis* in large ultra-oligotrophic lakes like Crooked Lake and Lake Druzhby is extremely low, usually less than five individuals per cubic metre, so the encounter rate of females with males would be very low and not conducive to supporting sexual reproduction.

The nauplii of the copepod *Acanthocyclops mirnyi* are seen periodically in plankton samples from the lower part of the water column in Crooked Lake, suggesting that it is a benthic species. It has been noted in samples from a range of other freshwater lakes in the Vestfold Hills (Dartnall 2000) and in Lake Figurnoye in the Bunger Hills (Borutzky 1962). In Lake Figurnoye it was mainly found close to the lake bottom, again suggesting that it is fundamentally benthic in its habits (Klokov et al. 1990).

Until recently it was assumed that the freshwater lakes of the McMurdo Dry Valleys lacked crustacean plankton. The only metazoans observed were rotifers (see Figure 1.30). However, a study that took large volume water samples of 20 l from each of four depths in the water column of Lake Hoare found fragments and one complete specimen of a copepod copepodite, which was tentatively identified as belonging to the genus *Boeckella*. (Hansson et al. 2011). Clearly the crustacean plankton of Lake Hoare is extremely sparse compared with other continental lakes since 80 l of water yielded only one intact crustacean specimen. Long-term sampling of Lake Hoare for microbial plankton has never revealed crustaceans, whereas in the lakes of the Vestfold Hills and in epishelf Beaver Lake, 5 l water samples regularly contain crustacean specimens.

Boeckella poppei (Copepoda) and *Branchinecta gainii* (Anostraca—fairy shrimp) are found in the Maritime Antarctic (see Figure 1.31d). *Branchinecta* feeds on the benthic cyanobacteria, bacteria, algae, and protozoa and debris. The resting phase of the life cycle is in the winter (June to September), at a time when lakes are ice covered. The nauplii hatch in October when the ice cover is still intact but beginning to thin. They are active for six months in the year (October to early May) and produce one generation each year (Pociecha and Dumont 2008). Around 2.5 months is sufficient for it to complete its life cycle (Hawes et al. 2008). *Boeckella poppei* nauplii appear in high concentrations in October maturing to copepodites in November. In Lake Wujka (South Shetlands, lake 68, Table 2.1) at least three cohorts of nauplii and copepodites occurred. Mature males and females appeared at the end of November to early December and the first gravid females noted in January (Pociecha and Dumont 2008). Low abundances of Crustacea are the norm. In Boeckella Lake, *Boeckella poppei* reached peak abundances of around 12 l^{-1} for all life cycle stages combined (Izaguirre et al. 2003). The cladoceran *Macrothrix ciliata* has been recorded in Lake Limnopolar (lake 53, Table 2.1); it is a nekto-benthic species associated with the aquatic moss *Drepanocladus longifolius*. *Daphniopsis studeri* and *Macrothrix ciliata* are the only cladocerans recorded in Antarctic lakes (Toro et al. 2007).

Planktonic rotifers show much greater species diversity than crustaceans in continental lakes. In the lakes of the Vestfold Hills and Larsemann Hills a number of species belonging to the genera *Philodina*, *Notholca*, and *Lecane* occur (Dartnall 1995; Ellis-Evans et al. 1998). However, their density in the water column is low. In Crooked Lake their numbers ranged from 0 in June up to 10.5 ± 4 l^{-1}, with the greatest abundance recorded in January and February (Laybourn-Parry et al. 1992). They are probably feeding on bacteria and nanoflagellates. A detailed study of zooplankton in Lake Hoare identified nine rotifer taxa, including species from the genera *Brachionus*, *Keratella*, *Kelliocottia*, *Notholca*, *Filinia*, *Lepadella*, and *Philodina* (Hansson et al. 2011). *Philodina alata*, a species common in the plankton of Lake Fryxell (a brackish lake—see Chapter 3), was the most abundant rotifer in Lake Hoare with up to 400 individuals l^{-1}. In the Maritime Antarctic, Lake Wujka (King George Island, lake 67, Table 2.1) supports populations of *Notholca squamula salina* which can reach concentrations of 150 l^{-1} in the winter (June). This rotifer is essentially a benthic species feeding on benthic algae, but it enters the plankton when the bottom waters are stirred up during storms, where it may remain for some time. The lake is freshwater but influx of sea water during storms can increase its conductivity, which favours *Notholca squamula salina* (Pociecha 2008). Boeckella Lake (Antarctic Peninsula) has a relatively low rotifer species diversity. The most common species was *Philodina gregarina* and other unidentified bdelloid rotifers with densities of up to 0.14 to 1.56 individuals l^{-1} (Izaguirre et al. 2003).

2.6 Carbon cycling in the planktonic environment

2.6.1 Primary production

Rates of primary production in Antarctic lakes across a gradient from the Dry Valleys to the Maritime Antarctic are low when placed in a global context (Table 2.2). Even when compared with the Arctic the rates are generally lower. A review of Arctic photosynthesis found rates ranging from 0 to 1500 µg C l^{-1} day^{-1}. A comparison of Arctic and Antarctic daily rates of primary production shows that Antarctic rates are more skewed (52%) to the lowest category rate of less than 10 µg C l^{-1} day^{-1} whereas in the Arctic where the data set included more lakes, 43% had values of more than 100 µg C l^{-1} day^{-1} (Figure 2.15) (Lizotte 2008). These data include saline and freshwater lakes in Antarctica. There are a range of factors that account for these differences between the polar regions, among which are that many Arctic lakes are situated in vegetated catchments and consequently received allochthonous nutrient inputs, whereas Antarctic lakes are nutrient limited. Moreover, summer water temperatures are generally higher in the Arctic. For example, Toolik Lake and the Barrow tundra ponds reach a maximum of 15 °C (Alexander et al. 1980; Whalen and Alexander 1984) while in the High Canadian Arctic and Svalbard maxima of 7 °C were recorded (Markager et al. 1999; Laybourn-Parry and Marshall 2003). As indicated in Section 2.3, Antarctic lake temperatures in the euphotic zone are usually well below 4 °C.

Continental Antarctic lakes tend not to have accumulated snow cover, because it is blown off by katabatic winds that sweep down from the continental ice sheet. Freshwater lakes have highly transparent ice for a significant part of the year, particularly during winter and spring, exceptions being the Dry Valley lakes (see Figure 1.33). Most data relate to the summer, but there are a small number of investigations that cover the winter (see Table 2.2 indicated by asterisks). Once the sun returns in July primary production commences in Crooked Lake (Vestfold Hills), so that by August there are relatively high rates throughout the photic zone which extends down to between 10 and 12 m (Figure 2.16)

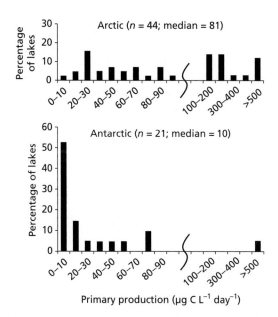

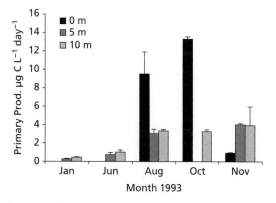

Figure 2.15 A comparative review of maximum photosynthetic rates in Arctic and Antarctic waters (saline and freshwater). Note differences in scales between graphs. From Lizotte (2008) with the permission of Oxford University Press.

Figure 2.16 Primary production throughout a year (1993) at different depths in the water column of Crooked Lake, Vestfold Hills. Data from Bayliss et al. (1997).

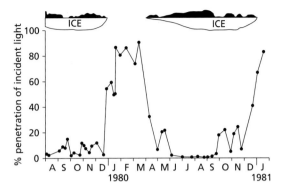

Figure 2.17 Seasonal variation in the light climate of Sombre Lake and Heywood Lake, Signy Island. Note the effect of snow-covered ice on light transmission. From Hawes (1985a) with the permission of Dr Junk Publishers, Dordrecht.

(Bayliss et al. 1997). In nearby Watts Lake, which is shallower and has higher chlorophyll a concentrations (Table 2.2), carbon fixation was measurable in all months except July. The highest rates of photosynthesis occurred in September with a maximum rate of 154 µg C l^{-1} day^{-1} (Heath 1988). A similar pattern was seen in Lake Verkhneye (Schirmacher Oasis) but here there was no detectable photosynthesis between mid-May and early August (Kaup 1994). Thus even in winter, when PAR levels are extremely low, the phytoplankton of these extreme environments are capable of fixing carbon.

Measurements of photosynthesis in the Maritime Antarctic are restricted to Signy Island for which there are excellent year-long data sets (Table 2.2). As one might anticipate photosynthetic rates are higher than in continental lakes. Even in oligotrophic Tranquil and Moss Lakes the reported rates reach maxima of 107 and 28 µg C l^{-1} day^{-1} respectively, which in most cases exceeds rates of photosynthesis reported for continental freshwater lakes (Table 2.2). Snow accumulates on the ice cover of Signy Island lakes during winter and has a profound impact on the transmission of PAR to the water column (Figure 2.17). Clear ice does attenuate PAR to the water column, but the presence of snow on the ice strongly reduces light transmission, so that when there was thick snow cover between June and September of 1980 virtually no light reached the underlying water. This in turn caused a reduction in photosynthesis, even when chlorophyll a levels were high (Figure 2.18) (Hawes 1985b).

As is apparent in Figure 2.16 there are differences in the rates of photosynthesis with depth in the water column, and the highest rates do not always occur at the water surface or immediately under the ice, nor do they necessarily correlate with concentrations of chlorophyll a, as Figure 2.19 shows. Lake Hoare phytoplankton is subject to nitrogen deficiency. In this and other Dry Valley lakes nutrients in the euphotic zone are replaced by upward diffusion from deep nutrient pools (Priscu 1995). The light climate of Lake Hoare is also challenging to phytoplankton. Through a thick, debris laden ice cover only 1.7%–3.3% of surface incident PAR penetrates

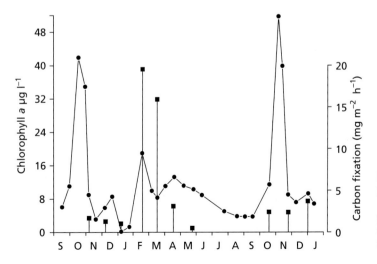

Figure 2.18 Chlorophyll a and primary production in Heywood Lake during 1979 and 1980. ● : chlorophyll a, ■ : vertical lines—primary production. See extent of ice and snow cover in Figure 2.17. Redrawn from Hawes (1985a).

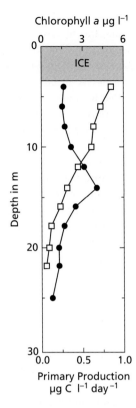

Figure 2.19 Primary production (□) and chlorophyll a (●) concentration in the water column of Lake Hoare during November and December 1993. Redrawn from Priscu (1995).

to the underlying water column (Howard-Williams et al. 1998), whereas in Crooked Lake, previously described, 18%–15% of surface incident PAR penetrates the ice to the water column.

The physiological basis of adaptation to shaded conditions is not well researched in Antarctic communities and has only been considered in detail in the Dry Valley lakes. The light climate is consistently low, allowing the evolution of a series of specific adaptations. It is evident that the phytoplankton of the Taylor Valley lakes (Hoare, Bonney, and Fryxell) possess a photosynthetic apparatus that has evolved to function under extremely shaded conditions (Neale and Priscu 1998). The phytoplankton has low irradiance for the onset of light-saturated photosynthesis and a high sensitivity to photoinhibition. Adaptations have been considered at the species level. *Chlamydomonas subcaudata* is a common species in Lake Bonney, one of the brackish Dry Valley lakes (See Table 3.1, Chapter 3). The strain isolated from Lake Bonney is a psychrophile growing only at 16 °C and below. This phytoflagellate possesses higher levels of the xanthophylls lutein, neoxanthin, and violaxanthin in comparison with mesophilic species (*Chlamydomonas reinhardtii*) (Neale and Priscu 1995; Morgan et al. 1998). Based on this observation it has been suggested that the phytoplankton of Lake Bonney have traded off photoprotection mechanisms against an enhanced ability for efficient light-harvesting and energy usage. Interestingly, when *C. subcaudata* is grown under relatively high light conditions of 150 µmol m^{-2} s^{-1}, the flagellate retains its ability to adjust the xanthophyll cycle and the capacity for dissipating excess energy as heat. The stoichiometry of the photosystem II: photosystem I: coupling factor is significantly altered in *C. subcaudata* when compared with *C. reinhardtii*, indicating that *C. subcaudata* is well adapted for growth at the low irradiances typical of Lake Bonney and other Dry Valley lakes (Morgan et al. 1998).

Mixotrophy is common among species of Antarctic lake phytoflagellates in both saline and freshwater systems (see Section 1.8.3, Chapter 1). Cryptophytes are a conspicuous element in the phytoplankton of Lake Hoare, where they practise mixotrophy, feeding on bacteria to supplement their carbon budget (Roberts and Laybourn-Parry 1999; Marshall and Laybourn-Parry 2002) (see Figure 1.26e). Thus the growth of the phytoplankton is not solely attributable to photosynthesis. The dependence on carbon derived from ingesting bacteria varies across the summer months and is lowest in December when grazing rates fall to a low level (Figure 2.20). The supposition is that the cryptophytes become more dependent on heterotrophy, as opposed to photosynthesis, as they move into winter. In 2008 the LTER field season was extended into early April to determine how the plankton responded to the onset of winter. Phototrophic nanoflagellates showed no marked decline and continued to graze bacteria up to 15 bacteria cell^{-1} day^{-1}. Their community grazing exceeded bacterial production (Thurman et al. 2012). The concentration of bacterial and phytoflagellate cells varies considerably from year to year (Table 2.7). This inter-annual variability is a characteristic of Antarctic continental

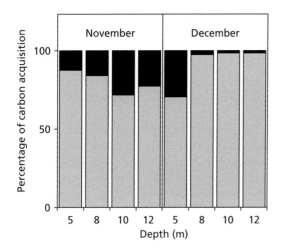

Figure 2.20 The balance between carbon acquisition from photosynthesis and grazing by mixotrophic cryptophytes during November and December 2000 in the Lake Hoare water column. Pale grey section of columns = photosynthesis, black section of columns = grazing. Redrawn from Marshall and Laybourn-Parry (2002).

lakes. Grazing rates also vary from year to year. In the summer of 1997/1998 the grazing rates varied between 4.8 and 24 bacteria cell^{-1} day^{-1}, while in the summer of 2000 the rates were between 5.3 and 10.1 bacteria cell^{-1} day^{-1}. These variations are difficult to explain but probably relate to concentrations of bacterial prey, competition from heterotrophic nanoflagellates, and the species composition of the bacterial community. Flagellates probably do not graze indiscriminately and are likely to be selecting prey on the basis of size and palatability. Ciliated protozoans have been shown to be highly selective and it is likely that this feeding behaviour also occurs in other protozoans (Thurman et al. 2012).

While there are no winter data for Lake Hoare, a remote sampling device in Lake Fryxell (brackish, see Table 3.3, Chapter 3) collected and preserved plankton samples over a winter (McKnight et al. 2000). While numbers were lower in winter, vegetative cells were the most abundant form of all species. One species, *Stichococcus*, was observed in winter and two of the cryptophyte species were more abundant than in summer. The supposition is that at least some of these species are surviving by adopting mixotrophy.

2.6.2 Bacterial production

Investigations of bacterial production in freshwater lakes are not extensive, but do cover a gradient from the McMurdo Dry Valleys through the Vestfold Hills to Signy Island in the Maritime Antarctic (Table 2.2). A number of the data sets cover an entire year (indicated by an asterisk in Table 2.2). The maximum rates of bacterial production range from 0.15 µg C l^{-1} day^{-1} to 12.5 µg C l^{-1} day^{-1}. In Heywood Lake which is undergoing eutrophication, rates achieved ranged between 4.8 and 26.4 µg C l^{-1} day^{-1}. Bacterial production represents 2.8%–44.6% of primary production across lakes from the Dry Valley to the Maritime Antarctic. A review of planktonic production across a wide range of marine and freshwater environments indicates that bacterial production averages 20% of primary production and on an areal basis of the entire water column around 30% (Cole et al. 1988). There are exceptions, for example in Arctic Toolik Lake (68°N) bacterial production represents 66% of primary production (O'Brien et al. 1997). This lake lies in a vegetated catchment and receives a large influx of labile terrestrial dissolved organic matter associated with snow melt (Crump et al. 2003).

Rates of bacterial production in Antarctic freshwater systems (Table 2.2) are comparable with rates reported for large temperate oligotrophic lakes, for example lakes in Wisconsin and Michigan (Pace and Cole 1994) and Loch Ness (Laybourn-Parry et al. 1994). Bacterial production in small proglacial lakes in Svalbard ranged between 0.96 and 8.9 µg C l^{-1} day^{-1},

Table 2.7 Abundances of bacteria and phototrophic nanoflagellates in lake Hoare to illustrate inter-annual variation.

Year	Bacteria × 10^8 l^{-1}	Phytoflagellates × 10^5 l^{-1}
November/January 1997/1998[1]	9.2–11.9	99.6
November 2004[2]	–	6.5–30.2
November/December 2000[3]	5.0–22.0	1, 2
March 2008[4]	2.7–5.5	1.0–13.5
November/January 1993/1997[5]	3–70	–

1: Roberts and Laybourn-Parry 1999, 2: Laybourn-Parry unpublished data, 3: Marshall and Laybourn-Parry 2002, 4: Thurman et al. 2012, 5: Takacs and Priscu 1998.

within the range seen in Antarctic lakes (Anesio et al. 2007; Mindl et al. 2007). These proglacial lakes are not surrounded by tundra vegetation.

As indicated in Section 1.8, heterotrophic planktonic bacteria use dissolved organic carbon as an energy source. The DOC pool of Lake Hoare, along with other Taylor Valley lakes, is derived from a number of sources, exudation of labile DOC from the phytoplankton and benthic algal mats, upward diffusion across sediments and in meromictic lakes (Fryxell and Bonney) across the chemocline, and inputs from glacial streams. Based on bacterial production and estimated bacterial respiration the carbon demand of the bacterial community can be calculated. The DOC pool was three to eight times smaller than the estimated bacterial demand, indicating that a major source of bacterial carbon was unaccounted for (Table 2.8) (Takacs et al. 2001). Some of the missing DOC may have been due to the dissolution of stream particulate organic matter (POC) and lake POC. This is indicated in Table 2.8 and makes little difference to the budgets. Even when bacterial growth efficiency is doubled, thereby reducing respiratory costs, the budget still does not balance. Growth efficiency is the proportion of the 'consumed' DOC that is converted to bacterial biomass. The major question is where does the 'missing' DOC come from? Other sources may be derived from the lysis of cells by grazers, which in these lakes are mainly heterotrophic and mixotrophic protozoans and rotifers, and the lysis of bacterial cells by viruses. A further important point is that this budget pertains to the summer, there are no winter data for the Dry Valleys so that bacterioplankton DOC demand may be balanced when measured on an annual basis (Takacs et al. 2001).

In Crooked Lake and Lake Druzhby (Vestfold Hills) bacterial production continues during the winter. Estimated bacterial growth efficiency was 2%–5% with high community respiration rates between 7.3 and 42.4 µg C l^{-1} day^{-1} (Säwström et al. 2007d). Seasonal lake carbon calculations suggest that bacterial respiration occasionally exceeded the estimated carbon input by 74%–97%, indicating that these lakes are net heterotrophic systems, like the Dry Valleys lakes. Not all potential sources of DOC were explored in Crooked Lake and Lake Druzhby. There are fluctuations of DOC over the year. Even in winter there are peaks of DOC that sustain bacterial growth, as evidenced by peaks in bacterial biomass (Figure 2.21). Primary production occurs for most of the year (see Figure 2.16) but is unlikely to supply

Table 2.8 A partial DOC budget for the photic zone of Lake Hoare (freshwater) and Lakes Bonney and Frxyell (saline, see Chapter 3) (Takacs et al. 2001). Bacterial respiration was calculated from bacterial production (del Giorgio and Cole 1998) and dissolution of lake and stream POC was estimated assuming a dissolution rate of 0.08% day^{-1} (Priscu 1992). 2 × BGE (bacterial growth efficiency) is bacterial respiration when BGE is doubled.

kg Dissolved organic carbon (sink)	Lake Fryxell	Lake Hoare	Lake Bonney East lobe	Lake Bonney West lobe
Phytoplankton exudation	304	24	668	365
Streams	154	15	136	94
Diffusion	97	27	29	12
Particulate organic carbon dissolution	803	201	449	185
INPUT	1358	267	1282	656
Bacterial production	1109	205	561	277
OUTPUTS				
Bacterial respiration	4615	2335	5915	1966
Bacterial respiration 2 × BGE	1753	1065	2677	845
Net (deficit)	(3257)	(2067)	(4633)	(1310)
Net (deficit) 2 × BGE	(395)	(789)	(1395)	(189)

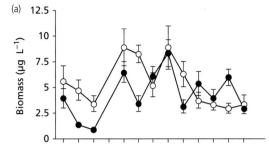

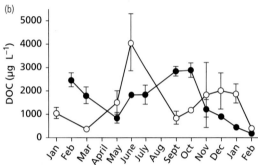

Figure 2.21 Changes in bacterial biomass and dissolved organic carbon (DOC) concentrations in Crooked Lake and Lake Druzhby, Vestfold Hills during 1999 to 2000. (●) = Crooked Lake, (○) = Lake Druzhby. From Laybourn-Parry et al. (2004) with the permission of Blackwell Publishing Ltd.

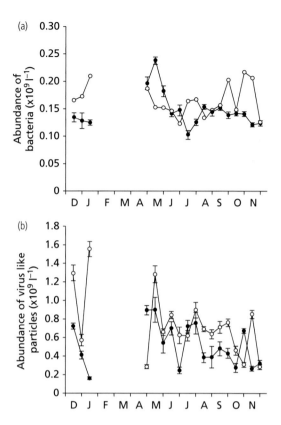

Figure 2.22 Virus and bacterial abundance in Crooked Lake and Lake Druzhby over an annual cycle (2003/2004). Panel A bacterial abundance and Panel B virus abundance; (●) = Crooked Lake, (○) = Lake Druzhby. From Säwström et al. (2007d) With the permission of Springer Science & Business Media Inc.

sufficient DOC to account for the winter peaks. Virus abundance remains relatively high during winter (Figure 2.22) and in these two lakes the lytic cycle predominates causing the lysis of bacterial cells of which 22.7%–34.2% are infected (see Section 2.5.2). During winter an estimated >60% of the carbon supplied to the DOC pool originates from viral lysis, while in summer the figure is lower at around <20% (Säwström et al. 2007d). Thus Lakes Hoare and Druzhby and Crooked Lake are net heterotrophic systems, as is also the case in some lower latitude oligotrophic lakes (Cole et al. 2002). It is suggested that 5 mg DOC l^{-1} is the threshold for the transition between net autotrophy and net heterotrophy in lakes, and that around half of lakes surveyed worldwide are probably net heterotrophic systems (Prairie et al. 2002; Sobek et al. 2007). DOC concentrations in Antarctic lakes are at around a maximum of 2 mg l^{-1} (Table 2.3). It appears that the freshwater Antarctic lakes, for which we have detailed data, do not fit into the pattern for net heterotrophy/net autotrophy seen at lower latitudes.

2.6.3 Heterotrophic grazing

The major grazers of the bacterioplankton are heterotrophic nanoflagellates, and in some lakes (e.g. Lake Hoare), mixotrophic phytoflagellates (see Figure 1.26). Ciliated protozoans and rotifers feed on bacteria, flagellates, and elements of the phytoplankton. Antarctic lakes have simple food webs, so it is relatively easy to disentangle trophic interactions (Figure 1.23). The grazing rates of protozoans are usually determined by monitoring the uptake of heat-killed fluorescently stained bacteria in short-term incubations. The fluorchrome used is DTAF

(5-(4,6–dichlorotriazinyl)aminofluorescein) and the stained bacteria termed fluorescently labelled bacteria or FLBs. The ingested FLBs are clearly visible within the food vacuoles of the protozoans under epifluorescent microscopy (Sherr and Sherr 1993). The bacteria are dead and the cells are covered with a fluorescent stain. The assumption is that the protozoans, both flagellates and ciliates, do not practise selective feeding and cannot discern the difference between live bacteria and stained dead bacteria. It is now possible to culture bacteria that express different colour (green, yellow, red, and blue) fluorescent proteins which are clearly visible inside the food vacuoles of protozoans under epifluorescent microscopy. This approach has several advantages, firstly the cells are live and their cell surface is unmodified and secondly, having bacteria of different colours enables one to determine whether there are feeding preferences. Such feeding selectivity has been seen in ciliated protozoans that select one bacterial species over another, even when the preferred bacterium is present in low concentrations relative to the other species (Thurman et al. 2010). The grazing rates of nanoflagellates are higher when fed on live bacteria with fluorescent proteins compared with FLBs (Laybourn-Parry, unpublished data). Thus while the development of FLBs has been very important in giving us an insight into the grazing of protozoans and their impact on their food sources, we have probably been underestimating grazing rates. Moreover, if nanoflagellates are also selective feeders, then the diversity of the bacterial community will be a factor in their grazing impact and selective feeding may affect the composition of the bacterial assemblage.

The grazing rates of heterotrophic nanoflagellates based on FLB uptake vary considerably across Antarctic lakes. They are lower than rates reported from lakes at lower latitudes, where rates range from 4 to 73 cells indiv^{-1} h^{-1} (Simek and Straskrabová 1992; Bennett et al. 1990; Vaqué and Pace 1992). Temperature is clearly an important factor in determining grazing rates, but food concentration is also likely to play a role. Low temperatures in Antarctic lakes undoubtedly depress grazing rates. Bacterial concentrations are shown in Table 2.4. The highest bacterial concentrations occur in Heywood Lake and the lowest in Lake Druzhby. Highest grazing rates were recorded in Lake Hoare during the summer, although they dropped considerably as winter approached. Grazing rates vary from year to year for mixotrophic cryptophytes in Lake Hoare, so it is likely that the same pertains for heterotrophic flagellates. These differences are difficult to explain, but may relate not only to food concentration, but the taxonomic make-up of the bacterial community and the flagellate community. For example in Heywood Lake the bacterial community has a diverse morphology with cocci, larger rods, and filamentous forms (Butler 1999). The larger bacteria are probably not accessible to filter feeding nanoflagellates, so while superficially there may appear to be an abundant bacterial food supply, only a portion of it can be exploited. The impact on bacterial production can be high, in some instances daily production is consumed by the flagellate and ciliate community, while in other lakes the impact is relatively small (Table 2.9).

Lake Hoare stands out among the freshwater lakes so far investigated because it possesses a significant community of mixotrophic phytoflagellates, made up of 3–4 species of cryptophyte (see Figure 1.26e). Over four years between 1997 and 2008 when the protozooplankton was investigated, phototrophic nanoflagellates always greatly exceeded heterotrophic nanoflagellates in abundance (see Table 2.4). Mixotrophic phytoflagellates have grazing rates that vary over the years between 0.2 and 1.0 bacteria cell^{-1} day^{-1} (Roberts and Laybourn-Parry 1999; Marshall and Laybourn-Parry 2002; Laybourn-Parry unpublished 2000; Thurman et al. 2012). Because the mixotrophs are more abundant than the heterotrophic flagellates their community grazing impact can be greater on occasion, despite lower ingestion rates. They removed more than 100% of daily bacterial production as winter encroached in 2008, at a time when photosynthesis was declining in response to declining PAR (Thurman et al. 2012). At this time chlorophyll a concentration continued to increase, indicating that part of the phytoplankton community was still active and switching to heterotrophy. During the summer (November/December) mixotrophic phytoflagellates removed between 1.14% and 4.3% of bacterial production.

In most continental freshwater lakes ciliate concentrations are low, so they are unlikely to impose

Table 2.9 Heterotrophic nanoflagellate grazing rates and grazing impact. *both ciliates and flagellates, ‡based on grazing rates from Laybourn-Parry et al. 1996.

Lake and location	Grazing rate bacteria indiv^{-1} h^{-1}	% Bacterial production grazed day^{-1}
42. Heywood Lake, Signy Island[1]	0.83 ± 0.12	up to 6%* in summer
51. Sombre Lake, Signy Island[1]	0.51 ± 0.13	up to 100%* in summer
56. Tranquil Lake, Signy Island[2]	0.51 ± 0.13‡	up to 100%‡
8. Lake Druzhby, Vestfold Hills[3]	0.28 ± 0.03 at 2 °C 0.49 ± 0.04 at 4 °C	2% in summer
6. Crooked Lake, Vestfold Hills[4]	0.21 ± 0.02 at 2 °C 0.20 ± 0.03 at 4 °C	1.1% –9.7% over the year
11. Lake Hoare, Dry Valleys[5]	6.2–6.7	1%–2% in summer
11. Lake Hoare, Dry Valleys[6]	0.33–1.28	7.2%–61% in March

1: Laybourn-Parry et al. 1996; 2: Butler et al. 2000; 3: Laybourn-Parry and Bayliss 1996; 4: Laybourn-Parry et al. 1995; 5: Roberts and Laybourn-Parry 1999; 6: Thurman et al. 2012.

any significant predation pressure on either bacteria or flagellates. However, in the Maritime Antarctic ciliates are much more abundant, for example in Heywood Lake, between 6000 and 40,000 l^{-1} and in Tranquil Lake, 11,000 l^{-1} (Laybourn-Parry et al. 1996; Butler 1999; Butler et al. 2000). Bacterivore ciliates grazed at rates of 119 ± 12 bacteria cell^{-1} h^{-1} in Heywood Lake and 70.6 ± 11 bacteria cell^{-1} h^{-1} in Sombre Lake. Together with heterotrophic flagellates they can remove up to 100% of daily bacterial production in summer (Table 2.9).

Zooplankton are sparse in Antarctic continental lakes and little is known about their feeding habits. *Daphniopsis studeri* occurs in the lakes of the Vestfold and Larsemann Hills, and possibly other eastern coastal Antarctic oases. In Crooked Lake it occurs in densities of around 5 individuals m^{-3} where it feeds on the flagellate phytoplankton, protozoa, and bacteria. It has estimated filtering rates of 0.061 ± 0.015 l individual^{-1} day^{-1} on bacteria and 0.048 ± 0.003 l individual^{-1} day^{-1} on phytoplankton. In temperate lakes *Daphnia* feeding on bacteria had filtering rates ranging between 0.024 and 0.067 l individual^{-1} day^{-1}. Thus *Daphinopsis studeri* is comparable to its lower latitude relatives. In the winter it may remove between 2.8% and 34% of bacterial production day^{-1} and in summer between 1.6% and 6.9%. The percentage of phototrophic biomass removed ranged between 0.6% and 3.5% over the year (Säwström et al. 2009).

These levels of zooplankton grazing are unlikely to impose any top-down effects on either the phytoplankton or bacterioplankton. Similarly heterotrophic nanoflagellate grazing removes between 1.1% and 9.7% of bacterial production and is also unlikely to impose major top-down effect. In common with other continental Antarctic freshwater lakes, Crooked Lake is controlled by bottom-up forces related to nutrient (nitrogen and phosphorus) and DOC availability. As previously indicated, recycling of carbon and other elements as a result of viral lysis of bacteria is an important component of carbon cycling in this system, and has a much greater impact on bacterial production than heterotrophic grazing.

Crustacean zooplankton are more abundant in the Maritime Antarctic where there is some evidence of top-down control. In Sombre Lake (Signy Island) *Boeckella poppei* (see Figure 1.31a) can reach densities of between 60 and 1700 m^3. Not a great deal is known about its feeding biology. It has been reported as both a pelagic and benthic feeder in a range of lakes. Its clearance rates on different food sources vary from up to 0.155 l individual^{-1} day^{-1} on ciliates to 0.003 l individual^{-1} day^{-1} on phototrophic nanoflagellates in March. Lower clearance rates on ciliates occurred in the summer (December), while higher rates were apparent for flagellates at that time. *Boeckella* fed on a range of food sources including ciliates, heterotrophic and phototrophic flagellates, *Ankistrodesmus falcatus*, *Gymnodinium*, and *Chlorella*, indicating that it is an omnivore in Sombre Lake. In March, adult *Boeckella* were estimated to clear 24% of the lake of ciliates each day. Allowing for other developmental stage grazing, it is likely that there is the potential for control of the ciliate community at particular times of the year (Butler et al. 2005). In common with Sombre Lake, Boeckella Lake at Hope Bay (Antarctic Peninsula) has a crustacean zooplankton dominated by *Boeckella poppei*. It has a mean density of 3500 m^3 during summer and can exert a significant grazing pressure on the nano-phytoplankton, imposing a top-down impact upon this food source. The *Boeckella* community did not impact on the picoplankton. Gut content analysis showed that this calanoid copepod also exploited the periphyton as an alternative food source, indicating that it can feed both pelagically and in the benthos (Almada et al. 2004).

2.7 The benthic communities

2.7.1 Phototrophic benthic communities

The benthos of Antarctic lakes is dominated by Cyanobacteria that form complex 'algal mats' (Figure 2.23). In spite of being thought of as warm latitude organisms, the Cyanobacteria are the predominant biota in terms of biomass of Antarctic lakes and ponds and other polar habitats such as glaciers. The mats can be the major primary producers in Antarctic lakes. They form highly pigmented layers over sediments or rocks which gradually accumulate as mucilaginous films and mats that range in thickness from a few centimetres to tens of centimetres in thickness (Vincent 2000). Their success relates to their ability to withstand freezing and their wide salinity tolerances as well as the ability of some species to fix atmospheric nitrogen (Hawes et al. 1992) (see Section 1.8.1). They have successfully colonized both freshwater and saline lakes in Antarctica. The algal mats are composed of filamentous forms of Cyanobacteria where species belonging to the genera *Phormidium*, *Oscillatoria*, *Lyngbya*, and *Nostoc* predominate (see Figure 1.24). The cyanobacteria mats form a habitat for other

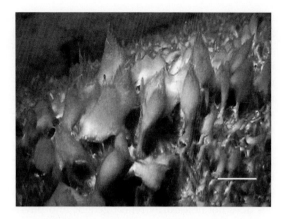

Figure 2.23 Photograph of benthic cyanobacterial mats in Lake Hoare; note the pinnacles. Photo courtesy of Ian Hawes. (See Plate 10)

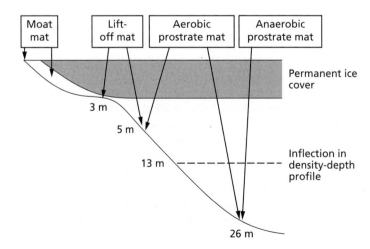

Figure 2.24 Diagrammatic representation of different benthic cyanobacterial mat forms in Lake Hoare. The extent of permanent ice cover and the summer moat are shown. The depth of a discontinuity in the depth/density profile, which divides the lake into upper and lower compartments, is indicated by a dashed line. From Hawes and Schwarz (1999), with the permission of John Wiley and Sons Inc.

phototrophs, among which diatoms are common and heterotrophic bacteria, protozoans, and a small number of invertebrate species. The oligotrophic nature of Antarctic lakes enables good light transmission through the water column, so that the photic zone often extends to the bottom of large sections of lake basins, enabling the development of extensive phototrophic benthic communities.

In recent years the taxonomy of mat-forming Cyanobacteria has been investigated with molecular techniques. Traditional taxonomic analysis that is based on morphology and ecological characteristics does not necessarily reflect genetic and physiological diversity. The molecular approach has produced a picture that differs from that derived from traditional taxonomic investigation, suggesting that novel and endemic taxa occur. Freshwater and saline lakes in eastern Antarctica (Larsemann Hills, Vestfold Hills, and Rauer Islands) revealed 17 morphospecies. Molecular analysis showed 28 gene-based operational taxonomic units (OTUs) belonging to 12 genera, of which 9 were Oscillatoriales, 3 Nostocales, and 5 Chroococcales (Taton et al. 2006a). A wider geographical investigation covered a total of 26 lakes which also included the Bølingen Islands (eastern Antarctica) and the lakes of the Dry Valleys (Taton et al. 2006b). Based on morphology, 12 species were identified which included four considered endemic to Antarctica (*Phormidium priestleyi, P. pseudopriestleyi, Leptolyngbya frigida* and *L. antarctica*). Much higher diversity was revealed by molecular analysis (16S rRNA) of 56 strains. For these, 21 OTUs were obtained which included 9 that were novel and 3 endemic to Antarctica. The divergence between morphological characteristics and molecular analysis was particularly evident in the Oscillatoriales. Seven morphospecies were identified, however, these concealed a relatively high degree of genotypic diversity (15 OTUs). The evolution of the Oscillatoriales is thought to be polyphilic, meaning that they were derived from more than one ancestor group. Cyanobacteria isolated from Arctic habitats contained three of the species listed above, that were thought to be endemic to Antarctica (*Phormidium priestleyi, Leptolyngbya frigida,* and *L. antarctica*) (Jungblut et al. 2010). Thus, there appear to be species that are common to both polar regions that are absent from lower latitudes. An approach that considers both the Arctic and Antarctic in terms of microbial biodiversity may provide a better insight into the evolution and adaptation of this important group.

Cyanobacterial mats that occur in shallow waters are subject to high levels of ultra-violet radiation (UVR), which in Antarctica has been exacerbated by stratospheric ozone depletion. There are a number of strategies that cyanobacteria can employ to avoid UVR damage. Some species are mobile and can migrate upwards and downwards within the mat to avoid UVR, and still acquire sufficient PAR to undertake photosynthesis. Other species possess protective pigmentation including scytonemin and mycosporine-like amino acids (MAAs) (Vincent 2000). These screen out harmful UVR (280–400 nm). A number of factors are implicated in determining the pigment content of cyanobacterial mats, among which light intensity and its spectral composition are major elements. Other factors include duration and extent of snow cover, species composition, water chemistry, and geographical location. In shallow lakes the cyanobacterial mats are subject to high levels of PAR and UVR so that scytonemins and carotenoid compound concentrations are high. Scytonemins provide UVR protection. Scytonemin is most abundant in lakes of less than 2 m depth in eastern Antarctic lakes. In deeper lakes, greater than 4 m in depth, scytonemin declines and both chlorophyll and xanthophyll dominate the mat pigment composition, suggesting a strategy balanced towards maximizing photosynthesis in a poorer light climate, as opposed to investment in UVR protection and repair as occurs in shallower lakes. It appears that the major factor in determining the pigment composition of mats is the depth of lakes, which clearly relates to light climate and levels of PAR and UVB radiation. However, turbidity, conductivity dissolved oxygen, sulphate, and geographical location are also factors implicated in some of the variance seen (Hodgson et al. 2004).

In perennially ice-covered Lake Hoare (Dry Valleys), exposure to UVR through thick debris containing ice is not a challenge, rather it is maintaining photosynthesis in a poor light climate. Different types of mats show zonation and extend into deep waters (Figure 2.24). Light penetrating the ice cover is rich in blue and blue-green wavelengths of the PAR spectrum, favourable for photosynthesis particularly by phycobolin-containing organisms like Cyanobacteria. In Lake Hoare, *Oscillatoria* and *Leptolyngbya* morphotypes predominated with the diatoms *Hantzschia amphioxys* var. *maior*, *Navicula*, and *Pinnularia* spp. (Hawes and Schwarz 1999). As well as possessing chlorophyll *a*, the Cyanobacteria contain accessory photosynthetic pigments among which the phycobilins are important in light-harvesting. These are the phycoerythrins that absorb light at 540–560 nm, and the phycocyanins that absorb light at 610–630 nm. These accessory pigments spread the absorption spectrum of the 'trap' for light energy or quanta (Reynolds 1984). Photosynthetic pigments show a marked vertical profile in Lake Hoare (Figure 2.25). At depth, irradiance was usually below a level at which photosynthesis is light saturated, consequently the mat communities are operating at high conversion efficiencies for most of the time (Hawes and Schwarz 1999).

It is clear that the physical structure of mats is affected by depth (Figure 2.24). This is also the case in shallow systems. The benthic mats from 62 freshwater and saline lakes in eastern Antarctica from the Larsemann Hills, the Bølingen Islands, and the Rauer Islands revealed a range of mat morphologies that showed a significant relationship with lake depth (Figure 2.26) (Hodgson et al. 2004). Finely

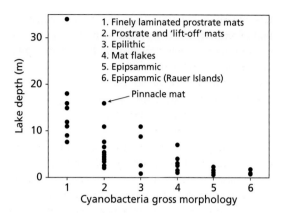

Figure 2.26 The distribution of benthic cyanobacterial mat gross morphology in relation to lake depth in lakes from the Larsemann Hills, the Bølingen Islands, and the Rauer Islands (Prydz Bay, eastern Antarctica). From Hodgson et al. (2004), with the permission of Inter-Research.

laminated mats were mainly found in deeper lakes (>7 m). Less structure prostrate mats and 'lift-off' mats are formed by cyanobacterial colonies competing for optimum light conditions at the mat surface, resulting in the disruption of the continuous filamentous structures seen in deeper lakes. These can form concentric colonies that resemble 'dinner plates'. Colonies can also lift off as a result of trapped gases and may accumulate on the underside of the

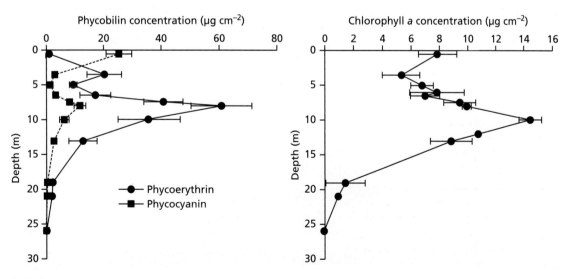

Figure 2.25 Area specific concentrations of photosynthetic pigments of benthic cyanobacterial mats in Lake Hoare. From Hawes and Schwarz (1999), with the permission of John Wiley and Sons Inc.

lake ice cover in winter. These detached portions of mat eventually move upward by melt through the ice and may be blown away. One Larsemann Hills lake had pinnacle mats with a calcite interior. Mat flakes consisting of small unattached fragments about 0.5 to 2 cm in size were found in lakes less than 2.3 m deep. Episammic cyanobacterial communities (colonies attached to sand grains or sandy sediments) were only observed in shallow ponds less than 1.8 m deep. These included saline ponds in the Rauer islands.

Lake Untersee (lake 4, Table 2.1) is one of the largest surface lakes in Antarctica. The floor of the lake lying above the anoxic region (see Section 2.3) is covered with cyanobacterial mats to a depth of 100 m. The mats form two distinct macroscopic structures. The first is common throughout Antarctica, being composed of cm-scale cuspate pinnacles dominated by *Leptolyngbya antarctica* and *L. angustissima*. The second mat type is not common. These mats are made up of laminated, conical stromatolites that rise to 0.5 m above the lake floor, and are made up mainly of *Phormidium autumnale*, with other species including *Leptolyngbya antarctica, L. angustissima, Oscillatoria koettlizii,* and *O. fracta*. In both types of mat two types of unidentified green unicells were also noted. Diatoms were very rare, which is unusual in cyanobacterial mat habitats. This may be a result of the isolation of the lake and its remarkable degree of closure. The ice cover of Lake Untersee and its transparent water column permits good light transmission, allowing 1% of

Figure 2.27 A photograph of benthic mosses from a freshwater lake in the Syowa Oasis. Photo courtesy of S. Kudoh. (See Plate 11)

surface irradiance to reach 42 m and 0.1% to reach 135 m (Andersen et al. 2011).

In some lakes benthic mosses form extensive cover where, like cyanobacterial mats, they provide a habitat for a range of other organisms (Figure 2.27). The lakes of the Syowa Oasis have well-developed moss pillars that can be about 40 cm in diameter and up to 60 cm in height. They are made up of species of *Leptobryum* with lesser quantities of the moss *Bryum pseudotriquetrum*, along with cyanobacteria and diatoms. The colonies are usually found from 3 to 5 m in depth and do not occur in water shallower than 1.5 m, which is the maximum depth of ice cover in winter (Figure 2.28) (Imura et al. 1999). Molecular analysis of mosses from the Syowa Oasis

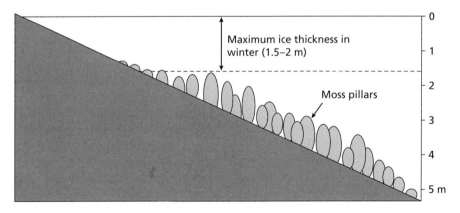

Figure 2.28 Schematic of benthic moss beds in freshwater lakes of the Syowa Oasis. Redrawn from Imura et al. (1999) with the permission of Springer-Verlag.

indicates that *Leptobryum* sp. is conspecific with *Leptobryum wilsonii* described from South America. There is virtually no genetic variation between the mosses from Antarctica and *L. wilsonii* from South America. The evidence suggests that *L. wilsonii* arrived in Antarctica from South America during the Holocene, indicating that it is a postglacial immigrant (Kato et al. 2013). *Bryum pseudotriquetrum* has been found in Lake Radok (Amery Oasis, lake 2, Table 2.1). Here it was found in water up to 81 m in depth. All the samples collected from a range of depths between 66 and 81 m were etiolated and the leaves were very small with relatively high interleaf distances (Wagner and Seppelt 2006). This moss species has also been reported from a depth of 80 m in Lake Figurnoya (Bunger Hills, lake 5, Table 2.1) (Korotkevivh 1972 quoted in Kaspar et al. 1982). Thus in spite of challenging light climate conditions this moss can survive at considerable depth in large ultra-oligotrophic lakes. Aquatic mosses can be extensive in the lakes of the Maritime Antarctic. In aptly named Moss Lake (Signy Island, lake 59, Table 2.1) around 40% of the sediment below 5 m is covered by mosses, mostly *Calliergon sarmentosum* and *Drepanocladus* sp. The mosses have an unusual growth form with large leaves and stems up to 0.4 m long (Priddle 1980a). *Drepanocladus longiformis* occurs in a range of lakes on the Byers Peninsula (Livingston Island, South Shetlands) where it forms benthic carpets on the lake bottom under the shallow littoral zone (Toro et al. 2007).

While diatoms are rare in the plankton of Antarctic freshwater lakes, they are abundant in the algal mats and among mosses, where they exhibit a high degree of biodiversity across the continent (see Figure 1.29a). Taxonomically they are the most abundant components of mats (Vinocur and Pizarro 2000). There appears to be a gradient in diversity, with lower biodiversity in continental lakes compared with the Maritime Antarctic, but there are differences between locations. For example the Larsemann Hills has lower diatom diversity than the neighbouring Vestfold Hills, which is attributed to a greater range of conductivity among the Vestfold Hills lakes and the fact that some of the Vestfold Hills lakes are meromictic (Sabbe et al. 2004). There appears to be a high degree of endemism. In the Larsemann Hills a survey of 66 freshwater and saline lakes and ponds revealed that 39% of the species found were endemic and a further 26% were uncertain, due to an inability to assess their distribution (Sabbe et al. 2003). However, there is debate about the issue of Antarctic diatom endemism due to a lack of taxonomic consistency and incomplete exploration of water bodies across the continent and Maritime Antarctic. There is no Antarctic diatom flora text, so workers have had to rely on literature pertaining to lower latitudes, which has resulted in investigators assigning the same morphotype to different species (Jones 1996). Molecular analysis will undoubtedly assist in resolving the issue of Antarctic diatom diversity and endemism.

Diatom distribution is controlled by physical and chemical parameters. Species abundance is largely controlled by nutrient and salinity gradients in the Maritime Antarctic (Jones et al. 1993). A detailed analysis of lakes on Signy Island and Livingston Island indicated that conductivity, potassium, chlorophyll *a*, sodium, and ammonium are significant factors in explaining differences in diatoms assemblages. Particular species are associated with nutrient-rich waters, for example *Fragilaria construens* var. *binodis*, *Achnanthes pinnata*, *Gomphonema angustatum*, and *Achnanthes subatomoides*, which have total phosphorus optima of >10 µg l^{-1} in a study of Canadian systems (Jones et al. 1993). This is supported by the identification of assemblages typical of oligotrophic, mesotrophic, and proglacial lakes on Signy Island (Oppenheim 1990). In eastern continental lakes, species richness was marginally higher in shallow lakes less than 2.5 m deep, where the communities were characterized by aerial species such as *Diadesmis* cf. *persusilla*, *Hantzschia* spp., *Luticola muticopsis*, *Pinnularia microstauron*, and *Pinnularia borealis*. In deeper lakes the attached species *Psammothidium abundans* dominated the species poor diatom community of finely laminated cyanobacterial mats. Some species dominated hypersaline water bodies, for example *Amphora veneta* and *Craticula* cf. *molesta* (Sabbe et al. 2004).

2.7.2 Heterotrophic benthic communities

There are limited data for heterotrophic bacteria in benthic mats. A study of lakes from the Dry Valleys (Lakes Hoare and Frxyell), a range of freshwater and saline lakes in the Vestfold Hills, and Lake Reid in the Larsemann Hills revealed that cultured strains belonged to the α-Proteobacteria, the β-Proteobacteria, the γ-Proteobacteria, the high and low guanine to cytosine (G + C) Gram-positive bacteria (Actinobacteria), and the Cytophaga-Flavobacterium-Bacteriodes groups and phyla. The nearest named phylogenetic neighbours of the isolated strains belonged to taxa that have been isolated from cold aquatic environments, for example *Shewanella baltica*, *Psychrobacter glacincola*, and *Flavobacterium frigidarium*. Of the strains 16 showed pairwise sequence similarities of less than 97% to their nearest validly named neighbours, which indicates that they are new taxa belonging to species for which no sequences were available. There appeared to be a high diversity, based on culture-dependent techniques, and apparently a degree of endemism (Van Trappen et al. 2002). Interestingly this study was undertaken as part of a European Community funded project with industry, called MICROMAT. Its aim was to characterize microbial diversity in Antarctic microbial mats and to screen isolates for novel bioactive chemicals. Some of the isolated strains (*Arthrobacter agilis*) produced potent antibiotics active against Gram-positive bacteria (Rojas et al. 2009).

The benthic environment is relatively rich in metazoans compared with the plankton. Mats and sediments harbour nematodes, tardigrades, rotifers, and some crustaceans such as *Acanthocyclops mirnyi* (see Section 2.2.5). Moraine ponds, thaw ponds, and nearshore ponds on King George Island in the Maritime Antarctic support nematodes, mostly belonging to the families Plectidae and Monhysteridae. In all water bodies the most common species were *Eumonhystera* sp. and *Plectus antarticus*. Tardigrades were also common, in moraine ponds *Dactylobiotus ambiguous* was common, but it did not occur in nearshore ponds where *Hypsibius arcticus* constituted 50% of the tardigrade community. In thaw ponds tardigrades were numerous where *Hypsibius papillifer* and *Hypsibius renaudi* dominated. Rotifers are common in the benthos and are also part of the pelagic community of Antarctic lakes. Both Monogononata and Bdelloidea groups were represented (see Section 1.5.7). In moraine ponds *Resticula gelida* dominated, in thaw ponds and nearshore ponds *Notholca squamula salina* was common, where it constituted up to 50% and of all rotifers (Janiec 1996).

In deeper water bodies in the Vestfold Hills a range of benthic species have been observed. In Watts Lake, which is slightly brackish but contains a freshwater biota, unidentified tardigrades and nematodes have been recorded. The rotifers *Encentrum mustela*, *Lepadella patella*, *Lindia torulosa*, *Notholca* sp. *Ptygura crystallina*, *Adineta grandis*, and *Philodina gregaria* occur. A similar range of species was noted in Crooked Lake and Lake Druzbhy (Dartnall 2000). Lakes in the Larsemann Hills, including Progress Lake and Discussion Lake (lakes 29 and 57, Table 2.1), harboured two species of the tardigrade *Isohypsibius*, nematodes, and two to three rotifers species (*Collotheca ornata cornuta*, *Lapadella patella*, *Notholca* sp.) (Dartnall 1995).

2.7.2 Carbon cycling in the benthos

Investigations of heterotrophic bacterial production have focused on the plankton, and as yet we have no information on heterotrophic bacteria that are clearly an integral part of benthic mat communities Given that there must be significant decay of mat material and probably high exudation of photosynthate from mats, the benthic environment is likely to support high bacterial production compared with the plankton. To date the focus has been on algal mats.

Rates of primary production in cyanobacterial mats are often higher than in the plankton (Table 2.10, see also Table 2.2), though there are exceptions. With such high photosynthetic biomass this is not surprising. In Lake Hoare estimated integrated plankton production derived from Priscu (1995) of 7.5 mg C m^{-2} day^{-1} in December are less than one tenth of the values measured at some depths in microbial mats (Figure 2.29) (Hawes and Schwarz 1999). While higher biomass and acclimation to low irradiances are prime factors in enabling high production, another factor is nutrient availability. The phytoplankton is nutrient limited (Priscu 1995) whereas in benthic mats it is likely that concentration gradients of a few mm below the upper layers of the mats are steep, with regeneration of fixed and sediments nutrients. The metabolically

Table 2.10 Freshwater benthic microbial mat primary production across Antarctica from the Dry Valleys to the Maritime Antarctic.

Lake	Depth m	Gross photosynthesis µg C cm² day⁻¹
Lake Hoare, Dry Valleys[1]	6	32.2 ± 3.6
	10	49.4 ± 6.9
	13	34.8 ± 2.4
Watts Lake, Vestfold Hills[2]	3 to 12	74
ephemeral pond, Livingston Island[3]	–	64.8
Lake Kitiesh, King George Island*[4]	9.8	66.2 (mean for Feb)

*benthic moss 1: Hawes and Schwarz 2000; 2: Heath 1988; 3: Fernández-Valiente et al. 2007; 4: Montecino et al. 1991.

active surface of the mats quickly takes up nutrients before they can diffuse into the water column, in effect it is a self-sustaining system for the recycling of nutrients (Hawes and Schwarz 1999; Quesada et al. 2008). Some cyanobacteria are capable of fixing nitrogen (see Section 1.8.1), a capability that confers an advantage in nutrient-poor conditions. This aspect will be covered in more detail for mats in saline environments in Chapter 3.

Most studies of mat photosynthesis are undertaken in the laboratory. Few studies are conducted in situ, but in situ investigations of mats in Lake Hoare have confirmed that at depths of up to 16.5 m the mats are net producers of oxygen during summer and that photosynthesis occurs at close to maximum photosynthetic efficiency (Vopel and Hawes 2006).

Further north in Watts Lake (Vestfold Hills), which has a similar maximum depth to Lake Hoare (Table 2.1) but clear annual ice with a thickness of around 2 m, estimated phytoplankton production exceeded benthic production (10.1 g C m⁻² year⁻¹ and 5.5 g C m⁻² year⁻¹ respectively) (Heath 1988). Annual mat production in Lake Hoare is calculated at 15–16 g C m⁻² year⁻¹ when ≥ 5% of ambient PAR was transmitted through the ice cover (Moorhead et al. 2005). It appears that the more extreme environment of the Dry Valleys is capable of sustaining higher benthic productivity. The chlorophyll specific rates of photosynthesis (assimilation number) were higher in phytoplankton of Lake Watts, ranging from 0.18 to 17.3 µg C µg chl a^{-1} h⁻¹ over a year compared with 0.001 to 3.29 µg C µg chl a^{-1} h⁻¹ in benthic mats. An assimilation number of 0.05 µg C µg chl a^{-1} h⁻¹ for mats was noted in mats covering the bottom of an ephemeral pond on the Byers Peninsula (Livingston Island) (Fernández-Valiente et al. 2007). Lake Kitiesh on King George Island is used as a water supply. Its benthic moss cover has an annual primary production rate similar to cyanobacterial mats (Table 2.9). Areal phytoplankton production averaged 95 µg C cm⁻² day⁻¹ during February, which is almost a third higher than benthic production (Montecino et al. 1991). The assimilation numbers of the phytoplankton and benthic mosses were similar (4.53 and 4.86 respectively), and are higher than reported for benthic cyanobacterial mats in Watts Lake and an ephemeral pond on Livingston Island.

Rates of primary production by benthic mats are similar across a gradient from the Dry Valleys to the Maritime Antarctic, with significantly less variation than is seen in planktonic rates (Tables 2.2 and 2.10), where there are several orders of magnitude

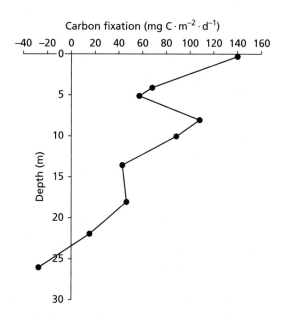

Figure 2.29 Estimated daily in situ primary production by benthic cyanobacterial mats in Lake Hoare. The estimates are derived from incident irradiance, attenuation by ice and water, and mat photosynthesis-irradiance relationships. From Hawes and Schwarz (1999), with the permission of John Wiley and Sons Inc.

difference between lakes. Arctic mats communities have rates of production comparable to those seen in Antarctic lakes (Quesada et al. 2008).

As indicated in Section 2.7.1, microbial mats in lakes of the coastal oases and the Maritime Antarctic support a metazoan and protozoan community that graze on bacteria, flagellates, and algae associated with the mat matrices. Benthic mats from Lakes Hoare and Frxyell support a thriving protozoan community of ciliates and heterotrophic flagellates, as well as rotifers (Laybourn-Parry unpublished). These organisms living in what is a diverse and productive environment undoubtedly have complex trophodynamics. This is an area of Antarctic limnology that demands investigation. We know a great deal more about the more complex benthic communities of Arctic lakes (Vincent et al. 2008; Thomas et al. 2008).

CHAPTER 3

Saline lakes

3.1 Introduction

The definition of saline as opposed to freshwater Antarctic lakes is arbitrary; however, it has been proposed that lakes with a salinity greater than 3‰ (around 9% seawater, seawater = 35‰) should be regarded as saline (Bayly and Williams 1966). Saline lakes fall into two broad categories based on their chemistry. The majority are dominated by chloride ions, either as sodium chloride or combined with another common element such as calcium. However, there are a small number of lakes that are dominated by carbonates, particularly sodium carbonate; these are the so-called soda lakes. They are most often associated with volcanic regions such as the African Rift Valley (e.g. Lake Natron, Lake Magadi, Lake Nakuru, Lake Elmenteita). Saline lakes occur worldwide, the largest being the Caspian Sea, which has an area of 386,000 km² and a mean depth of 187 m. It has a salinity of around 12‰. There are numerous other large saline lakes, for example the Great Salt Lake (North America) and the Dead Sea (between Israel and Jordan). Some global saline lakes can vary considerably in area. Lake Eyre in the northern South Australia desert is a good example and is one of many saline lakes in Australia. Its huge shallow basin is 144 km long by 77 km wide. It fills around every eight years and has only filled to full capacity three times since its discovery by Edward John Eyre in the 1800s. Saline lakes, both within Antarctica and elsewhere, are usually closed or endorheic basins. They have inflows but lack outflows. Water loss is by evaporation, which concentrates the salts dissolved in the water. The water evaporates after the lake fills and this volume can drive significant changes in salinity. For example, Lake Natron in the African Rift Valley shows considerable differences in its area, and hence volume, depending on variations in annual rainfall. Between 1960 and 1980 the area varied between 36 and 49 km², with corresponding 17-fold changes in salinity (Burgis and Morris 1987).

Antarctica has many saline lakes (Table 3.1). Those listed in Table 3.1 are lakes that have been investigated in detail and for which there are morphometric data. There are many more that are unnamed or have not been subject to any significant investigation. Inevitably lakes within easy distance of research stations have been subject to the most detailed research, for example the lakes of the McMurdo Dry Valleys and the Vestfold Hills. Ponds of various sizes have also been subject to investigation, for example the upland ponds of the Taylor Valley, which show changes in size and geochemistry over periods of years (Lyons et al. 2012). Most of Antarctica's saline lakes are dominated by sodium chloride, indicating a marine origin, but there are exceptions, most notably Lake Vanda and Don Juan Pond (Wright Valley) which are dominated by calcium chloride (Tomiyama and Kitano 1985; Green and Lyons 2009).

The Arctic has few saline lakes by comparison. The best known are Lake Sophia on Cornwallis Island and Garrow Lake on Little Cornwallis Island. Lake Sophia is a meromictic lake, which is stratified with strong physical and chemical gradients. The upper waters to around a depth of 10 m are virtually freshwater, but below the chemocline they increase to a maximum of 58‰ (Ouellet et al. 1987). Garrow Lake is also meromictic with salinity in the upper mixolimnion at around 4.5‰–5.7‰, reaching a maximum in the lower monomolimnion of 91.2‰ (Ouellet et al. 1989). Regrettably Garrow Lake has been subject to major human impact

Antarctic Lakes. Johanna Laybourn-Parry and Jemma L. Wadham.
© Johanna Laybourn-Parry and Jemma L. Wadham 2014. Published 2014 by Oxford University Press.

Table 3.1 Brackish, saline, and hypersaline lakes of Antarctica. The lakes listed have published limnological data. There are other saline lakes, both named and unnamed, for which there are no data. Some publications do not give depth, area, or coordinates and have been excluded from the table. Lakes are listed by decreasing surface area.

Lake and notes	Location	Surface area km²	Maximum depth m	Salinity ppt (‰)
1. Lake Fryxell (M)[1]	McMurdo Dry Valleys (77°37'S, 163°09'E)	7.08	20	*6.2
2. Lake Vanda (M)[1]	McMurdo Dry Valleys (77°32'S, 161°34'E)	5.2	69	*124
3. Lake Wilson (M)[2]	McMurdo Dry Valleys (79°49'S, 159°33'E)	~4.4	95	*4
4. Lake Bonney West (M)[1]	McMurdo Dry Valleys (77°43'S, 162°26'E)	3.32	40	*149
5. Burton Lake (M)[3] (has periodic marine incursions)	Vestfold Hills (68°62S, 78°62'E)	1.46	18	*45
6. Lake Bonney East (M)[1]	McMurdo Dry Valleys (77°43'S, 162°26'E)	0.99	37	*239
7. Lake Joyce (M)[1]	McMurdo Dry Valleys (77°43'S, 161°37'E)	0.83	35	*3.5
8. Ekho Lake (M)[3]	Vestfold Hills (68°52'S, 78°27'E)	0.44	43	*165
9. Suribati Ike (M)[4]	Syowa Oasis (69°29'S, 39°39'E)	0.41	31.2	219
10. Laternula Lake (M)[3]	Vestfold Hills (68°64'S, 77°97'E)	0.36	9	*200
11. Clear Lake (M)[3]	Vestfold Hills (68°63S, 77°98'E)	0.36	62	*14
12. Shield Lake (M)[3]	Vestfold Hills (68°53'S, 78°26'E)	0.20	39	*172
13. Highway Lake[5]	Vestfold Hills (68°30'S, 78°14'E)	0.20	15	4
14. Ace Lake (M)[3]	Vestfold Hills (68°28'S, 78°11'E)	0.18	25	*43
15. Oblong Lake (M)[3]	Vestfold Hills (68°62'S, 78°23'E)	0.18	15	*233
16. Oval Lake (M)[3]	Vestfold Hills (68°53'S, 78°28'E)	0.18	16	*228
17. Rookery Lake[5] (receives allochthonous inputs from a penguin rookery; has marine incursions)	Vestfold Hills (68°31'S, 78°00'E)	0.18	2	87
18. Lake Williams (M)[3]	Vestfold Hills (68°48'S, 78°15'E)	0.17	7	150
19. Pendant Lake[5]	Vestfold Hills (68°46'S, 78°24'E)	0.16	12	19
20. Hunazoko Ike[4]	Syowa Oasis (69°26'S, 39°33'E)	0.142	9.2	212
21. Lake Polest[6]	Bunger Hills (66°15'S, 100°56'E)	0.13	6.0	79*

continued

Table 3.1 Continued

Lake and notes	Location	Surface area km²	Maximum depth m	Salinity ppt (‰)
22. Lake Anderson (M)[3]	Vestfold Hills (68°60′S, 78°17′S)	0.12	21	160
23. Lake McCallum (M)[3]	Vestfold Hills (68°64′S, 78°12′E)	0.11	28	25
24. Deprez Lake (M)[3]	Vestfold Hills (68°50′S, 78°20′E)	0.10	10	120
25. Fletcher Lake (M)[3]	Vestfold Hills (68°45′S, 78°25′E)	0.093	12	110
26. South Angle (M)[3]	Vestfold Hills (68°64′S, 77°91′E)	0.079	16	235
27. Lake Abraxas (M)[3]	Vestfold Hills (69°48′S, 78°28′E)	0.077	24	24
28. Lake Johnstone (M)[3]	Vestfold Hills (68°49′S, 78°40′E)	0.075	12	216
29. Lake 1 Filla Island[7]	Rauer Islands (68°48′S, 77°52′E)	0.070	7	71
30. Deep Lake[8]	Vestfold Hills (68°34′S, 78°11′E)	0.064	36	350
31. Lake Bayly (M)[3]	Vestfold Hills (68°44′S, 78°25′E)	0.064	8	120
32. Lake Franzmann (M)[3]	Vestfold Hills (68°48′S, 78°25′E)	0.063	9	140
33. Lake Reid[9]	Larsemann Hills (69°23′S, 76°23′E)	0.055	4	14
34. Lake Farrell (M)[3]	Vestfold Hills (68°52′S, 78°31′E)	0.053	10	224
35. Glider Lake (M)[3]	Vestfold Hills (68°S46′S, 78°28′E)	0.052	9	168
36. Organic Lake (M)[3]	Vestfold Hills (68°45′S, 78°19′E)	0.047	7	230
37. Ephyra Lake (M)[3]	Vestfold Hills (68°57′S, 78°23′E)	0.047	9	33
38. Lake 5 Torckler Island[7]	Rauer Islands (68°53′S, 77°49′E)	0.04	4	55
39. Scale Lake (M)[3]	Vestfold Hills (68°58′S, 78°17′E)	0.038	10	34
40. Crater Lake[10]	Larsemann Hills (69°24′S, 76°11′E)	0.032	12	2.5
41. Nurume Ike (M)[4]	Syowa Oasis 69°14′S, 39°40′E)	0.031	16.6	174
42. Holl Lake[11]	Windmill Islands (66°24′S, 110°24′S)	0.03	7	30
43. Beall Lake[11]	Windmill Islands (66°17′S, 10°29′E)	0.026	6.2	55

continued

Table 3.1 Continued

Lake and notes	Location	Surface area km²	Maximum depth m	Salinity ppt (‰)
44. Lake Anna[10]	Larsemann Hills (69°23′S, 76°17′E)	0.025	7.6	3
45. Lake 3 Filla Island[7]	Rauer Islands (68°49′S, 77°49′E)	0.025	2.5	82
46. Lake 2 Filla Island[7]	Rauer Islands (68°48′S, 77°51′E)	0.02	3	77
47. Lake Alanna[9]	Larsemann Hills (69°28′S, 75°55′E)	0.016	4	2
48. Lake Warrington[11]	Windmill Islands (66°20′S, 110°27′E)	0.011	1.9	30
49. Sarah Tarn[10]	Larsemann Hills (69°23′S, 76°23′E)	0.01	2.5	34
50. Jaw Lake[12]	Bunger Hills (66°12′S, 101°58′E)	Unknown 300 m long	~7	22.7

* taken at the lake bottom, (M) meromictic
1: Spigel and Priscu 1998; 2: Webster et al. 1996; 3: Gibson 1999; 4: Tominaga and Fukui 1981; 5: Laybourn-Parry et al. 2002; 6: Kaup et al. 1993; 7: Hodgson et al. 2001b; 8: Ferris and Burton 1988; 9: Ellis-Evans et al. 1998; 10: Sabbe et al. 2004; 11: Roberts et al. 2001; 12: Roberts et al. 2000).

since 1981 when a mining project started discharging high levels of lead and zinc mine tailings into the lake at a depth of 20 m below the chemocline. This has led to major irreversible biological changes. Other saline lakes have been described from Ellesmere Island in the high Arctic, including Lake Tuborg and Lakes A, B, and C (Ludlam 1996 and reference therein). A detailed investigation of meromictic Lake A's geochemistry suggests that it was originally connected to an epishelf lake when relative sea level was higher, and subsequently became cut off to form an isolated lake (Gibson et al. 2002b).

Productivity in Antarctic brackish and saline lakes is generally higher than that seen in the freshwater lakes (Table 3.2, see also Table 2.2 for comparison). In the hypersaline lakes there is reduced biodiversity with an increased dominance of prokaryotes along the continuum from around seawater to ten times seawater—the most saline lakes (see Table 3.1). Throughout this chapter salinity will be given in ‰ (parts per thousand, ppt). There is a range of ways to express salinity, of which conductivity and ppt are the most widely used in limnology, but marine scientists more commonly use PSU (practical salinity units).

3.2 Distribution of saline lakes in Antarctica

While Table 3.1 lists those saline lakes for which there are data, it does not give an indication of the wider distribution of saline lakes and ponds for which there are few data. The Syowa Oasis (see Figure 1.1) has a significant number of hyposaline, saline, and hypersaline lakes, some of which are meromictic (see Section 1.6.1 and Figure 1.14). Lake Hunazoko-ike (lake 20, Table 3.1) is an example of a hypersaline lake with a salinity of 95‰ in the top two metres of the water column, increasing to 200‰ in the lower 2–7 m. Other hypersaline lakes include Suribati-ike, which is up to 32 m in depth. This lake is meromictic with a strong chemocline above which salinity is 48‰, increasing to 200‰ in the lower monimolimnion. Meromictic hyposaline lakes include Kobati-ike with a mixolimnion salinity of around 8‰ and a monomolimnion salinity of 16‰–17‰ and Nurume-ike where the upper waters are 15‰ and the lower waters 32‰ (Kimura et al. 2010). Along with the Vestfold Hills, the Syowa Oasis possesses some of the most saline lakes in Antarctica. There is little information about the geomorphological origin of the lakes in the

Table 3.2 Chlorophyll *a* concentration, primary production, and bacterial production in saline Antarctic lakes.

Lake and location Note: lake numbers correspond with those in Table 3.1	Chl *a* µg l^{-1}	Primary production µg C l^{-1} day^{-1}	Bacterial production µg C l^{-1} day^{-1}
1. Lake Fryxell (McMurdo Dry Valleys 77°37′S, 163°09′E)[1,2,3,4]	0.2–21 over 2 summers, between 4 and 18 m	Max. 73[2] over summer 0–14.4[3] over summer, between 4.5 and 15 m	0–10 over 4 summers between 5 and 18 m
2. Lake Vanda (McMurdo Dry Valleys 77°32′S, 161°34′E)[5]	0.5–0.92 between 3.25 and 57.5 m	0.1–1.6 between 3.25 and 57.5 m	–
3. Lake Wilson (McMurdo Dry Valleys 79°49′S,159°33′E)[6]	0.05–0.4 between 5 and 45 m	0–7.2 at 10 m, 45 m, and 50.5 m	–
4. Lake Bonney West (McMurdo Dry Valleys 77°43′S, 162°34′E)[1,7,4]	0.1–7.5 over 2 summers between 4 and 39 m	0–10 between 4 and 25 m	0–1.1 over 4 summers between 4 and 38 m
6. Lake Bonney East (McMurdo Dry Valleys 77°43′S, 162°26′E)[1,7,4]	0.1–1.8 over 2 summers between 4 and 39 m	0–1.6 between 4 and 39 m	0–4.9 over 4 summers between 4 and 30 m
7. Lake Joyce (McMurdo Dry Valleys 77°43′S, 161°37′E)[1,8]	0.05–3.0 over 2 summers between 5 and 40 m	0–2 between 5 and 28 m	–
5. Burton Lake (Vestfold Hills 68°52′S, 78°27′E)[9] (experiences occasional marine incursions)	33.2 ± 5.8	–	–
8. Ekho Lake (Vestfold Hills 68°52′S, 78°27′E)[9]	4.4 ± 2.7	–	–
9. Suribati Ike (Syowa Oasis 69°29′S, 39°39′E)[10]	1–10 between 0 and 30 m	–	–
12. Shield Lake (Vestfold Hills 68°53′S, 78°26′E)[9]	0.7 ± 0.2	–	–
13. Highway Lake (Vestfold Hills 68°30′S, 78°14′E)[11,12]	0–12.6 over year between 2 and 8 m	0–436.8 over year between 2 and 8 m	12.0–53.5 over summer, between 0 and 4 m
14. Ace Lake (Vestfold Hills 68°28′S, 78°11′E)[12,13]	0.7–5.74 over year between 2 and 10 m	0–345.6 over year between 2 and 10 m	18–72 over year between 2 and 8 m
16. Oval Lake (Vestfold Hills 68°53′S, 78°28′E)[9]	3.2 ± 2.8	–	–
17. Rookery Lake (Vestfold Hills 68°31′S, 78°00′E)[11] (receives allochthonous inputs from neighbouring penguin rookery, and has occasional marine incursions)	2.0–35.0 over summer between 0 and 1 m	3.6–229.4 over summer between 0 and 1 m	177.6–7848 over summer, between 0 and 1 m
17. Lake Williams (Vestfold Hills 68°48′S, 78°15′E)[11]	2.5–25.0 over summer between 0 and 1 m	7.9–583 over summer between 0 and 1 m	5.5–146.6 over summer between 0 and 1 m
19. Pendant Lake (Vestfold Hills 68°46′S, 78°24′E)[11,13]	7.5–29.0 over summer between 2 and 10 m	26.4–148.8 over year between 2 and 12 m	12–79.2 over year between 2 and 12 m
20. Hunazoko Ike (Syowa Oasis 69°26′S, 39°33′E)[10]	0.1–2.3 between 0 and 7 m	–	–
22. Lake Polest (Bunger Hills 66°15′S, 100°56′E)[14]	0.25–0.6 Feb to March between 0.5 and 5.5 m	2.5–160 Feb to March between 0.5 and 5.5 m	–
27. Lake Abraxas (Vestfold Hills 69°48′S, 78°28′E)[9]	0.2 ± 0	–	–
33. Lake Reid (Larsemann Hills 69°23′S, 76°23′E)[15]	0.5	–	–
36. Organic Lake (Vestfold Hills 68°45′S, 78°19′E)[9]	1.4 ± 0.7	–	–
42. Nurume Ike (Syowa Oasis 69°14′S, 39°40′E)[10]	0.15–1 between 0 and 9 m	–	–
49. Sarah Tarn (Larsemann Hills 69°23′S, 76°23′E)[15]	0.72	–	–

1: Roberts et al. 2004a; 2: Vincent 1981; 3: Priscu et al. 1987; 4: Takacs and Priscu 1998; 5: Vincent and Vincent 1982; 6: Webster et al. 1996; 7: Priscu 1995; 8: Parker et al. 1982; 9: Perriss and Laybourn-Parry 1997; 10: Tominaga and Fukui 1981; 11: Laybourn-Parry et al. 2002; 12: Laybourn-Parry et al. 2005; 13: Laybourn-Parry et al. 2007; 14: Kaup et al. 1993; 15: Ellis-Evans et al. 1998

Syowa Oasis, but they are likely to have the same history as coastal saline lakes elsewhere in Antarctica (see Sections 1.7 and 3.3).

Saline lakes occur in the Bunger Hills, 12 of which are listed by Kaup et al. (1993) (see Figure 2.3). The lakes range from close to freshwater (3.4‰) to hypersaline. The largest and most saline is Lake Polest (Lake 21, Table 3.1) which displayed salinity stratification with a halocline in February, while it had a uniform salinity in the water column in March (65‰). The lakes of the Bunger Hills are less saline than those found in the Syowa Oasis and the Vestfold Hills. They are relatively shallow, Lake Polest is only 5.5 m deep, and consequently permanently stratified meromictic lakes have not developed in this region.

The Schirmacher Oasis lacks saline lakes and this is probably due to it being 80–100 km distant from the sea. The total salt content of the majority of the lakes in the region is below 500 mg l^{-1}. Salt concentrations above this are rare, but do occur in shallow endorheic pools that have been subject to evaporation. These can dry down completely and may freeze to their bases in winter (Wand 1995).

Saline lakes are not reported from the Maritime Antarctic. There are some small waters bodies that have elevated conductivities but these cannot really be classified as brackish. Some ponds close to the Polish Station on King George Island (South Shetlands), that clearly experience a marine influence have conductivities of around 2.6 to 2.8 mS cm^{-1} (Kawecka et al. 1998), which equates to a salinity of around 2‰, fitting into the classification of freshwater. Another pond on the Potter Peninsula, King George Island had a reported conductivity of 3.9 mS cm^{-1}, equating to a salinity of around 3‰ (Vinocur and Pizarro 2000), again on the borderline between freshwater and brackish. Lake 37, a shallow pond on Livingston Island, has a conductivity of 2.9 mS cm^{-1} (Jones et al. 1993). The Maritime Antarctic limnology is typified by small freshwater systems (see Table 2.1).

The Vestfold Hills has a large number of brackish (hyposaline), saline, and hypersaline lakes (Figure 2.1). The presence of Davis Station, which has overwintering research programmes, has resulted in the intensive investigation of a number of these lakes, notably Ace Lake, Deep Lake, Ekho Lake, Organic Lake, Pendant Lake, and Highway Lake (lakes 14, 30, 36, 19, and 13, Table 3.1). Meromictic lakes are common, around 25 have been profiled (Gibson 1999). These range in maximum salinity (measured in the monimolimnion) between 14‰ and 235‰. Hyposaline amictic lakes such as Highway Lake, Lake Vereteno, and Lake Collerson are also common. The Vestfold Hills has a number of extremely hypersaline monomictic lakes that undergo thermal stratification in summer and are mixed in winter, and which are so saline they rarely develop winter ice covers. The most saline is Deep Lake which has a salinity of around ten times that of seawater (lake 30, Table 3.1). Without the thermal protection of an ice cover the temperature in Deep Lake can fall to between –17 °C and –18 °C in the winter, while for brief periods in the summer temperatures may reach 7 °C–11.5 °C (Ferris and Burton 1988).

The lakes of the nearby Larsemann Hills are generally smaller and shallower than those of the Vestfold Hills (see Tables 2.1 and 3.1 and Figure 2.2). The majority are freshwater, but some are hyposaline (Table 3.1). The Rauer Islands lie in close proximity to the Vestfold Hills. They carry a suite of small, shallow saline lakes that range from hyposaline to hypersaline (around three times seawater) (Hodgson et al. 2001b). The Windmill Islands in Eastern Antarctica (66°S, 110°E) also have a number of saline lakes that range from 4‰ to 66‰ (Roberts et al. 2001).

Among the largest saline lakes in Antarctica are those of the McMurdo Dry Valleys (Table 3.1 and Figure 2.7). The lakes of the Taylor Valley have been subject to detailed investigation as part of the United States National Science Foundation Long Term Ecosystem Research Program on the Dry Valleys (LTER). The Taylor Valley saline lakes are Lake Bonney and Lake Fryxell (lakes 4/6 and 1, Table 3.1). Lake Vanda in the Wright Valley, Lake Wilson in the Darwin Valley, and Lake Joyce in the Pearce Valley have received less attention (lakes 2, 3, and 7 Table 3.1). All of these lakes show chemical and thermal stratification. They range from hyposaline (Lakes Wilson, Joyce, and Fryxell) to hypersaline (Lakes Bonney and Vanda).

Saline ponds are common in Antarctica. Many freeze to their bottom in winter but may have a layer of highly saline slush underlying the ice. This type of freeze/thaw regime poses considerable challenges to the organisms that have successfully colonized them. The most saline and unusual of these ponds is Don Juan Pond in the Wright Valley (McMurdo Dry Valleys). It is extremely saline up to 671‰ (around 19 times seawater) being a near saturated calcium chloride solution. It is 100 m × 300 m and only around 10 cm deep. On occasion it completely evaporates. Its high salt concentration prevents it from freezing even at temperatures as low as −55 °C (Marion 1997; Samarkin et al. 2010). There are numerous upland ponds in the Dry Valleys, some of which have been subject to long-term studies. The Marr Ponds and Parera Ponds complexes in the Taylor Valley show annual variations in size and ionic concentration. These variations relate to glacial melt during summer. The ponds in each complex are in close proximity but they have a diverse geochemistry ranging from Na-rich to Ca-rich and Cl-rich and HCO_3-rich (Lyons et al. 2012). Sodium chloride and Na-HCO_3 ponds occur in the Victoria Valley and Bull Pass region (Webster et al. 1994) and in the Labyrinth region of the Wright Valley (Healy et al. 2006). Shallow coastal ponds are also found in the McMurdo region. They usually freeze to their bases in the winter. An example is Pony Pond on Cape Royds, Ross Island. Its salinity varies considerably as it thaws and freezes. At midsummer it is effectively freshwater with a salinity of around 1.8‰–2.3‰, but as summer ends and freezing progresses, salts in the water become concentrated making the pond a hyposaline environment (3.9‰) (Dieser et al. 2012). Forlidas Pond in the Dufek Massif in the Pensacola Mountains at the northern end of the TransAntarctic Mountains is another example of a saline pond. It is the remains of what was once a much more extensive water body that has evaporated (Peeters et al. 2011). Ponds, both freshwater and saline, are a common feature of the ice-free areas of Antarctica. They have attracted much less attention from scientists than lakes, so the database on their biology and chemistry is not extensive.

3.3 Formation of saline lakes

A broad outline of the formation of Antarctic lakes, including saline lakes, is given in Section 1.7. As outlined in Section 1.4 box 1.1 stable isotopes are a valuable tool in determining the sources of lake waters and interpreting lake history. The saline lakes of the Dry Valleys have a complex history. Chloride is a major ion in these systems. It is extremely soluble and is not controlled by biochemical or geochemical processes in aquatic ecosystems. Two stable isotopes of chlorine, ^{37}Cl and ^{35}Cl, can undergo fractionation and hence, the $\delta^{37}Cl$ can be used to interpret cycling of Cl^- in saline lake waters, enabling inferences regarding lake evolutionary history. The precipitation of chloride salts is the major process that fractionates chlorine isotopes at low temperatures. The halite that precipitates is enriched in ^{37}Cl, while the residual brine simultaneously becomes depleted in ^{37}Cl. Consequently brine from which halite has been precipitated should be depleted in ^{37}Cl and a solution into which halite has been dissolved should be enriched in ^{37}Cl (Lyons et al. 1999).

Lake Bonney (lake 4/6, Table 3.1) is located about 30 km from the Ross Sea at an elevation of 57 m (for map see Figure 2.7). It has two lobes or basins where the mixolimnion of each lobe is in contact at above around 13 m (see Table 3.1). The geochemistry of the two lobes differs and this reflects different evolutionary histories. A lake has existed in the Lake Bonney basin for about 300,000 years, which has waxed and waned in response to climatic change in the region. It is suggested that the mixolimnia of both lobes of Lake Bonney and Blood Falls, a subglacial discharge from the snout of the Taylor Glacier, are the remnants of a marine fjord that was overlying the Taylor Valley during the Tertiary period (Hendy 2000; Green and Lyons 2009). The main source of Cl^- to the Lake Bonney lobes mixolimnia is the discharge from Blood Falls. This has a red-yellow colour resulting from abundant iron oxides. The hypothesis is that Blood Falls and probably saline deposits beneath the Taylor Glacier are the frozen portion of a larger saline lake known as Greater Lake Priscu (Lyons et al. 2005; Green and Lyons 2009). The falls discharge sporadically, the discharge becomes coloured when the reduced iron oxidizes on contact with the air, giving it a red-yellow colour (Mikucki et al. 2004). During the 1994–1995 flow season, Blood Falls and the stream draining Blood Falls contributed ~95% of the total Cl^- to Lake Bonney, although in other flow years it may be lower (Lyons et al. 1998b; Green and Lyons 1999). Assuming this has been the case over the recent past, it explains why the surface waters of Lake Bonney are less depleted in $\delta^{37}Cl$ representing the isotopic signature of Blood Falls. The value of –0.37‰ for $\delta^{37}Cl$ implies that the chloride in Blood Falls may originate from chloride salts close to the final stages of halite precipitation from seawater (Lyons et al. 1999; Lyons et al. 1998b). The monomolimnia of the two lobes are very different (Figure 3.1). The evidence suggests that the source of Cl^- in the west lobe is marine salts that have been unaffected by fractionation processes. In contrast the monomolimnion of the east lobe is supersaturated with respect to NaCl, indicating that halite and dihydrohalite in the surficial sediments are derived from the precipitation of the lake waters. The depletion of ^{37}Cl in the east lobe indicates a brine from which NaCl has been removed (Lyons et al. 2005).

Lake Fryxell (lake 1, Table 3.1) lies lower down the Taylor Valley around 8 km from the Ross Sea at an elevation of 18 m (Figure 2.7). It has a large number of short melt streams flowing into it. The $\delta^{37}Cl$ of Lake Fryxell waters within the mixolimnion indicates that it is influenced by marine-aerosol-borne NaCl. At various times since the early Pliocene the Lake Frxyell basin has experienced marine incursions as sea level has changed with glacial advances and retreats. Thus the Cl^- in the monomolimnion may have been of marine origin. An isotopically depleted $\delta^{37}Cl$ in the monomolimnion (Figure 3.1) suggests previous fractionation via NaCl precipitation (Lyons et al. 2005). The present day Lake Fryxell and Lake Bonney are derived from Glacial Lake Washburn that occupied the Taylor Valley during the LGM (see Section 1.7). Both of these lakes dried down to hypersaline ponds at around 1000–1200 yrs B.P. and have subsequently refilled (Lyons et al. 1998a). Analysis of a 9.14 m sediment core from Lake Fryxell indicates that a lake was present in the Frxyell basin at about 48,000 yrs B.P., before the Taylor Valley became occupied by proglacial Lake Washburn (see Section 1.7). Lake Washburn lowered at the end of the LGM leaving a lake in the Fryxell basin. The lake experienced a series of high and low stands. From about 4000 yrs B.P. conditions were similar to those experienced today. Subsequently the lake evaporated to a hypersaline pond and has since refilled (Wagner et al. 2006).

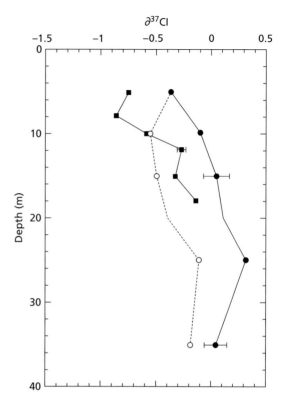

Figure 3.1 Profiles of $\delta^{37}Cl$ in relation to depth in Lakes Bonney and Fryxell, Taylor Valley, McMurdo Dry Valleys. ■ Lake Fryxell, ○ Lake Bonney East, ● Lake Bonney West. Redrawn from Lyons et al. (2005).

Lake Vanda (lake 2, Table 3.1) in the Wright Valley is fed by the Onyx River, which is its sole water input, derived from seasonal melt of the Lower Wright Glacier (Figure 2.7). Unlike many of the other Dry Valley lakes it possesses a smooth, transparent perennial ice cover around 4 m thick. It also differs from the other lakes in having a calcium chloride brine as opposed to a sodium chloride brine. It is suggested that the salts in the lake are derived from two main sources, firstly the deep-ground water reservoir of the Don Juan Basin and the Onyx River. The water above 52 m (the mixolimnion) has Ca and Cl concentrations of 2.18 and 2.16 mM at 12 m respectively. The lake is highly stratified. Below 52 m salinity rises to 124‰ and temperature to close to 25 °C. It is suggested that the Lake Vanda brine is the result of an older deeper lake that evaporated into a hypersaline pool, possibly at about 2000 to 3000 years ago (Wilson 1964;

Green and Lyons 2009). With the advent of warmer climatic conditions the Onyx River flowed into the Lake Vanda basin and covered, without mixing, the denser existing brine. These lower hypersaline waters are very stable over long-term timescales (Canfield et al. 1995). In common with other Dry Valley lakes, Lake Vanda has risen in the recent past. Since the early 1980s the lake has risen 10 m due to increased inflow (Chinn 1993).

Lake Wilson (lake 3, Table 3.1) in the Darwin Valley has risen around 25 m since 1975 (Figure 2.7). The increased flow has augmented the mixolimnion waters and has not mixed with the monomolimnion. Like the other Dry Valley lakes, Lake Wilson experienced a major evaporative phase and dried down to a saline pool at about 1000 yrs B.P. (Webster et al. 1996).

Coastal saline lakes mostly postdate the LGM. As described in Section 1.7 these lakes were formed by glacioisostatic uplift or rebound. In some cases the water in the closed basins became highly concentrated by evaporation to form hypersaline lakes, while in others the water became diluted by inflow and became hyposaline lakes. Some lakes have complex histories having experienced successive marine incursions and phases of meromixis. One of the most studied lakes is Ace Lake in the Vestfold Hills (for map see Figure 2.1). The lake's palaeolimnology has been particularly well documented (Burton and Barker 1979; Bird et al. 1991; Fulford-Smith and Sikes 1996; Zwartz et al. 1998; Roberts and McMinn 1999; Cromer et al. 2005a). Studies based on diatom, protozoan and animal fossils, $\delta^{13}C$ and $\delta^{18}O$ analysis, sulphur chemistry, and radiocarbon dating have allowed a detailed picture of the evolution of Ace Lake (lake 14, Table 3.1) to be constructed.

Following the retreat of the polar ice sheet at the end of the LGM, the exposed Ace Lake basin gradually filled with glacial meltwater, starting at around 13,000 yrs B.P. Thus the original lake was freshwater, and this is evident from the sediment diatom assemblages, which are benthic hyposaline–freshwater, and the loricae of the rotifer *Notholca* sp. affin. *verae*. Relative sea level rise then occurred allowing incursion of seawater across the sill that separated the lake from the sea at around 8000 to 9000 yrs B.P. These incursions were probably seasonal, occurring during the summer melt phase. During the winter the lake ice cover, which ranges between 1.5 and 2.0 m, would have precluded inflow (Figure 3.2). It is suggested that the freshwater would have been replaced fairly rapidly, perhaps in a matter of years. The sediments laid down during this phase contain diatoms of marine planktonic and sea-ice origin. The entering seawater would have brought with it other marine organisms, many of which leave no identifiable remains. However, copepod eggs, skeletal remains and faecal pellet, the testate marine ciliate *Magnifolliculina*, and tintinnid ciliate tests all occur in the sediments laid down during this period. At this time the lake was probably meromictic. Increasing relative sea level gradually transformed the lake into a marine basin (Figure 3.2) with typical planktonic marine and sea-ice diatoms and marine animal and protist remains being laid down in the sediment. The presence of diatoms and other fossil remains is reduced during this period (approximately 9000–5700 yrs B.P.). However, the diversity of animal and protist remains is higher, perhaps suggesting lower productivity, or alternatively greater grazing pressure resulting in a paucity of remains. A subsequent reduction in relative sea level once again transformed the lake into a seasonally isolated marine basin, which eventually became isolated from the sea into the present day Ace Lake at about 5100 yrs B.P. (Figure 3.2). The marine planktonic and sea-ice diatom assemblage was replaced by the benthic hypersaline taxa that dominate the lake today. The more recent sediments show a reduction in the abundance of ciliate, rotifer, and copepod remains. Ace Lake has probably undergone several phases of meromixis interspersed with phases of mixing. Anderson Lake (lake 22, Table 3.1) also started its existence as a freshwater lake at around 8500–7000 yrs B.P. It then became a marine inlet due to relative sea level rise and at about 6000 yrs B.P. was isolated from the sea as a lake (Roberts and McMinn 1998). Today Anderson lake is a meromictic, hypersaline lake.

Carbon-14 dating of Highway Lake (lake 13, Table 3.1) and Organic Lake (lake 36, Table 3.1) sediment cores indicate that they have a less complex history than Ace Lake. Highway Lake was isolated from the sea at least 4600 yrs B.P. The sediment core shows a clear transition from marine to lacustrine. Subsequently the lake's waters have become

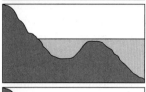
Unit 1: Freshwater lake
ca. 13,000–9400 cal yr B.P.
Anoxia: absent

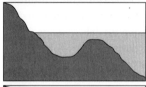
Unit 2: Seasonally isolated marine basin
9400–*ca.*9000 cal yr B.P.
Anoxia: present

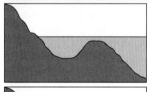
Unit 3: Open marine basin
ca. 9000–5700 cal yr B.P.
Anoxia: absent

Unit 4: Seasonally isolated marine basin
5700–5100 cal yr B.P.
Anoxia: present

Unit 5, 6: Saline lake
ca. 5100 cal yr B.P. - Present
Anoxia present

Figure 3.2 The evolution of saline Ace Lake, Vestfold Hills. Starting from the top panel where the lake was freshwater to the bottom panel which represents the present lake. Redrawn from Cromer et al. (2005a).

diluted by snow melt so that the present day lake is hyposaline. Organic Lake, in contrast, was isolated at about 2700 yrs B.P. The sediment core shows a 13 cm transition from marine to lacustrine material, suggesting that Organic Lake existed as a lake with seasonal marine incursions for several hundred years, similar to what is seen today in Burton Lake (Bird et al. 1991; Zwartz *et al.* 1998). Organic Lake has evolved into a meromictic, hypersaline lake.

Analysis of a sediment core from Lake Abraxas (lake 27, Table 3.1) suggests that this lake predated the LGM. Radiocarbon dating of the basal sediments gives a date of 22,370 ± 1280 yrs B.P. which predates the coastal Antarctic glaciation. However, the authors (Gibson et al. 2009) caution that ^{14}C dates from Antarctic lake sediment can be misleading because of input of old carbon trapped in glaciers and permafrost, reduced gas exchange with the atmosphere due to perennial ice coverage, recycling of old carbon into meromictic lakes, and incorporation of old carbon into sediments through reworking. Biological sedimentation is evident from the second oldest ^{14}C date at around 12,860 ± 620 yrs B.P. The most notable fossil is the formaminiferan *Paratrochammina minutissima*. Foraminiferans are exclusively marine. The lake has been saline throughout its history. There is no evidence of Holocene marine incursion, as seen in Ace Lake. Based on a range of evidence, Gibson et al. (2009) conclude that Lake Abraxas must have acquired its salt waters prior to the LGM and may possibly be over 120,000 years old.

Based on diatom inferred salinities, Holl Lake and Beall Lake (lakes 42 and 43, Table 3.1) in the Windmill Islands have experienced a decrease in salinity during the late Holocene (Hodgson et al. 2006b). Beall Lake was isolated from the sea at about 4800 yrs B.P. The base of a core taken from Beall Lake is dominated by *Navicula glaciei*, a species typical of hypersaline conditions. Above this layer is a sediment containing hyposaline diatoms. A rapid short-term increase in salinity then occurred with a dominance of the hypersaline diatom *Navicula cf. cryptotenella*. A hyposaline phase then ensued with a relatively high species diversity, including freshwater and saline and euryhaline taxa. The uppermost section of the sediment shows a return to the hypersaline conditions seen today. In contrast, Holl Lake started as a hyposaline lake at about 6639–6941 yrs B.P., with freshwater and euryhaline diatom taxa. It has never experienced direct marine input and was a proglacial lake. The lake then became freshwater when the freshwater diatom *Psammothidium abundans* dominated. This was followed by a hyposaline period and finally by a hypersaline environment. Both lakes show a pattern of decreasing salinity during the late Holocene indicating a positive moisture balance, probably from an annual input of snow meltwater. The rapid increase in salinity over the last few decades is indicative of water bodies that have switched from a positive to a negative water balance, where evaporative water losses and sublimation from ice cover exceeds meltwater inputs from the catchment. These changes are likely to be related to local climate change.

The Bunger Hills were probably ice-free during the LGM, as was the Broknes Peninsula in the Larsemann Hills (Gore et al. 2001; Hodgson et al. 2001a), see Section 2.2. Lake Reid (lake 33, Table 3.1) is situated on the Broknes Peninsula. An analysis of a sediment core indicates that this lake has existed since at least the late Pleistocene (Hodgson et al. 2001a).

Today Lake Reid is a hyposaline lake, among a suite of lakes that are predominantly freshwater.

3.4 Patterns of stratification and temperature

Saline lakes show three broad types of stratification. Those indicted by the letter M in Table 3.1 are meromictic. They are permanently stratified with strong physical and chemical gradients (see Figure 3.3, Figure 1.14, and Section 1.6.2). Many of the hyposaline lakes are amictic and are mixed both under ice and when they are ice-free. Some of the most hypersaline lakes are monomictic, with thermal stratification in summer and mixis in winter.

As Figures 3.3 and 1.14 show, meromictic lakes possess strong gradients in temperature, salinity, and oxygen. The upper waters, the mixolimnion, are usually colder, lower in salinity, and well oxygenated, while the lower waters (the monomolimnion) are anoxic, higher in salinity, and may in some cases be warmer. Lake Abraxas, Echo Lake, and a number of other meromictic lakes in the Vestfold Hills have monomolimnion temperatures higher than those in the mixolimnion (Gibson 1999). Among the Dry Valley lakes, Lake Vanda has temperatures in its monomolimnion that are around 25 °C. There has been conjecture as to the sources of these warm waters. Geothermal sources were suggested, but there is now unequivocal evidence from the Dry Valley Drilling Project 1973–1974, that solar energy

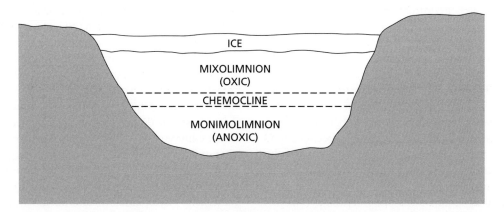

Figure 3.3 A schematic showing the structure of a permanently stratified meromictic lake. The upper mixolimnion encompasses the euphotic zone where the waters are oxygenated. The chemocline is a zone of strong gradients in physical/chemical conditions (temperature, salinity and nutrients). The lower waters or monimolimnion are permanently anoxic.

penetrates the ice and is stored at depth within the water column being stabilized by salinity gradients (Wilson et al. 1974; Spigel and Priscu 1998). Lake Vanda has clear ice cover and extremely transparent waters, which allow penetration of solar energy to great depth. The same phenomenon of solar heating operates in other meromictic lakes.

The water column of these thermally and chemically stratified lakes effectively has two clear biological zones, the upper mixed oxygenated waters within the photic zone, and lower permanently anoxic waters that are the domain of anaerobic microorganisms. As we shall see in Sections 3.6.1 and 3.6.2, the chemocline that separates the mixolimnion and monimolimnion is a region of high biological activity because nutrients diffuse upwards from the nutrient-rich monimolimnion. The chemocline is usually within the photic zone, albeit with poor levels of PAR. Consequently deep chlorophyll maxima are often associated with the chemocline.

Surface water temperatures in coastal meromictic and amictic lakes can be lower than neighbouring freshwater lakes, particularly when ice does not break out completely or is late in so doing. For example in the summer of 1999–2000, the ice on Pendant, Highway, and Ace Lakes did not melt, although moating occurred. Maximum summer water temperatures ranged between –1.1 °C and 2.1 °C (Laybourn-Parry et al. 2002). However, in summers when there is a lengthy ice-free phase, temperatures can climb to well above those seen in the freshwater lakes. For example, in February 1996 the temperature in the mixolimnion of Ace Lake ranged from 1 °C at the surface to 8 °C above the chemocline (Bell and Laybourn-Parry 1999a). A similar summer temperature was recorded for hypersaline Nurume Ike in the Syowa Oasis (Torii et al. 1988). In the more extreme perennially ice-covered lakes of the Dry Valleys, surface water temperatures ranged between 0.0 1 °C and 2.7 °C in Lake Fryxell, and in Lake Bonney between 0 °C and 3.6 °C; in all cases increasing with depth in the mixolimnion (Spigel and Priscu 1998; Roberts and Laybourn-Parry 1999).

Extremely saline lakes like Deep Lake (lake 30, Table 3.1), which is around ten times the salinity of seawater, are so saline that they rarely develop ice covers. Without the thermal protection of an ice cover the surface water is in contact with the extremely cold atmosphere, and is subject to extreme cooling. Vertical circulation normally occurs for a period of two to three months between July and September (Figure 3.4). At this time water temperatures can fall as low as –17 °C to –18 °C. During the summer the lake becomes thermally stratified (Figure 3.4) when surface temperatures can rise to between 7 °C and 11.5 °C. In late February to April the stratification breaks down and mixis occurs (Figure 3.4). The changes in density in Deep Lake water are 11 times greater than those that occur in freshwater in water warmed from 4 °C to 9 °C. Thus Deep Lake will tend to stratify as the result of relatively small thermal gradients. It cools at a rate less than half its rate of heating and is consequently slow to cool when compared with tropical and temperate lakes (Ferris and Burton 1988). Ike Hunazoko in the Syowa Oasis has a similar salinity to Deep Lake. It is shallower than Deep Lake, 9.2 m compared with 36 m. In the winter the water temperature of Ike Hunazoko falls to around –15°C (Torii et al.1988). For more details of thermal stratification see Section 2.3.

3.5 Water chemistry

The chemistry of saline lakes varies considerably and is particularly complex in meromictic systems that possess permanently anoxic lower waters. A detailed introduction to geochemistry is given in Chapter 1, Section 1.6. Table 3.3 shows concentrations of major nutrients in the water columns of a range of saline lakes, both meromictic and amictic. Concentrations of ammonium (NH_4^+) are elevated in the anoxic monomolimnion of meromictic lakes, whereas concentrations in the mixolimnion range from undetectable to 10.5 µM l^{-1} (Figure 3.5). Nitrate (NO_3^-) in contrast is either lacking or relatively low compared with the mixolimnion (Figure 3.5). Orthophosphate (PO_4^{3-}) may also be elevated in the monomolimnion, particularly so in the meromictic lakes of the Syowa Oasis. The concentration of ammonium and SRP in stagnant lake bottom waters often arises from the degradation of settling organic matter (e.g. phytoplankton), and the associated accumulation of nutrients over very long time periods and in the absence of strong water circulation in lakes (Matsumoto 1993).

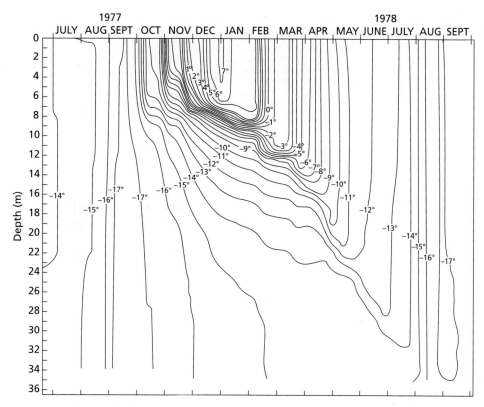

Figure 3.4 Graph showing the thermal stratification of hypersaline Deep Lake, Vestfold Hills over an annual cycle. During the summer months the lake is thermally stratified; during autumn the stratification breaks down resulting in winter mixis. The high salinity of Deep Lake inhibits the development of an ice cover. Redrawn from Ferris and Burton (1988).

Nutrients diffuse upwards across the chemocline into the mixolimnion. PAR at low levels reaches this region of the water column allowing high levels of photosynthesis in the upper section of the chemocline where oxygen is still present. Consequently deep chlorophyll maxima develop on the chemocline. Figure 3.6 shows chlorophyll a concentrations across the chemocline of Ace Lake (lake 14, Table 3.1) at 4 cm intervals. Above the chemocline at 9.5 m there is virtually no chlorophyll, as evidenced by a lack of colour in the first tube. The chlorophyll a concentration increases strongly over the next 12 cm (a deep chlorophyll maximum), dropping to a lower concentration in the anoxic monomolimnion. The latter is bacterial chlorophyll from photosynthetic purple sulphur and purple non-sulphur bacteria (see Section 3.6.2). The pattern in chlorophyll a concentration is mirrored in nutrient concentrations (Figure 3.6), as indicated by higher ammonium concentrations over a similar depth interval. The phytoplankton that accumulate within the upper portion of the chemocline are effecting a trade-off between a decreased light availability but access to increased concentrations of nutrients.

In amictic saline lakes nutrient concentrations tend to be low. In summer both nitrogen and phosphorus may be below the limit of detection for standard analytical methods, indicating that primary production may be limited by nutrient availability. Where year-long studies have been conducted concentrations of these nutrient species tend to be higher in the winter when biological demand is lower.

As one might anticipate, DOC concentrations in the monimolimnia of meromictic lakes are relatively high compared with those in the upper waters (Table 3.3, see Section 1.6.3). High DOC is also reported from monomictic hypersaline Hunazoko Ike

Table 3.3 Nutrients (ammonium, nitrate, soluble reactive phosphorus), dissolved organic carbon (DOC), and particulate organic carbon (POC) in saline Antarctic lakes. ND – not detectable Data are for the summer with the exception of data with an asterisk indicating that the data relate to an entire year. Lake numbers correspond with those in Table 3.1.

Lake	Zone in water column	NH_4-N µM l^{-1}	NO_3-N µM l^{-1}	PO_4-P µM l^{-1}	DOC mg L^{-1}	POC mg L^{-1}
1. Lake Fryxell (Dry Valleys)[1]	Mixolimnion	0.05 ± 0.08	ND	0.08 ± 0.01	4.4 ± 1.6	0.47 ± 0.3
	Monomolimnion	227.4 ± 232.4	ND	13.3 ± 13.0	14.7 ± 4.9	0.67 ± 0.2
2. Lake Vanda (Dry Valleys)[2,3]	Mixolimnion	0.39–0.83	6.05–10.9	0.003–0.028		0.050–0.075
	Monomolimnion	up to 1330	0	3.20		
3. Lake Wilson (Dry Valleys)[4]	Mixolimnion	ND–0.11	1.33–8.0	0.01–0.05		
	Monomolimnion	ND–1.05	11.8–79.4	0.03–0.06		
4. Lake Bonney West (Dry Valleys)[1]	Mixolimnion	1.9 ± 3.7	9.9 ± 3.7	0.05 ± 0.01	1.2 ± 0.6	0.23 ± 0.13
	Monomolomnion	191.3 ± 87.9	12.0 ± 10.6	0.27 ± 0.19	11.4 ± 6.1	0.23 ± 0.11
6. Lake Bonney East (Dry Valleys)[1]	Mixolimnion	1.9 ± 3.3	13.4 ± 9.5	ND	1.5 ± 0.6	0.17 ± 0.03
	Monomolimnion	112.2 ± 64.8	125.5 ± 50.5	0.20 ± 0.18	18.1 ± 8.7	0.25 ± 0.11
7. Lake Joyce (Dry Valleys)[1]	Mixolimnion	0.39 ± 0.26	36.1 ± 0.19	0.05 ± 0.02	2.2 ± 1.7	0.30 ± 0.15
	Monomolimnion	104.1 ± 109.4	30.6 ± 0.39	1.59 ± 1.49	3.7 ± 0.8	0.24 ± 0.03
13. Highway Lake (Vestfold Hills)[5,6]	unstratified	2.05–6.65	0.64–1.77	0.15–0.31	3.0–32.5*	
14. Ace Lake (Vestfold Hills)[7]	Mixolimnion	ND–10.5*	ND*	ND–0.76*	5.3–10.7*	200–350*
	Monomolimnion	ND–27.7*	ND*	0.10–3.5*	6.5–14.2*	215–350*
17. Rookery Lake (Vestfold Hills)[5]	unstratified	0.55–26.3	0.88–13.1	0.50–7.48	4.9–39.0	
18. Lake Williams (Vestfold Hills)[5]	Mixolimnion	2.66–8.31	0.67–2.41	0.15–5.13	2.5–30.1	
19. Pendant Lake (Vestfold Hills)[5,6]	unstratified	3.36–6.92	0.96–2.01	0.26–0.47	4.1–43.8*	
33. Lake Reid (Larsemann Hills)[8]	unstratified	5.43	0.27	ND	2.65	1.12
49. Sarah Tarn (Larsemann Hills)[8]	unstratified	3.10	0.24	ND	2.06	0.79
45. Lake 3 Filla Island (Rauer Is.)[9]	unstratified	0.04	ND	ND	ND	
46. Lake 2 Filla Island (Rauer Is.)[9]	unstratified	0.12	ND	0.03	ND	
21. Lake Polest (Bunger Hills)[10]	unstratified	ND	0.02–0.06	0.05–0.13		
9. Suribati Ike (Syowa Oasis)[11,12]	Mixolimnion	0–1.7	ND	0.05–0.40	20.6–85.5	
	Monomolimnion	458–575	ND	54.0–68.1	106–128	
20. Hunazoko Ike (Syowa Oasis)[11,12]	monomictic	1.7–6.0	1.0–1.9	0.25–0.45	103–186	
41. Nurume Ike (Syowa Oasis)[11,12]	Mixolimnion	ND–0.5	< 1.0	0.04–0.17	1.7–9.0	
	Monomolimnion	557–2420	1.7–1.8	70–156	15.7–39.7	

1: Roberts et al. 2004a; 2: Priscu et al. 1989; 3: Canfield and Green 1985; 4: Webster et al. 1996; 5: Laybourn-Parry et al. 2002; 6: Madan et al. 2005; 7: Bell and Laybourn-Parry 1999a; 8: Ellis-Evans et al. 1998; 9: Hodgson et al. 2001b; 10: Kaup et al. 1993; 11: Tominaga and Fukui 1981; 12: Fukui et al. 1985.

in the Syowa Oasis (Table 3.3). The amictic saline lakes have a wide range of DOC concentrations. Where annual studies have been conducted the range can be wide, as seen in Highway and Pendant Lakes (Table 3.3, lakes 13 and 19, Table 3.1). These pools of DOC are in most instances derived from autochthonous sources. There are exceptions, such as Rookery Lake (lake 17, Table 3.1) which has faecal inputs from a nearby Adelie Penguin Rookery. In the Dry Valley lakes the major components of the DOC pool are fulvic and hydrophilic acids (McKnight et al. 1993). The minor components vary between lakes and between their water and sediment. This reflects differences in their microbial communities and lake history (Lyons and Finlay 2008).

Stable isotopic compositions of carbon and nitrogen in organic matter can be used to trace biogeochemical processes in aquatic environments (see Box 1.1 Section 1.4). This approach in the Dry Valley lakes, which assessed the $\delta^{13}C$ compositions of

In perennial ice-covered lakes, like those of the Dry Valleys, physical mixing is absent precluding carbon dioxide exchange with the atmosphere. This restricts the redistribution of total dissolved inorganic carbon (DIC), which is the sum of CO_2, HCO_3^-, and CO_3^{2-}. The surface waters of these lakes can be depleted in inorganic carbon relative to the atmosphere with low partial pressures of CO_2. The depletion is due to uptake of CO_2 by photosynthetic organisms without replenishment from the atmosphere (Lyons and Finlay 2008). The photosynthetic demand for CO_2 is diminished in winter, when it is replenished. There are a number of sources for replenishment: firstly from CO_2-rich meltwater streams that flow only for a short period during the summer and advect water directly under the ice cover; secondly from the respiration of organic matter in the upper aerobic zone; thirdly from upward diffusion of CO_2 from the lower CO_2-rich zones of the lakes; and lastly from calcite precipitation ($Ca^{2+} + 2HCO_3^- \leftrightarrow CaCO_3 + CO_2 + H_2O$). The annual uptake of CO_2 varies among the Dry Valley lakes (Figure 3.7). In Lake Bonney West an uptake of 250% of CO_2 dissolved in the water occurs, suggesting that every molecule of CO_2 is recycled 2.5 times each year, while in Lake Fryxell the figure is 10%, indicating that each molecule is recycled every tenth year (Neumann et al. 2001).

3.6 The planktonic biota of saline lakes

The wide range in salinity found in Antarctic lakes leads to very considerable variations in the biota. At one end of the salinity continuum, extremely hyposaline lakes have been invaded by freshwater species such as the cladoceran *Daphniopsis studeri*, while in extremely hypersaline lakes biodiversity is severely reduced. In hypersaline Deep Lake (lake 30, Table 3.1) the only phytoplankter is the halophile phytoflagellate *Dunaliella* sp. (Ferris and Burton 1988). Younger coastal lakes are marine derived and their biota is effectively marine. However, the diversity is reduced compared with the neighbouring marine environment. Only species able to adapt to changes in salinity, reduced nutrient regimen, and wider temperature ranges have successfully colonized. These lacustrine conditions have imposed very considerable evolutionary pressure

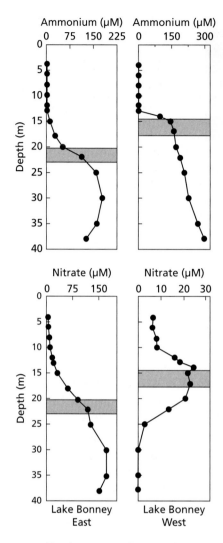

Figure 3.5 Profiles of ammonium and nitrate in the east and west lobes of Lake Bonney, Taylor Valley, McMurdo Dry Valleys. Redrawn from Voytek et al. 1998.

benthic organic matter and particulate organic matter, showed that Lake Fryxell and freshwater Lake Hoare are dominated by benthic production from their extensive algal mats, whereas in Lake Bonney pelagic production dominates the carbon cycle. In all cases the moat and the under ice zones had distinctly different isotopic compositions. The moat organic matter was significantly enriched in ^{13}C compared with the deeper environments. (Lawson et al. 2004).

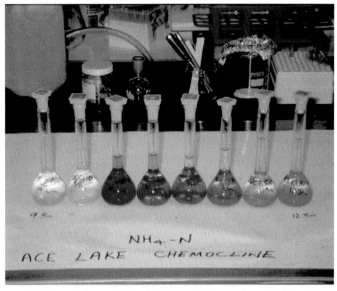

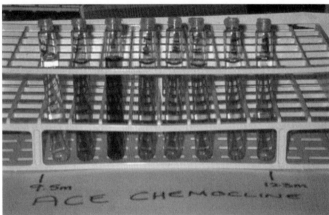

Figure 3.6 Photograph of ammonium and chlorophyll *a* concentrations over the chemocline of Ace Lake at 4 cm intervals. Ammonium analysis is by a colorimetric technique, thus the intensity of the blue colour is related to the concentration of ammonium (NH_4-N). Photo J. Laybourn-Parry. (See Plate 12)

on organisms, so that new species and subspecies have evolved in what is a relatively short time span. Moreover, in the Vestfold Hills the wide diversity of conditions among a relatively large number of hyposaline, saline, and hypersaline lakes has effectively provided a natural laboratory in which to study evolutionary divergence (Rengefors et al. 2012; Logares et al. 2012).

3.6.1 Heterotrophic Bacteria and Archaea

The concentrations of heterotrophic planktonic bacteria in a range of Antarctic saline lakes are given in Table 3.4. The range is wide, from 1.0 to 72.8×10^8 l^{-1}. In tropical soda lakes bacterial concentrations ranged between 2.4×10^{10} l^{-1} and 3.6×10^{11} l^{-1} (Kilman 1981), several orders of magnitude higher than in Antarctica. However, the saline lakes of Africa have chlorophyll *a* concentrations of up to 616 µg l^{-1} and are clearly highly productive. Chlorophyll *a* concentrations in Antarctic saline lakes range between 0.5 and 29 µg l^{-1}, and in lakes which still have periodic connection with the sea the range is between 5.8 and 35 µg l^{-1} (Table 3.2). There is a clear positive relationship between bacterial and chlorophyll *a* concentrations in both saline and freshwater African lakes (Zinabu and Taylor 1997) and in freshwater lakes worldwide (Bird and Kalff 1984).

Detailed depth analysis of bacterial communities from Lake Bonney (Dry Valleys, lake 4/6, Table 3.1)

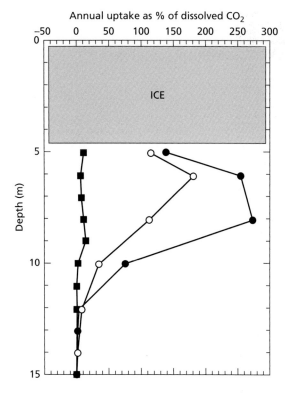

Figure 3.7 Annual CO_2 uptake as a percentage of CO_2 dissolved in the water. ■ Lake Fryxell, ○ Lake Bonney West, ● Lake Bonney East. The value of 250% in Lake Bonney West indicates that every CO_2 molecule is recycled two and a half times each year. In contrast lower values in Lake Fryxell suggest that every CO_2 molecule is recycled every tenth year. From Neumann et al. (2001), with the permission of Springer.

and Ekho Lake (Vestfold Hills, lake 8, Table 3.1) reveal a predonderance of alpha-Proteobacteria and gamma-Proteobacteria, Cytophaga/Flexibacter/Bacteriodetes (or CFB) and Actinobacteria (Labrenz and Hirsch 2001; Glatz et al. 2006). The two lobes of Lake Bonney differed in their bacterial community composition, undoubtedly reflecting the different chemical and physical conditions that prevail in the two lobes of the lake. Bacteriodetes constituted the major fraction of clones isolated from the west lobe at depths of 13 m (43%) and 16 m (46%), while in the east lobe they contributed only 18%. In contrast, sequences within the gamma-Proteobacteria dominated the east lobe at 16 m, 19 m, and 25 m (38%, 43%, and 47% respectively), while in the west lobe they accounted for 10% at 10 m and 26% at 16 m.

Halomonas sequences were common (Glatz et al. 2006). *Halomonas* spp. are common in the hypersaline saline lakes of the Vestfold Hills. In Ekho Lake, *Halomonas meridian* and *Halomonas subglaciescola* contributed 23% and 16% of peak abundance. They were less common in Organic Lake (2%–3%). These species showed seasonal patterns in abundance, in most cases peak abundance occurred in spring/summer with lowest abundance in winter. Highest concentrations occurred above the chemocline (James et al. 1994).

In extremely hypersaline Deep Lake (salinity 350‰) the only viable prokaryotes recovered from water samples were *Halobacterium* spp. Both *Halomonas subglaciescola* and *Halobacterium* isolated from Organic Lake (230‰) and Deep Lake are obligate halophile Archaea. *Halomonas* isolated from Organic Lake is capable of growing down to –5.4 °C and *Halobacterium* from Deep Lake grew down to –4 °C. In both cases liquid water is present in these lakes well below these growth temperatures. Selective pressures for lower temperatures have only operated in these lakes for 4000–7000 years. Eventually selection might enable these halophiles to fully exploit their extreme environment (McMeekin and Franzmann 1988). More recently phylogenetic reconstructions using SSU rRNA gene sequences cluster four isolates from Deep Lake within the family Halobacteriaceae and confirm that overall community diversity is very low in this extremely hypersaline lake. Moreover, the genomic characteristics of the four Haloarchaea indicate a high level of horizontal gene transfer (see Section 1.8.1). It is argued that in this extreme ecosystem there is accentuated gene exchange between archaeal genera (DeMaere et al. 2013).

Some degree of endemism does appear to exist among the bacteria isolated from Antarctic lakes. For example, new species of *Flavobacterium*, *Carnobacterium*, *Saligentibacter*, *Colwellia*, *Methylosphaera*, and *Methanogenium* have been isolated from saline lakes in the Vestfold Hills (McCammon et al. 1998; McCammon and Bowman 2000; Franzmann and Rohde 1991; Franzmann et al. 1997; Bowman et al. 1997, 1998). Similarly in the Dry Valleys (Lakes Fryxell and Bonney), dilution to extinction culturing of psychrotolerant bacteria indicated that the majority of the sequences were not related to

Table 3.4 Abundances of bacteria, heterotrophic nanoflagellates (HNAN), and phototrophic nanoflagellates (PNAN) in saline lakes. * includes a *Stichococcus* bloom.

Lake and notes; numbers correspond with numbers in Table 3.1	Bacteria × 10^8 l^{-1}	HNAN × 10^4 l^{-1}	PNAN × 10^4 l^{-1}
1. Lake Fryxell (Dry Valleys)[1,2,3] Nov–Dec 2000, Nov–Jan 1997–1998, Nov–Dec 1996 and 1997	11.0–47.5	12.2	227.9 ± 264
2. Lake Vanda (Dry Valleys)[4] in top 55 m	1.0–10.0		
4. Lake Bonney West (Dry Valleys)[3,5] Nov–Dec 1996 and1997 in mixolimnion March–April 2008 between 5 and 13 m	1.96 ± 0.75 1.48–4.73	9.3 ± 8.1 6.0–22.2	11.3 ± 16.7 350–1050
6. Lake Bonney East (Dry Valleys)[3,5] Nov–Dec 1996 and1997 in mixolimnion March–April 2008 between 5 and 20 m	1.77 ± 0.75 1.87–5.22	9.8 ± 7.1 7.5–24.0	7.1 ± 11.4 75–970
7. Lake Joyce (Dry Valleys)[3] Nov–Dec 1996 and 1997 in mixolimnion	1.92 ± 0.43	3.7 ± 3.0	3.8 ± 5.3
8. Ekho Lake (Vestfold Hills)[5] October 1994, between 0 and 10 m	3.1	41.0	21.0
12. Shield Lake (Vestfold Hills)[5] October 1994, between 0 and 10 m	4.0	62.2	59.0
13. Highway Lake (Vestfold Hills)[6] Nov–Feb 1999–2000 between 2 and 6 m	2.4–3.5	40–72	82–190
14. Ace Lake (Vestfold Hills)[7] Dec 1995–Feb 1997, between 0 and 12 m	1.26–72.8	92–402	50–500
17. Rookery Lake (Vestfold Hills)[6] Nov–Feb 1999–2000 between 2 and 6 m	24.5–48.9	12–884	500–2500
18. Lake Williams (Vestfold Hills)[6] Nov–Feb 1999–2000 between 0 and 2 m	1.6–30.0	34–190	125–590
19. Pendant Lake (Vestfold Hills)[6] Nov–Feb 1999–2000 between 2 and 6 m	5.5–23.0	32–58	2600–15,530*
27. Lake Abraxas (Vestfold Hills)[5] October 1994, between 0 and 10 m	4.1	62	11
33. Lake Reid (Larsemann Hills)[8] Dec–Feb 1992–1993	4.08	42	25
49. Sarah Tarn (Larsemann Hills)[8] Dec–Feb 1992–1993	4.99	86	22

1: Marshall and Laybourn-Parry 2002; 2: Roberts and Laybourn-Parry 1999; 3: Roberts et al. 2004a; 4: Takii et al. 1986; 5: Thurman et al. 2012; 5: Perriss and Laybourn-Parry 1997; 6: Laybourn-Parry et al. 2002; 7: Bell and Laybourn-Parry 1999a; 8: Ellis-Evans et al. 1998.

previously described species (Stingl et al. 2008). There is currently no consensus on the degree of endemism in Antarctic bacteria. In effect we have only scratched the surface of bacterial diversity in Antarctica. More particularly we need much greater insights into the functional role of microbe groups in Antarctic lakes.

The many marine-derived brackish, saline, and hypersaline lakes of the Vestfold Hills offer a natural laboratory in which to study the evolution of bacterial communities and the pressures that drive that process. A detailed molecular and taxonomic analysis of 13 lakes ranging in salinity from 4‰ to 100‰ and two freshwater lakes (Crooked Lake and

Lake Druzhby) showed that the taxonomic composition of the communities was strongly correlated with salinity, and weakly correlated with geographical distance between the lakes (Logares et al. 2012). A few of the taxa found were shared with the neighbouring marine environment, but many of the taxa found in the lakes were not present in the sea. Both habitat specialists and generalists were detected among abundant and rare taxa, with the specialists being relatively more abundant at the extremes of the salinity gradient. It was evident that long-term changes in the salinities of the lakes during development to their present day condition has promoted the diversification of the bacterial communities derived from the ancestral marine communities. A test of the neutral model (NM) of community assembly explained only 25% of the variation in community composition, indicating that strong environmental gradients may favour filtering over neutral dynamics. A comparison with a similar study of 15 freshwater Scandinavian lakes showed that NM accounted for 50% of the variability. These Scandinavian lakes have relatively homogeneous physico/chemical conditions that favour neutral dynamics. Some of the saline lakes share similar taxa to saline lakes elsewhere in the world, which supports the view that long distance dispersal occurs (Marshall 1996). This was also true of freshwater Crooked Lake and Lake Druzhby which shared taxa with lakes at lower latitudes, as well as possessing what might be Antarctic endemics (Logares et al. 2012).

In the recent past, extremophile bacteria from saline lakes have been a source of considerable interest to industry, because of their biotechnological potential for low temperature enzymes, antifreeze proteins, and other molecules. For example *Marinomonas prioryensis*, isolated from Ace Lake, has a powerful antifreeze protein, which is Ca^{2+} dependent and can produce over 2 °C of freezing point depression (Gilbert et al. 2005). Cold active enzymes such as alkaline phosphatase, lipase, and β-galatosidase have also been isolated from Antarctic bacteria (Dhaked et al. 2005; Yang et al. 2004; Coker et al. 2003).

The chemocline and anoxic lower waters of meromictic lakes provide a distinct environment for particular physiological groups of bacteria. Methanotrophic bacteria have been isolated from Vestfold Hills lakes. These bacteria are a subset of a physiological group known as the methylotrophs. They are unique in their ability to use methane as a sole carbon source, and are able to function both aerobically and anaerobically. A new methanotrophic bacterium, *Methylosphaera hansonii*, belonging to a Group 1 methanotrophic genus (gamma-Proteobacteria) was isolated from Ace Lake and Burton Lake (Vestfold Hills). Methane increases across the chemocline of Ace Lake from 0 to around 4.9 mM at 23 m in the water column. *M. hansonii* was not detected in the mixolimnion, while on the chemocline it occurred in concentrations of 4.0×10^2 l^{-1}, increasing to between 1.0×10^3 l^{-1} and 3.6×10^5 l^{-1} between 11 and 13 m. It was absent between 14 and 23 m. Methanotrophs represented only a small fraction of the total bacterial community in Ace Lake, in the order of <0.1 to 1%. Methane production in Ace Lake is slow in the region of <2.5 µmol kg^{-1} day^{-1}, due to low temperatures and slow input of organic matter; nevertheless there is sufficient methane to support methanotrophs (Bowman et al. 1997). Possible methanotrophs have been identified from Lake Fryxell. A number of phylotypes of *Euryarhaeota* were found within the highly suphidic and methanotrophic deep waters of the lake (Karr et al. 2006).

Most methanogenic Archaea can reduce CO_2 with H_2 to methane, as occurs in the psychrophilic methanogenic Archaea *Methanogenium frigidum* which has been isolated from the monimolimnion of Ace Lake. It grows most rapidly at 15 °C, failing to grow between 18 °C and 20 °C. However, growth is slow with a doubling time of 2.9 days. Growth is fastest on H_2 plus CO_2, with slower growth on formate. Acetate, methanol, and trimethylamine are not catabolised, but acetate is required as a carbon source (Franzmann et al. 1997). *Methanococcoides burtonii*, another methanogenic Antarctica archaeon, has been isolated from anoxic waters in Ace Lake, where it uses methylamines and methanol as precursors to methanogenesis (Franzmann and Rohde 1992). A single phylotype of the methanogen *Crenarchaeota* was located in Lake Fryxell close to the oxycline (Karr et al. 2006).

In the cold, maintaining the fluidity of cell membranes is crucial. *M. burtonii* controls membrane fluidity by manipulating the lipid composition, such that the level of unsaturated lipids is higher in cells grown at 4 °C compared with those grown at 23 °C. This phenomenon also occurs in Bacteria (Nichols et al. 2004). *M.burtonii* grows most rapidly at 23 °C. It expresses a heat shock protein (DnaK) when grown at its optimal temperature, indicating that it is stressed (Goodchild et al. 2004).

The biomass of methanogenic Archaea, based on the identification and quantification of phospholipid-derived ether lipids (PLEL), has been determined in Ace Lake. Methanogenic Archaea possess unique cell membranes that consist of lipids formed with ether linkages and isoprenoid branching. A conversion factor can be applied to concentrations of PLEL to derive an estimate of biomass or cell numbers. In the monimolimnion this equates to $<1.0 \times 10^8$ l^{-1} to 7.4×10^8 cells l^{-1} between 17 and 23 m. Much higher cell concentrations were found in the upper sediment layer (maximum 17.7×10^9 cells g^{-1} dry weight of sediment) (Mancuso et al. 1990).

Sulphate-reducing bacteria are also common in the sulphate-rich lower waters of meromictic lakes where they reduce sulphate to sulphide by means of a number of electron donors, including H_2, fatty acids, alcohols, and aromatic compounds. Some ester-linked phospholipid-derived fatty acids (PLFA) can act as signatures for known taxa. The sulphate-reducing bacteria *Desulfobacter* and *Desulfovibrio* cell numbers in Ace Lake were determined by this method. *Desulfobacter* in the monimolimnion increased from 0.5×10^8 cells l^{-1} at 10 m to 4.8×10^8 cells l^{-1} at 23 m, while *Desulfovibrio* was much less abundant, ranging between 0.1 and 0.9×10^8 cells l^{-1} (Mancuso et al. 1990). Analysis of environmental DNA and isolates cultured from Lake Fryxell (Dry Valleys) has revealed a diverse group of sulphate-reducing bacteria including members of the genera *Desulfovibrio* and *Desulfobacter*. Other genera included *Desulfobacterium*, *Desulfosarcina*, *Desulfotignum*, and *Desulfofaba*. There was a clear localization of some groups in the water column in relation to chemical and physical conditions (Figure 3.8). Sulphate reaches its maximum at 13 m, decreasing with depth, dropping sharply at 17 m to just 0.34 mM at 17.5 m, suggesting that maximum sulphate reduction rates are occurring close to the sediments. The groups that occur near the water/sediment interface are probably the most abundant and physiological active members of the community (Karr et al. 2005).

Sulphur oxidizing bacteria obtain their energy by the oxidation of reduced inorganic sulphur compounds. Some use sulphide, the sulphide being supplied as hydrogen sulphide, or as a metal sulphide such as iron or copper sulphide. Others use elemental sulphur (S^0), or thiosulphate ($S_2O_3^{2-}$). Some can use all three forms of sulphur. In all cases the end product is sulphuric acid, which leads to the lowering of pH during the growth of these microorganisms. The oxygen and sulphide gradients in many meromictic lakes provide an ideal environment for sulphur oxidizing bacteria. Three strains of *Thiobacillus thioparus* were isolated from 9 m, 10 m, and 11 m in Lake Fryxell, which were able to use thiosulphate, sulphide, elemental sulphur, and thiocyanate. Most Probable Number (MPN) culture analysis showed a peak in concentration at 9.5 m of 2×10^5 cells l^{-1}, within the narrow zone where sulphide and oxygen coexist in the water column. However, sulphur oxidizing bacteria were still detectable at 13 m, well into the anoxic water. The strains were able to grow at temperatures of up to 30 °C, as is the case for many bacteria that have been isolated from Antarctic lakes (Sattley and Madigan 2006). Many of the organisms that have colonized these extreme environments are not true psychrophiles.

3.6.2 Photosynthetic bacteria

The Cyanobacteria undertake oxygenic photosynthesis and filamentous forms are a conspicuous feature of the benthos where they form extensive algal mats (see Section 3.9). However, some filamentous Cyanobacteria are found in the plankton of Lake Fryxell and Lake Joyce (Dry Valleys) (Spaulding et al. 1994; Roberts et al. 2004a). These included *Oscillatoria* and *Phormidium*, which were probably derived from the benthic mats.

Picocyanobacteria occur in plankton of freshwater maritime lakes (see Section 2.5.4). They were first observed in the plankton of the Dry Valleys lakes by Goldman et al. (1967) and were noted as a dominant component of the deep chlorophyll maximum of Lake Vanda (Vincent and Vincent 1982). *Synechococcus* was recorded in the plankton of Lake Fryxell in 1990 and 1991, but not in the summers of the preceding three years (Spaulding et al. 1994). It

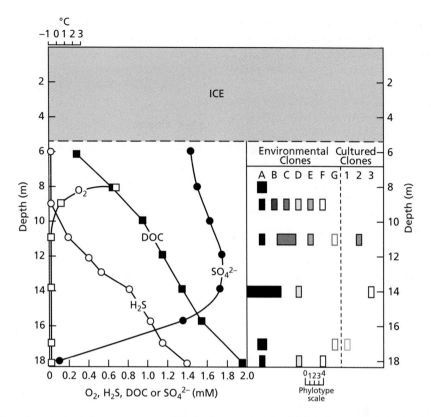

Figure 3.8 Left panel—main chemical parameters (oxygen (O_2), dissolved organic carbon (DOC), hydrogen sulphide (H_2S), and sulphate (SO_4)) in relation to depth in Lake Fryxell, Taylor Valley, McMurdo Dry Valleys. Right panel shows different phylotypes at particular depths in clones from environmental sample analysis and from cultured clones. The width of the boxes is proportional to the number of phylotypes residing at different depths. From Karr et al. (2005) with the permission of the American Society for Microbiology.

occurs in the plankton of Ace Lake (Vestfold Hills) (Rankin et al. 1997). *Synechococcus* sp. was most abundant on the oxycline and contributed to the deep chlorophyll maximum (DCM) seen at this level (see Figure 3.6). Picocyanobacteria showed a clear seasonal pattern in Ace Lake with lowest numbers in winter (May to September), increasing in summer, ranging between $<10^7$ cell l^{-1} increasing to 8×10^9 cells l^{-1} in the DCM. *Synechococcus* has also been reported from Lake Abraxas and Pendant Lake in abundances up to 1.5×10^{10} cells l^{-1} and 1.5×10^{10} cells l^{-1} respectively. However, it is absent from hypersaline lakes such as Ekho Lake and Organic Lake and from lower salinity lakes such as Clear Lake (Powell et al. 2005).

In the anoxic monomolimnia of meromictic lakes bacterial chlorophyll is clearly evident (see Figure 3.6). These are bacteriochlorophylls and are contained in the cells of the photosynthetic Proteobacteria. There are two subgroups: the purple sulphur bacteria and the purple non-sulphur bacteria. These photosynthetic bacteria undertake anoxygenic photosynthesis. In the purple sulphur bacteria the process involves the following two reactions (which are not balanced):

$$H_2S + CO_2 \rightarrow (CH_2O)_n + S^0 \quad (3.1)$$

$$S^0 + CO_2 + H_2O \rightarrow (CH_2O)_n + H_2SO_4, \quad (3.2)$$

where $(CH_2O)_n$ represents organic carbohydrate carbon.

The process of photosynthesis in non-sulphur bacteria is represented by the following equation:

$$H_2A = CO_2 \rightarrow (CH_2O)_n + A, \quad (3.3)$$

where H_2A = an organic compound such as acetic acid.

It is evident from these equations that oxygen is not formed during this type of photosynthesis (Perry et al. 2002).

There is only limited information available on the photosynthetic bacteria of Antarctic lakes. In thirteen meromictic lakes in the Vestfold Hills the only bacteriochlorophyll detected was bacteriochlorophyll c, in concentrations usually below 200 µg l^{-1}. The green sulphur bacterium *Chlorobium* spp. was found in seven of the lakes; these are strict anaerobes. *Chromatium* spp., a purple sulphur bacteria, occurred in four of the lakes. Under anaerobic light conditions these bacteria grow as photolithoautotrophs using sulphide or elemental sulphur as an electron acceptor and CO_2 as a carbon source. Members of the purple non-sulphur bacteria *Rhodopseudomonas palustris* and *Thiocapsa roseopersicina* were cultured from five of the lakes. Members of this group can grow aerobically as heterotrophs, but under anoxic light conditions can carry out anoxygenic photosynthesis, much like the purple sulphur bacteria (Burke and Burton 1988). The concentration of purple photosynthetic bacteria is at its lowest between May and July when light levels are low, increasing in August, and peaking in December/January when numbers were found to reach 10^7 cells l^{-1} (Rankin et al. 1999).

Culture-dependent and culture-independent molecular techniques have revealed a wide diversity of purple non-sulphur bacteria in Lake Fryxell. However, there was no genetic evidence of purple sulphur bacteria. This is something of an enigma as there is a high abundance of sulphide in Lake Fryxell. As previously indicated, in spite of their name the purple non-sulphur bacteria use sulphide as an electron donor and species have specific sulphur tolerances. Different phylotypes occurred at different levels in the water column of Lake Fryxell, where H_2S concentration increased from 0 to 1.1 mM with depth from 9 to 18 m. Some of the species cultured from the lake had gas vesicles, which suggests that these microorganisms have evolved a means of maintaining their position at specific depths in the water column (Karr et al. 2003).

Given the high concentrations of bacterial chlorophylls found in the lower waters of meromictic Antarctic lakes, it is likely that these photosynthetic bacteria make a significant contribution to the carbon budget within these ecosystems. They have been much more widely researched in lower latitude saline environments (Ollvier et al. 1994; McGenity and Oren 2012).

3.6.3 Viruses

As outlined in Section 2.5.2, information on lacustrine viruses is much more widely available for lower latitude systems. Nevertheless we do have some information on viruses and their role in both freshwater and saline Antarctic lakes. Abundances of viruses in the water columns of saline lakes are shown in Table 3.5. A comparison with Table 2.5 reveals that virus numbers and the virus-to-bacterium ratio (VBR) are generally higher in saline lakes compared with freshwater systems. Saline lakes are usually more productive than freshwater lakes, as indicated by chlorophyll a concentrations, and consequently the continental lakes support larger bacterial communities and populations of other potential hosts (see Tables 2.4 and 3.4). Electron microscopy has revealed a range of viral types from large and small icosahedral viruses, with and without tails, in Lakes Fryxell and Highway and Pendant Lakes (Kepner et al. 1998; Laybourn-Parry et al. 2001a). Metagenomic analysis of samples from different depths in the water column of Ace Lake indicated that in the mixolimnion all viral size fractions were dominated by the *Phycodnaviridae*, which are large algal viruses. Other viral taxa found included the *Siphoviridae, Myoviridae,* and *Podoviridae* (Lauro et al. 2011). These analyses were based on two samples taken in December 2006. The data suggest that the dominant hosts at this time were algal, specifically *Mantoniella*. However, Antarctic lakes show considerable inter-annual variation in the abundances of species and care needs to be exercised in drawing broad conclusions from what is effectively a small snapshot.

Some viruses are themselves subject to viral attack. A virophage predator of phycodnaviruses that infect phytoflagellates, specifically prasinophytes, has been isolated from hypersaline Organic Lake (Yau et al. 2011). The first such virophage described was named the Sputnik virophage. *Acanthoamoeba polyphaga* mimivirus (APMV) is the largest virus so

Table 3.5 Virus abundance and virus-to-bacterium ratio (VBR) in saline Antarctic lakes. Numbers correspond with those in Table 3.1.

Lake and location	Month (s) sampled	Viruses × 10^9 l^{-1}	VBR	Chloro a µg l^{-1}
1. L. Fryxell (Taylor Valley)[1]	Dec 1999*	34.4–55.6	13.3–46.3	3.6–12.0
2. L. Vanda (Wright Valley)[1]	Dec 1999*	0.32–1.9	7.5–12.1	0–0.01
4. L. Bonney W. (Taylor Valley)[1]	Dec 1999*	1.67–3.83	2.1–31.9	0.79
6. L. Bonney E. (Taylor Valley)[1]	Dec 1999*	0.33–0.70	0.9–3.3	0.1–1.15
7. L. Joyce (Pearse Valley)[2]	Dec 1996–Jan 1997	4.2	2.9	0.05–3.0[3]
13. Highway L. (Vestfold Hills)[4]	Dec 2002–Jan 2004	12.4–96.6	18.6–126.7	0.32–6.74
14. Ace L. (Vestfold Hills)[4]	Dec 2002–Jan 2004	8.9–61.3	30.6–80.0	0.37–1.76
18. L. Williams (Vestfold Hills)[5]	Dec 1999	36.5	9.17	2.5–25.0[6]
19. Pendant L. (Vestfold Hills)[4]	Dec 2002–Jan 2004	11.5–120.1	30.5–96.7	0.79–7.46

* samples taken at a range of depths in the water column on one occasion.
1: Lisle and Priscu 2004; 2: Kepner et al. 1998; 3: Roberts et al. 2004a; 4: Madan et al. 2005; 5: Laybourn-Parry et al. 2001a; 6: Laybourn-Parry et al. 2002.

far described. Sputnik is a small icosahedral virus (50 nm) found associated with APMV. It is unable to multiply within an amoeba host by itself, but multiplies rapidly within APMV that has infected an amoeba. Infection by Sputnik is damaging to APMV. It results in the creation of abortive APMV forms and abnormal capsid assembly of the host virus (La Scola et al. 2008). The virophage found in Organic Lake was identified by metagenomic analysis of samples collected in December 2006 and of further samples collected in November and December 2008 (Yau et al. 2011). The discovery of virophage adds a further dimension to microbial community dynamics. The evidence indicates that in the Sputnik/APMV model infection by virophage causes a 70% decrease in infectivity of APMV particles, and a significant decrease in host cell lysis (La Scola et al. 2008).

Where annual studies have been conducted, viral and bacterial numbers remain high throughout the year, in Ace Lake, Pendant Lake, and Highway Lake. There was no clear seasonal pattern apart from a consistent decline in November across all three lakes (Figure 3.9) (Madan et al. 2005). The lytic phase predominated in summer when little or no lysogenic phage occurred. However, high rates of lysogeny were recorded in winter and spring: up to 32% in Pendant Lake and up to 71% in Ace Lake (Laybourn-Parry et al. 2007). In contrast, the percentage of lysogenic phage in the Dry Valley saline lakes during summer (December) ranged between 0% and 52.8%, the highest values occurring in Lake Bonney (Lisle and Priscu 2004). There are conflicting data on the seasonal occurrence of lysogeny in lower latitude ecosystems and the factors that drive it. For example, in a subtropical estuary at temperatures above 19 °C in summer up to 41% of lysogenic phage occurred (Cochran and Paul 1998). In Lake Superior a slightly higher proportion of the bacterial community contained lysogenic prophage in July and August compared with October (Tapper and Hicks 1998). However, in the Gulf of Mexico the percentage of lysogenic phage was not correlated with temperature (Weinbauer and Suttle 1996, 1999) and in the Baltic and Mediterranean seas levels of lysogeny in excess of 50% occurred at temperatures below 15 °C (Weinbauer et al. 2003). In the freshwater lakes of the Vestfold Hills there was an apparent lack of any significant lysogeny in waters where temperatures are continuously close to freezing (Säwström et al. 2007d). Thus temperature and its impact on productivity do not appear to be major factors in driving the lysogenic cycle in Antarctic lakes. Trophic status has also been implicated as a factor, however it has been argued that the relationship between lysogeny and the trophic status of an environment remains to be properly elucidated (Weinbauer 2004).

In Ace Lake a positive correlation was found between primary production and virus abundance

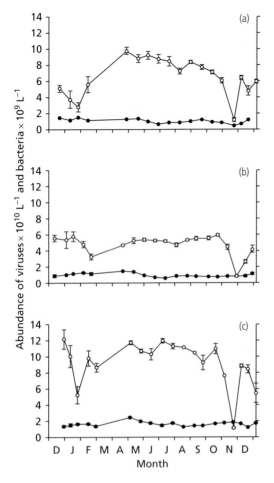

Figure 3.9 Annual patterns of virus and bacterial abundance in three saline lakes in the Vestfold Hills. A: Highway Lake, B: Ace Lake, and C: Pendant Lake. ○ viruses, ● bacteria. From Madan et al. (2005), with the permission of Blackwell Scientific Publications.

over a year, but not with bacterial numbers or bacteria production. In contrast, in Pendant Lake there was no significant correlation between virus abundance and primary production, but there was a highly significant correlation between bacteria numbers and production and virus particle concentration (Madan et al. 2005). Significant positive correlations between viral abundance and bacterial abundance, chlorophyll *a*, bacterial production, and soluble reactive phosphorus were evident in Lakes Fryxell, Bonney, and Vanda during December (Lisle and Priscu 2004). The implication is that the production of viruses is controlled by parameters that control the production of their hosts.

Virus lytic production rates in Antarctic saline lakes lie at the lower end of the range reported from lower latitude marine systems (0 to 9.58×10^6 ml^{-1} h^{-1}) (Fuhrman and Noble 1995; Hewson et al. 2001; Steward et al. 1992, 1996). In Ace Lake virus production ranged between 0.202 and 0.823×10^6 ml^{-1} h^{-1}, and in Pendant Lake, between 0.176 and 0.424×10^6 ml^{-1} h^{-1} during December (Laybourn-Parry et al. 2007). By comparison, in ultra-oligotrophic freshwater Crooked Lake and Lake Druzhby (Vestfold Hills), calculated lytic virus production rates were very much lower at up to 0.029 viruses ml^{-1} h^{-1} and 0.0098 viruses ml^{-1} h^{-1} respectively (Säwström et al. 2007d). The rates seen in saline Antarctic lakes are comparable with those reported from temperate oligotrophic/mesotrophic and eutrophic freshwater lakes in the French Massif Central, where lytic viral production reached 0.079×10^6 and 0.424×10^6 viruses ml^{-1} h^{-1} (Bettarel et al. 2004). Overall rates of lytic viral production in Antarctic lakes are low, particularly so in ultra-oligotrophic freshwater systems.

In polar freshwater environments, including cryoconite holes on glaciers, rates of bacterial infection by viruses are significantly higher than in lower latitude ecosystems (see Section 1.8.2). However, burst sizes are much lower (Säwström et al. 2008). At present we do not have data on rates of viral infection or burst sizes in saline lakes, however, it is likely that these will reflect what occurs in freshwater Antarctic systems.

3.6.4 Protozoa

Ciliated protozoa

Ciliated protozoans are a conspicuous feature of hyposaline and saline lake plankton. However, with increasing salinity into hypersaline environments their diversity and abundance decline. A survey of Vestfold Hills lakes ranging in salinity from 4‰ up to 170‰ found no ciliates in Oblong, Oval, and Organic Lakes, all of which have upper waters with salinities above 170‰. Ciliates were found in Shield Lake (mixolimnion salinity of around 70‰). One of the most ubiquitous ciliates in the Vestfold Hills lakes is the marine autotrophic ciliate *Mesodinium rubrum* (see Section 1.8.3 and Figure 1.28c). *M. rubrum* is common in Antarctic seas and in the marine environment worldwide. It has a wide salinity tolerance, also occurring in estuaries. In the saline lakes of the Vestfold Hills it occurs in lakes that have upper or mixolimnon waters ranging in salinity from 4‰ to around 70‰ (Perris and Laybourn-Parry 1997). Its success undoubtedly lies in its being pre-adapted to the wide ranges of salinity that it encounters in the lacustrine environment and its photosynthetic capacity. It is highly motile and is consequently able to migrate in the water column to facilitate nutrient uptake and exploit the best light climate for its cryptophycean endosymbiont. In some lakes it achieves abundances in the summer in excess of 200 × 10^3 cells l^{-1} and in a number of cases, for example Highway Lake, contributes between 83% and 99% of the ciliate community (Laybourn-Parry and Perriss 1995).

Apart from *M. rubrum*, other genera recorded in the Vestfold Hills saline lakes include *Euplotes*, *Strombidium*, *Monodinium*, and scuticociliates (see Figure 1.28). Lakes within the Vestfold Hills that experience periodic marine incursions, for example Rookery Lake (lake 17, Table 3.1), support a ciliated community that contains species found in the marine neighbouring environment, including several *Strombidium* species, *Tontonia* sp., *Lohmanniella oviformis*, *Tiarina* sp., *Laboea* sp., *Euplotes* spp., and *Codonellopsis* sp. (Bell and Laybourn-Parry 1999b). As lakes were isolated from the sea and evolved differing salinities and physico-chemical conditions, the original marine community became truncated to a simple ciliate community with low diversity and with *Mesodinum rubrum* predominating in many of the hyposaline and saline lakes. Hyposaline Lake Reid in the Larsemann Hills contains a ciliate community made up of *Strombidium*, *Askenasia*, and *Holophyra* (Ellis-Evans et al. 1998).

In contrast, the ciliated community of the considerably older Dry Valleys saline lakes shows much greater diversity (Table 3.6). Lake Fryxell possesses the highest diversity and Lake Vanda the lowest. However, caution should be exercised as the protozooplankton of Lake Fryxell has been studied more frequently and in more detail than the other lakes listed in Table 3.6. The communities of these lakes contain a wide range of nutritional types. Some are mixotrophs containing symbiotic zoochlorellae or sequestered plastids (see Figure 1.28a), some such as *Chilodonella* and *Vorticella* are bacterivores, others such as *Halteria* can feed on bacteria and small

Table 3.6 Ciliated protozoa in the Dry Valleys saline lakes.

Genus or species	L. Bonney	L. Fryxell	L. Joyce	L. Vanda
Askenasia	•	•	•	•
Blepharisma		•	•	
Bursaria	•	•	•	
Chilodonella		•		
Cyclidium		•		
Didinium	•	•		
Enchelydon		•		
Euplotes	•	•	•	•
Euplotes c.f. aediculatus		•		
Frontonia		•		
Halteria	•	•	•	
Lacrymaria		•		
Mesodinium		•		
Monodinium large species	•	•		•
Monodinium small species	•	•	•	
Nassula		•		
Plagiocampa	•	•	•	
Sphaerophrya	•	•		
Strombidium	•	•	•	•
Strombidium viride		•		
Uronema		•		
Urotricha		•		•
Urostyla	•			
Vorticella mayeri		•		
Vorticella	•	•		

Data taken from Laybourn-Parry et al. (1997), James et al. (1998), Kepner et al. (1999), Roberts et al. (2004a), and Thurman et al. (2012).

flagellates. *Plagiocampa* feeds on cryptophytes and other PNAN in Lake Fryxell and probably does so in the other lakes. In Lake Fryxell, the distribution of *Plagiocampa* in the water column closely mirrors that of its cryptophyte prey (Roberts and Laybourn-Parry 1999). *Didinium* is a highly motile predator of other ciliates at lower latitude environments. Of particular interest is the presence of the suctorian ciliate *Sphaerophrya*, a sedentary ambush predator of ciliates. It traps its prey on tentacles through which it extracts the cell contents. Thus the ciliate communities of the Dry Valleys lakes are nutritionally complex with several trophic levels. As indicated in Section 2.5.3 the greater ciliate diversity of the Dry Valleys lakes undoubtedly reflects their age. One can speculate as to the origin of these diverse communities. The Dry Valleys lakes were derived from large proglacial lakes and many subsequently experienced a drying down to small hypersaline water bodies. At this time ciliates may have been cryopreserved in the encysted state in the dried lake sediment, and subsequently recolonized the lakes as they refilled and their hypersaline waters became diluted. We have no idea how long cryopreserved protozoan cysts survive. The original large proglacial lakes are likely to have provided the ancestral communities, which may subsequently have been augmented by aerially derived propagules.

While *Mesodinium rubrum* can achieve high abundances in the lakes of the Vestfold Hills, where it is effectively part of the phytoplankton, abundances of heterotrophic ciliates are relatively low, at around 50–100 cells l^{-1} (Laybourn-Parry et al. 2002). Low ciliate abundances are also found in Lake Bonney, where the range is 24–240 cells l^{-1} in the mixolimnion (Roberts et al. 2004a; Thurman et al. 2012). Ciliates increased in number during the summer of 2007/2008, with a peak in April as winter encroached (Thurman et al. 2012). Lake Fryxell supported a larger ciliate community with up to 7720 cells l^{-1} in Januarys 1992 and 1994 (Laybourn-Parry et al. 1997), and up to 738 cells l^{-1} in the mixolimnion and 11,000 cells l^{-1} on the chemocline in the summers of 1996/1997 and 1997/1998 (Roberts et al. 2004a). These differences relate to the productivity of the lakes; Lake Fryxell is more productive than Lake Bonney (see Table 3.2). The ciliate community of Lake Vanda ranged between 7 and 6540 cells l^{-1} over three summers (James et al. 1998). Overall the data indicate considerable inter-annual variation.

Ciliates undoubtedly occur in the lakes of the Syowa Oasis, the Schirmacher Oasis, and the Bunger Hills, but at present we have no data sfor these environments.

Flagellated Protozoa

The typical concentrations of phototrophic nanoflagellates (PNAN) are given in Table 3.4. Among the PNAN found in saline lakes, cryptophytes are particularly common and often dominate the community (Burch 1988; Roberts et al. 2000; Laybourn-Parry et al. 2005; Bielwicz et al. 2011) (see Figure 1.26e). In many cases cryptophytes are mixotrophic in Antarctic lakes (see Section 1.8.3). While mixotrophic cryptophytes have been reported at lower latitudes, they do not commonly exploit this nutritional strategy. In Lake Fryxell (Lake 1, Table 3.1) three cryptophyte species dominated the community, contributing 79% of all PNAN biomass. Of these, two species were tentatively identified as *Cryptomonas* or *Rhinomonas*, and the third less common species resembled *Cryptomonas undulata* (Roberts et al. 2000). In Lake Bonney (lake 4/6, Table 3.1) cryptophytes contributed between 114 and 136 of the clones isolated from the water column; the vast majority were related to the marine species *Geminigera cryophila* (Bielewicz et al. 2011). Throughout an annual cycle in Ace Lake and Highway Lake (lakes 13 and 14, Table 3.1), the PNAN community was dominated by two morphological forms of a cryptophyte species (Laybourn-Parry et al. 2005). In both the Dry Valleys and Vestfold Hills lakes cryptophytes are mixotrophic, feeding on bacteria and in some instances on DOC (Roberts and Laybourn-Parry 1999; Marshall and Laybourn-Parry 2002; Laybourn-Parry et al. 2005; Thurman et al. 2012).

Another mixotroph, *Pyramimonas gelidicola*, is also common in the hyposaline and saline lakes of the Vestfold Hills (see Figure 1.26f). It occurs throughout the year in Ace Lake, where it peaked in January and February (Burch 1988; Bell and Laybourn-Parry 1999a; Laybourn-Parry et al. 2005). The PNAN community of Highway Lake also contains large numbers of *Pyramimonas*, feeding on DOC as well as bacteria (Laybourn-Parry et al. 2005). A species of *Pyramimonas* is common in Lake Fryxell, contributing 19% of the PNAN biomass at 8 to 10 m in the water column (Roberts et al. 2000). *Pyramimonas* does not appear to be mixotrophic in lower latitude ecosystems, only appearing to exploit this capability in extreme Antarctic systems.

Chlamydomonas is commonly seen in saline lakes such as Lake Bonney (Bielewicz et al. 2011; Thurman et al. 2012). *Chlamydomonas raudensis*, isolated from the water column of Lake Bonney, is an obligate psychrophile. It occupies a discrete layer in the lake at between 10 and 17 m, which is a transition zone between an oxygen-rich layer where oxygen production exceeds respiratory oxygen uptake, and a deeper anoxic region. At this level PAR levels are very low: <0.1% of surface incident PAR (Pocock et al. 2004). *Chlamydomonas subcaudata*, isolated from a layer at between 17 and 20 m in the east lobe of Lake Bonney, is also a psychrophile growing at temperatures of 16 °C or lower. Like *C. raudensis*, it is adapted to life in poor light conditions, but nonetheless retains the capacity to adjust the total xanthophyll pool size and consequently the ability to dissipate energy non-radiatively (Morgan et al. 1998). *Chlamydomonas* spp. are also reported from the water column of Lake Wilson (lake 3, Table 3.1) where its highest biomass occurred at 45 m decreasing to 51 m (Webster et al. 1996), and in Lake Fryxell where it showed no vertical stratification in the euphotic zone over a number of summers between 1987 and 1991 (Spaulding et al. 1994).

The mixotroph *Ochromonas* occurs in the phytoplankton of a number of saline lakes including Lake Wilson, Lake Frxyell, and Lake Bonney (Webster et al. 1996; Spaulding et al. 1994; Thurman et al. 2012). One of the most detailed and long-term studies of PNAN and algae in the Dry Valleys was conducted over five summers (1987–1991) in Lake Fryxell (Spaulding et al. 1994). While it confirmed the dominance of cryptophytes, it detailed the less common PNAN species including *Dinobryon* and *Mallomonas*, and a large number of Algae and Cyanobacteria. One of the interesting features of this long-term study is the considerable inter-annual variation seen in the species composition and abundances of various PNAN. Such marked variations are also evident from an analysis of other data sets from the Dry Valleys lakes (Thurman et al. 2012) and in the saline lakes of the Vestfold Hills (Laybourn-Parry et al. 2002).

PNAN diversity is low in hypersaline lakes. The only phototrophic flagellate that can tolerate extremely hypersaline conditions is *Dunaliella* spp. The genus has a wide salinity tolerance, from 20‰ to 350‰, with an optimum at around 120‰ (Borowitzka 1981). It apparently only occurs in hypersaline Antarctic lakes and has been recorded in Deep Lake (Lake 30, Table 3.1, salinity 350‰), Organic Lake (lake 36, Table 3, salinity 230‰), and Hunazoko Ike (Lake 20, Table 3.1, salinity 212‰) (Wright and Burton 1981; Franzmann and Rohde 1992; Tominaga and Fukui 1981). *Dunaliella*-like flagellates were seen in extremely hypersaline Don Juan Pond, where salinities of up to 671‰ have been recorded (Siegal et al. 1979).

Dinoflagellates are a conspicuous element of the phototrophic flagellate community of the Vestfold Hills suite of saline lakes (see Figure 1.26 d). To date the most saline lake in which they have been recorded is Lake Williams (lake 18, Table 3.1) and the least saline, Highway Lake (lake 13, Table 3.1), where they ranged in concentration between 0.4–2.7×10^5 cell l^{-1} and 0.02–0.45×10^5 cells l^{-1} respectively over a summer (Laybourn-Parry et al. 2002). The species diversity is low when compared with the Ellis Fjord and the neighbouring marine environment from which the lake communities were originally derived (Grey et al. 1997).

The communities in the lake are typically dominated by only three to four species including heterotrophic taxa. In Highway Lake (salinity 4‰) there were two dominant species, both of which were phototrophic. Molecular analysis showed one of these species to be a *Gymnodinium* sp. and the other as *Scrippsiella* aff. *hangoei*. A third common species was the heterotrophic or possibly mixotrophic *Gyrodinium glaciale*. Pendant Lake (salinity 19‰) had four dominant species including *G. glaciale*, *S.* aff. *hangoei*, a *Gymnodinium* species, and *Polarella glacialis*. The latter species has also been isolated from Lake Abraxas (24‰) and Ekho Lake (mixolimnion 60‰–150‰) (Rengefors et al. 2008). *P. glacilis* is common in the sea ice of Prydz Bay, off the Vestfold Hills, and elsewhere in Antarctica (Thomson et al. 2006; Patterson and Laybourn-Parry 2012). Life in the sea ice requires considerable physiological plasticity, particularly in relation to wide variations in salinity and extremely low temperatures, thus it is hardly surprising that *P. glacialis* has successfully colonized hyposaline to hypersaline lakes. In Ace Lake the dinoflagellate community was dominated by *G. glaciale*, *Gymnodinum* spp., with low numbers of *P. glacialis* (Rengefors et al. 2008). These various species show seasonal variations in Highway, Pendant, and Ace Lakes (Figure 3.10). In Pendant and Ace Lakes, *G. glaciale* is common during the winter months, while in much more hyposaline Highway Lake it is common in summer. Its heterotrophic/mixotrophic capabilities enable it to flourish in the

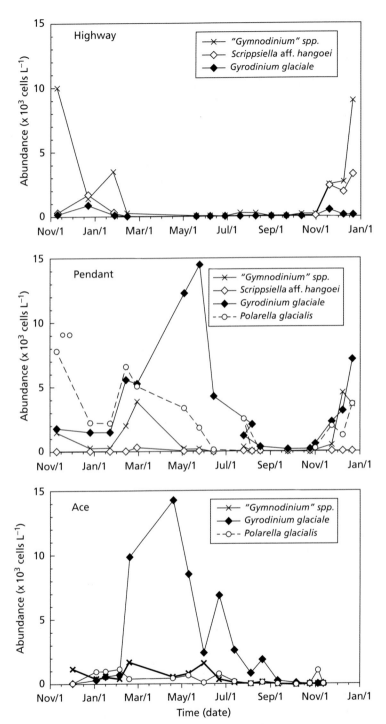

Figure 3.10 Abundances of dinoflagellate species in three saline lakes in the Vestfold Hills. From Rengefors et al. (2008), with the permission of Wiley and Sons.

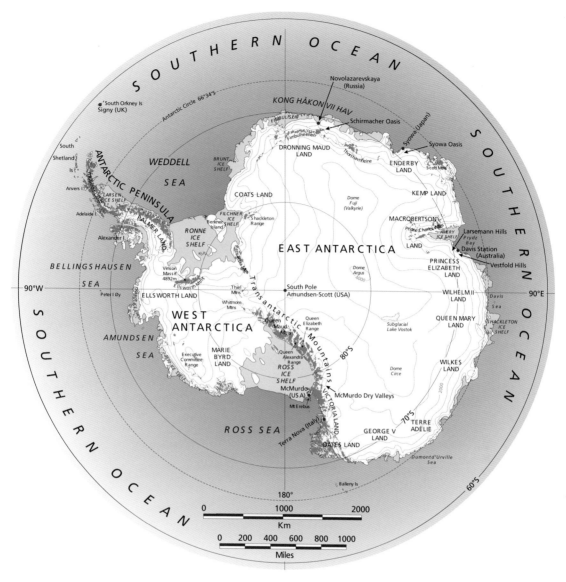

Plate 1 Map of Antarctica showing the position of coastal oases, the Dry Valleys, and the Maritime Antarctic. (See Figure 1.1)

Plate 2 Cryolake on Canada Glacier, Taylor Valley. Photo courtesy of M. Tranter. (See Figure 1.6)

Plate 3 Ice shelf lakes on the McMurdo Ice Shelf. Photo courtesy of W.F. Vincent. (See Figure 1.7)

Plate 4 Lake Nottingham (unofficial name), Vestfold Hills. Perennially ice-covered freshwater lake that abuts the continental ice sheet. Photo J. Laybourn-Parry. (See Figure 1.8)

Plate 5 (a) Rafted freshwater ice caused by tidal action on the shore of Beaver Lake, Amery Oasis. Beaver Lake camp in the background. (b) The perennial ice surface of Beaver Lake with twin otter used for deployment from Davis Station in the Vestfold Hills. Photos J. Laybourn-Parry. (See Figure 1.10)

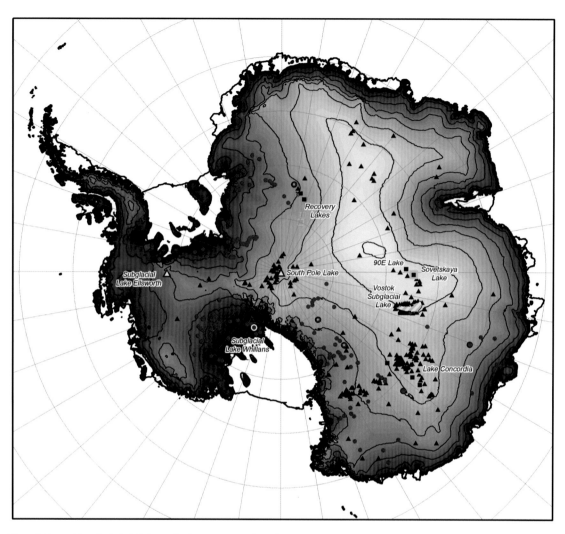

Plate 6 The position of subglacial lakes under the Antarctic Ice Sheet, from Wright and Siegert (2012) with the permission of Cambridge University Press. Colours and shapes indicate the type of investigation for each site: black triangle = radio-echo sounding, yellow = seismic sounding, green = gravitational field mapping, red circles = surface height change measurement, square = shape identified from ice surface feature. Lake Vostok is shown in outline. Original illustration provided by M. Siegert. (See Figure 1.11)

Plate 7 Lake Frxyell, Taylor Valley with the Canada Glacier in the background. Note polar haven erected over the sampling hole to stop it from freezing up. Photo J. Laybourn-Parry. (See Figure 1.12)

Plate 8 Arial photograph of a suite of saline lakes in the Vestfold Hills. From the left bottom seaward: Lake Jabs, Club Lake, Deep Lake (salinity X 10 seawater), Lake Stinear, and Lake Dingle. Photo J. Laybourn-Parry. (See Figure 1.13)

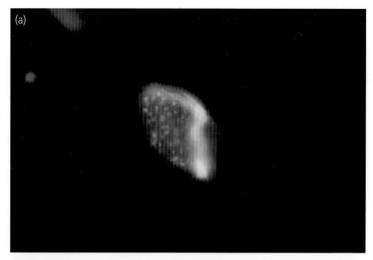

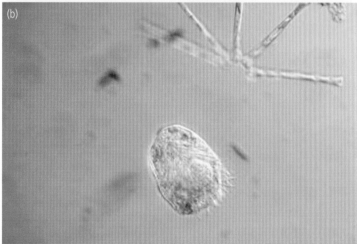

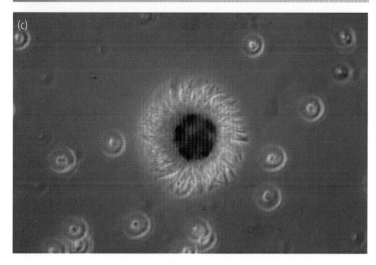

Plate 9 (a) *Euplotes* sp. containing symbiotic zoochlorellae from Lake Hoare. Stained with DAPI and viewed under epifluorescence microscopy; the long blue area is the ciliate macronucleus, the red areas are autofluorescing zoochlorellae with their individual blue DAPI stained nuclei; (b) *Strombidium* sp. containing sequestered (green) plastids from a temperate lake; (c) *Mesodinium rubrum* fixed in Lugol's iodine from Ace Lake; note deeply stained cryptophycaen endosymbiont within the cell. Photos J. Laybourn-Parry. (See Figure 1.28)

Plate 10 Photograph of benthic cyanobacterial mats in Lake Hoare; note the pinnacles. Photo courtesy of Ian Hawes. (See Figure 2.23)

Plate 11 A photograph of benthic mosses from a freshwater lake in the Syowa Oasis. Photo courtesy of S. Kudoh. (See Figure 2.27)

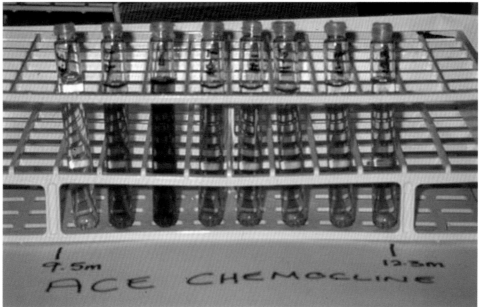

Plate 12 Photograph of ammonium and chlorophyll *a* concentrations over the chemocline of Ace Lake at 4 cm intervals. Ammonium analysis is by a colorimetric technique, and thus the intensity of the blue colour is related to the concentration of ammonium (NH_4-N). Photo J. Laybourn-Parry. (See Figure 3.6)

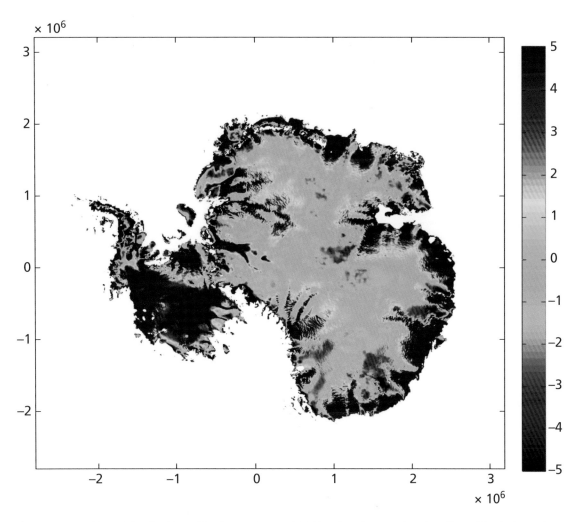

Plate 13 Estimated Antarctic basal melting and freezing rates in mm yr^{-1}. Melting, shown in warm colours, is positive and freezing is negative and shown in cold colours. Redrawn from Tulaczyk and Hossainzadeh (2011) with the permission of High Wire Press. Values that are higher or lower than ±5 mm yr^{-1} have been truncated. Calculations were performed in a similar manner to (Pattyn 2010). (See Figure 6.1)

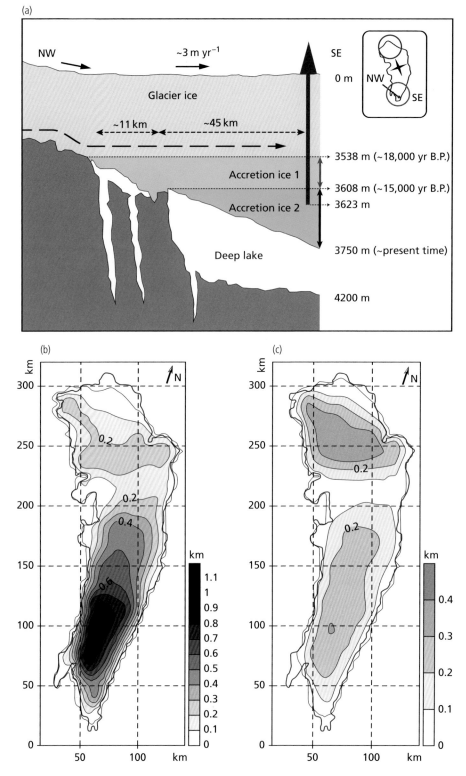

Plate 14 (a) Conceptual model of Subglacial Lake Vostok, showing the lake cross-section along a flow line (ice flow is NE to SW) (De Angelis et al. 2004). The two main areas of accretion ice are shown. Type 1 contains debris and Type 2 is clean, the thick vertical arrow denotes the site where the accretion ice has been cored. (b) Predicted water depths (resolution = 0.1 km). (c) Sediment thicknesses (contour interval = 0.1 km) in Subglacial Lake Vostok as inferred from airborne gravity data (by inversion), where the red line indicates the lake coast line from radar data (Filina et al. 2008). With the permission of Elsevier. (See Figure 6.3)

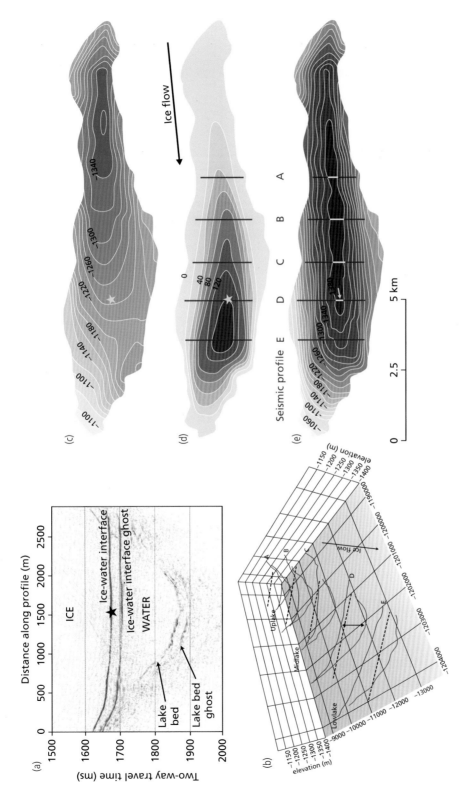

Plate 15 Seismic reflection data from Subglacial Lake Ellsworth (adapted from (Woodward et al. 2010). (a) An example of a seismic reflection data profile, with the main reflectors and ghosts identified (profile D). (b) An up-lake (into ice flow) 3D representation of the lake surface (red lines) and bed (blue lines) identified from five seismic profiles, with the drill site indicated by a black line in profile D. Black dashed lines indicate the critical boundary pressure for each seismic line (ice thickness = 3170 m). (c) Ice-water surface. (d) Water column thickness (relative to WRS-84 ellipsoid). (e) Lake bed topography (relative to WRS-84 ellipsoid), where yellow stars indicate drill site locations. Red lines in (d) and (e) indicate the measured positions of the lake bed (and water column thickness) from the seismic data and the parts highlighted in white are parts of the lake where the bed elevation is −1380 m. With the permission of the American Geophysical Union. (See Figure 6.4)

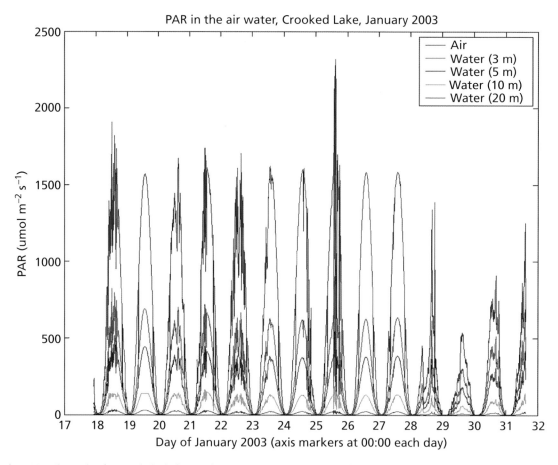

Plate 16 Daily PAR data from Crooked Lake for part of January 2003. Data collected every five minutes. The curves show marked differences in the patterns of PAR, for example the 19th, 26th, and 27th were cloudless sunny days, while the 18th, 20th, and other days were sunny with some cloud, resulting in a more jagged set of curves and the 29th was extremely overcast. (See Figure 7.2)

winter. However, some of the phototrophic species are also common in winter, begging the question of whether they might also be capable of mixotrophy.

Genotyping of clonal strains of *Scrippsiella* aff. *hangoei* isolated from four Vestfold Hills lakes using amplified fragment length polymorphisim (AFLP) has revealed high genetic differentiation among the lake populations. The clones were derived from Lakes Abraxas, Highway, McNeil, and Veretino with salinities ranging from 4‰ to 24‰. The lakes are within 10 km of each other. The hypothesis is that the lakes act as ecological islands, even though they are in close proximity (Rengefors et al. 2012). Using the Bayesian-based clustering method (STRUCTURE) two genetically distinct populations were identified, one population included all the strains isolated from Lake McNeill and the other mainly consisted of strains from the other three lakes. The genetic differentiation values (F_{ST}) ranged from 0.117 in Highway Lake up to 8.75 in Lakes Abraxas and Verentino. These values are high and indicate significant genetic variation among the lake populations of *Scrippsiella* aff. *hangoei*. The lakes differ in their physico-chemical conditions and their ages, imposing different evolutionary pressures on the organisms that have colonized them. Moreover, in dinoflagellates asexual reproduction is the norm, with sexual reproduction occurring in conjunction with resting cyst formation. While both asexual and sexual reproduction have been observed in Baltic Sea populations of *S. hangoei*, the life cycle of the Antarctic lake populations has not been elucidated. Given the high genetic diversity in the samples, occasional sexual reproduction is likely. It is also possible that different year cohorts of resting cysts excyst in different years, helping to maintain genetic diversity.

While dinoflagellates are a significant component of the microbial communities of saline lakes in the Vestfold Hills they are not commonly reported from the lakes of the Dry Valleys. The very detailed multi-summer survey conducted by Spaulding et al. (1994) did not record dinoflagellates. Parker et al. (1982) noted a colourless dinoflagellate in Lake Bonney East and Cathey et al. (1981) reported *Gymnodinium* and *Glenodinium* from Lake Fryxell. In neither case were concentrations for individual components of the plankton provided. However, a study that extended into April and the onset of winter darkness found both small phototrophic and heterotrophic (10–15 μm) dinoflagellates in both lobes of Lake Bonney in concentrations up to 100 l^{-1} in early April (Thurman et al. 2012). There are few data on the taxonomic make-up of the plankton in saline lakes elsewhere in Antarctica. Dinoflagellates were not recorded in a plankton survey of Nurume Ike, Hunazoko Ike, or Suribati Ike in the Syowa Oasis (Tominaga and Fukui 1981).

Concentrations of heterotrophic nanoflagellates (HNAN) in Antarctic saline lakes are given in Table 3.4. While we have quite a lot of data on their abundances, their taxomony has been poorly researched compared with the PNAN. A number of marine choanoflagellates have been identified in the saline lakes of the Vestfold Hills. *Acanthoecopsis unguiculata* occurs in Organic Lake (salinity 230‰) in concentrations up to 50×10^4 l^{-1}. It differs from the marine population in having a smaller protoplast size, although the lorica size is similar (Van der Hoff and Franzmann 1986). *Diaphanoeca grandis* dominated the HNAN community of Highway Lake (see Figure 1.26b). It occurred throughout the year with peak abundance in the winter. As well as feeding on bacteria it exploited the DOC pool as an energy source (Laybourn-Parry et al. 2005). *Diaphanoeca multiannulata* has been identified in the plankton of Clear Lake (Marchant 1985). The wide range of salinities in which choanoflagellates are found suggests that these flagellates have a strong tolerance to salinity variation. An aloricate collared flagellate contributed to the HNAN communities of Lake Fryxell (Laybourn-Parry unpublished data). A new species of the amoeboflagellate *Tetramitus* has been isolated from Pendant Lake (Murtagh et al. 2001) (see Figure 1.26c).

Growth rates of heterotrophic flagellates are generally slower than those seen in lower latitude lakes from temperate regions. Specific growth rates of natural assemblages of HNAN from Ace Lake ranged between 0.0032 and 0.017 h^{-1} giving doubling times of 119 to 37 hours at temperatures between 0° C and 8 °C, while in comparison in freshwater Lakes Constance, Michigan, and Loch Ness the range was between 0.0017 and 0.09 h^{-1}, with doubling times between 37 and 8 hours (Laybourn-Parry et al. 2000; Bjørnsen et al. 1988; Weisse et al. 1990;

Laybourn-Parry and Walton 1998). These water bodies have relatively low temperatures; for example, Loch Ness ranges from 5.7 °C to 13.5 °C at the surface and Lake Constance from 3.1 °C to 14.8 °C at 10 m, with a surface maximum at around 22 °C. Ace Lake has a temperature range from a few degrees below zero up to 8 °C when the ice cover is lost. At a range of temperatures between 0.0 °C and 8.0 °C, Ace Lake dinoflagellate community specific growth rates ranged between 0.0057 and 0.0216 h^{-1}, with the highest rate at 5 °C.

Grazing rates of mixotrophic PNAN and HNAN are shown in Table 3.7. At first glance there are very considerable differences in the grazing rates achieved by both PNAN and HNAN across the various lakes. This variation can be explained in relation to the productivity of the lakes and their bacterial community cell concentrations. Lake Bonney is less productive than Lake Fryxell (see Table 3.2) and has much lower concentrations of bacteria (Table 3.4). Consequently the grazing rates of both HNAN and mixotrophic PNAN are lower than in Lake Frxyell, despite relatively high clearance rates. The flagellates in Lake Bonney have to work harder to acquire food resources because the volume of water they filter contains less food. Lake Fryxell showed differences in cryptophyte grazing rates in the summers of 1997/1998 and 2000. This probably relates in part to inter-annual variations in the concentration of bacteria, because grazing rate correlates positively with bacterial cell abundance (Roberts and Laybourn-Parry 1999). In 1997/1998

Table 3.7 Heterotrophic and mixotrophic flagellated protozoa grazing rates. Numbers in brackets after location correspond with lake numbers in Table 3.1.

Functional group	Year and location	Temp. °C	Clearance rate nl cell^{-1} h^{-1}	Grazing rate bact. cell^{-1} day^{-1}
HNAN[1]	March/April 2008 **Lake Bonney E.** (6)	2	0.32 ± 0.18– 1.94 ± 0.60	2.88–20.80
PNAN[1]	March/April 2008 **Lake Bonney E** (6)	2	0.18 ± 0.07– 2.28 ± 0.18	1.68–12.48
HNAN[1]	March/April 2008 **Lake Bonney W.** (4)	2	0.82 ± 0.19– 2.34 ± 0.33	4.80–16.24
PNAN[1]	March/April 2008 **Lake Bonney W** (4)	2	0.16 ± 0.08– 1.16 ± 0.48	1.08–8.16
HNAN[2]	Nov/Jan 1997/1998 **Lake Fryxell** (1)	0.01–2.7	1.8–7.9	96–129
Cryptophytes[2]	Nov/Jan 1997/1998 **Lake Fryxell** (1)	0.01–2.7	0.1–1.6	38.4–86.4
Cryptophytes[3]	Nov/Dec 2000 **Lake Fryxell** (1)	0.3–2.2	0.006–0.32	5.28–10.08
HNAN[4]	Feb 2001–Feb 2002 **Ace Lake** (14)	1–2	0.04 ± 0.01 – 0.37 ± 0.01	0.24–34.8
Cryptophytes[4]	Feb 2001–Feb 2002 **Ace Lake** (14)	1–2	0.01 ± 0.005– 0.21 ± 0.04	0.31–24.9
Pyramimonas[4]	Feb 2001–Feb 2002 **Ace Lake** (14)	1–2	0.005 ± 0.001	0.15
HNAN[4]	Apr 2001–Feb 2002 **Highway Lake** (13)	1–2	0.02 ± 0.001– 1.80 ± 0.30	0.48–65.7
Cryptophytes[4]	Apr 2001–Feb 2002 **Highway Lake** (13)	1–2	0.004–1.05	0.17–38.2

1: Thurman et al. 2012; 2: Roberts and Laybourn-Parry 1999; 3: Marshall and Laybourn-Parry 2002; 4: Laybourn-Parry et al. 2005.

bacterial concentration ranged from 1.35×10^9 cells l^{-1} at 6 m to 4.07×10^9 l^{-1} at 9 m, while in 2000 it was lower, ranging from 1.90 to 3.40×10^9 l^{-1}. In both studies the species composition of the flagellate community was not determined. This may have varied across years contributing to the observed differences in grazing rate. Cryptophyte species and different morphotypes of the same species possess differing grazing rates. In Lake Frxyell cryptophytes never become entirely photosynthetic, although the balance between carbon derived from grazing and photosynthesis changes over the summer, with photosynthesis contributing its highest in December (Marshall and Laybourn-Parry 2002). We have no conclusive data on what happens in the winter, but based on automated phytoplankton sample collection in winter the supposition is that mixotrophy plays an important role in sustaining populations of vegetative cells throughout winter darkness (McKnight et al. 2000).

Year-long investigations at Ace Lake and Highway Lake in the Vestfold Hills showed that PNAN and HNAN remained active in the winter and that both cryptophytes and *Pyramimonas gelidcola* in Ace Lake grazed on bacteria throughout the year (Table 3.7). Bacterial concentrations were similar in both lakes, but the cryptophytes in Highway Lake were on average larger than those in Ace Lake, which may explain the higher grazing rates in this lake. The Highway Lake population of *Pyramimonas* did not take up bacteria but took up dextrans in the range 4 kDa to 500 kDa, indicating that they exploited the DOC pool as an energy source. The cryptophyte community in Highway Lake also took up dextrans to a lesser extent. The HNAN community in this lake was dominated by the collared flagellate *Diaphanoeca grandis* (Figure 1.27b) which ingested bacteria and dextrans with a preference for 4 kDa and 500 kDa sizes (Laybourn-Parry et al. 2005).

Clearly nutritional versatility is important to the success of dominant PNAN groups in saline Antarctic lakes and for some HNAN. Sustaining active populations through the winter allows a ramping up of production in the spring, so that populations enter the short austral summer actively growing with a sustained biomass. However, some groups do enter resting cysts, as is obviously the case for dinoflagellates in Highway Lake (see Figure 3.10).

In Ace Lake the photosynthetic ciliate *Mesodinium rubrum* overwinters with part of the population in the active form surviving on endogenous energy reserves, while a portion of the population encysts. In the spring excysting individuals augment the actively growing population (Bell and Laybourn-Parry 1999a).

3.6.5 Algae

Together with the Cyanobacteria and PNAN (see Sections 3.6.2 and 3.6.4), the algae make up the phytoplankton. In common with the freshwater lakes, diatoms are not a common feature of the plankton where motile organisms predominate. The halophilic diatom *Tropidoneis laevissima* is reported from Suribati Ike and the euryhaline *Achnanthes brevipes* from Hunazoko Ike in the Syowa oasis (Tominaga and Fukui 1981). In a year-long study of Ace Lake (Vestfold Hills) diatoms including *Nitzschia* spp. only appeared in the plankton in the summer, occasionally attaining abundances of 2.5×10^5 l^{-1}. It is thought that these diatoms were derived from fringing littoral benthic mats as a result of flushing of meltwaters and/or wind-driven turbulence as the lake lost its ice cover (Bell and Laybourn-Parry 1999a). *Cyclotella* sp., *Navicula* spp., and *Nitzschia* sp. were noted in low numbers in Lake Fryxell during the summer of 1987. In the summers of 1988–1991 only one species of *Navicula* was recorded (Spaulding et al. 1994). It is likely that these diatoms were derived from algal mats, as was the case in Ace Lake.

Occasionally blooms of algae occur, for example in Pendant Lake. In most years this lake has chlorophyll *a* concentrations ranging between 3.7 µg l^{-1} and 7.46 µg l^{-1} during the summer and has a phytoplankton dominated by phytoflagellates (Perriss and Laybourn-Parry 1997; Madan et al. 2005). In the summer of 1999/2000 the lake supported chlorophyll *a* concentrations of between 5µg l^{-1} and 29 µg j^{-1}. The bloom was largely made up of the chlorophyte *Stichococcus bacillaris* which reached concentrations of 1.5×10^7 l^{-1} (Laybourn-Parry et al. 2002). There is no obvious explanation for this unusual bloom. In the summer of 1999/2000 the ice held in for the summer with only a moat melting out, although this was also the case in other years for which data are available. Possibly there was some unexplained perturbation causing the release of nutrients from the well-developed sediments in Pendant Lake. This unusual bloom and the variations seen in the long-term investigation of Lake Fryxell by Spaulding et al. (1994), serve to illustrate the considerable inter-annual variation seen in Antarctic lakes. *Stichococcus* was also collected in an automatic sampling device in Lake Fryxell during the winter of 1990. It was absent in the following winter of 1991 and, interestingly, has not been recorded in the summer (McKnight et al. 2000).

3.6.6 Zooplankton

In common with the freshwater lakes, zooplankton are sparse in saline lakes and they are absent from the hypersaline lakes. The freshwater cladoceran *Daphniopsis studeri* has invaded the most hyposaline lakes, such as Highway Lake (salinity 4‰), where it has replaced the marine copepod *Paralabidocera antarctica* which occurs in more saline lakes such as Ace Lake, Lake Abraxas, and Pendant Lake (lakes 14, 27, and 19 respectively, Table 3.1) (Wright and Burton 1981). The planktonic harpacticoid *Amphiascoides* sp. occurs with *P. antarctica* in Lake Abraxas, where there appears to be spatial separation of the two species in the water column. The bulk of copepodid and adult stages of the *P. antarctica* population resided at a depth of around 5 m, whereas the post-naupliar *Amphiascoides* population occurred lower in the water column, with a peak at around 16–17 m (Bayly and Eslake 1989). This separation suggests that they may have been exploiting different food resources. Lake Abraxas appears to be the only saline lake in the Vestfold Hills that supports a planktonic harpacticoid, as indicated from a survey of 150 lakes (Burton and Hamond 1981). This is probably one of the oldest lakes in the Vestfold Hills; it is believed to predate the LGM and has always been saline (see Section 3.3).

P. antarctica remains active throughout the year in Ace Lake (Figure 3.11). Both copepodid (including adults) and naupliar stages occur in the population suggesting that development continues through the winter. Peak abundance occurs in the summer (Bell and Laybourn-Parry 1999a). The lacustrine populations are smaller than neighbouring marine

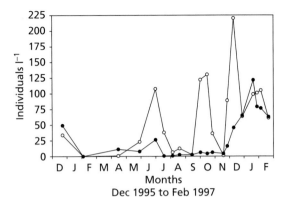

Figure 3.11 Abundance of *Paralabidocera antarctica* individuals in Ace Lake, Vestfold Hills. ○ naupliar stages, ● copepodid stages (including adults). Redrawn from Bell and Laybourn-Parry (1999).

populations. Adult lengths in Ace Lake were 40% of the marine population (Swadling et al. 2000). Calanoid copepods can feed both raptorially and by filter feeding. It is likely that the larger stages and the adults feed both the phototrophic and heterotrophic flagellates and ciliated protozoans, while the smaller naupliar stages may exploit bacteria and smaller flagellate species. Grazing experiments using a radiotracer indicate that the *P. antarctica* population can clear up to 3% of the water column each day (Swadling and Gibson 2000).

As stated in Section 2.5.5, large water volume sampling of freshwater Lake Hoare revealed evidence of a very sparse population of a crustacean thought to be *Boeckella*. The only saline Dry Valleys lake in which Crustacea have been found is hyposaline Lake Joyce (lake 7, Table 3.1, salinity 3.5‰). Five-litre water samples collected at 26 m contained a few copepod nauplii on two occasions in December 1997 (Roberts et al. 2004a). The lake has a maximum depth of 35 m so it is possible that the nauplii were of benthic origin. Crustacea have not been reported from the saline lakes of either the Syowa Oasis or elsewhere in Antarctica. However, this is largely because no one has looked for them. It is quite likely that they occur in the hyposaline lakes of the Bunger Hills.

Rotifers are primarily a freshwater group with few marine species, so that their limited occurrence in saline lakes is not surprising. Planktonic rotifers are sparse in the saline lakes of the Dry Valleys; a few specimens, mostly *Philodina*, have been noted in samples from Lake Bonney and Lake Joyce (Roberts et al. 2004a, Thurman et al. 2012). Higher numbers were recorded in Lake Fryxell, between 8 and 41 l^{-1}. Two species of *Philodina* occurred, the smaller of the two was identified as *Philodina alata*. The morphology of *Philodina* suggests that it is primarily a benthic species. It may have been washed into the plankton from the littoral algal mats along with filamentous cyanobacteria (Laybourn-Parry et al. 1997). Rotifers also occur in the hyposaline lakes of the Vestfold Hills. A small number of *Notholca* were noted in Ace Lake (Bell and Laybourn-Parry 1999a) and they have been observed in samples from Highway Lake and Pendant Lake (Dartnall 2000; Laybourn-Parry unpublished data).

3.7 Carbon cycling in the plankton

3.7.1 Primary production

Rates of primary production in saline Antarctic lakes are given in Table 3.2. The data are largely limited to the McMurdo Dry Valleys and the Vestfold Hills. A comparison with freshwater lakes (Table 2.2), for which there are many more data covering a transect from the Dry Valleys to Signy Island in the Maritime Antarctic, shows that the production of freshwater and saline Dry Valleys lakes are similar, while the saline lakes of the Vestfold Hills and Lake Polest in the Bunger Hills are much more productive than the freshwater lakes in continental coastal locations. In Antarctic lakes the initiation of the spring phytoplankton growth in non-turbulent waters under ice is entirely a function of seasonal increase in PAR. Where data cover an annual cycle, as is the case for Highway Lake, Pendant Lake, and Ace Lake (see Table 3.2), lowest rates of photosynthesis occur in the winter, increasing in the spring to a maximum in the summer in relation to solar radiation. This has also been shown in Lake Bonney where measurements of phytoplankton biomass and solar radiation in the water column showed that biomass increases when solar radiation first penetrates the ice, reaching its maximum in January (Priscu et al.1999b; Morgan-Kiss et al. 2006).

The lack of turbulence, combined with strong vertical gradients in nutrients and temperature in the water column of ice-covered meromictic lakes, creates conditions for the stratification of the phototrophic community. The development of three distinct phytoplankton communities in Lake Bonney East was followed during the winter/spring transition. Between the bottom of the ice cover down to 8 m the community was dominated by cryptophytes with *Chlamydomonas*; below this to 16 m the community was largely made up of cryosophytes including *Ochromonas* and a small coccoid alga (<4 μm). From 16 m to 20 m the assemblage was made up of *Chlamydomonas subcaudata*, *Ochromonas* and the coccoid alga. Photosynthesis had already started prior to the investigation on 9th September (Figure 3.12) increasing to highest rates in November and December. Highest photosynthetic efficiency occurred in the spring, decreasing over the growing season (Lizotte et al. 1996 Lizotte and Priscu 1994). A similar zonation of taxa occurred in Lake Vanda and Lake Fryxell (Vincent and Vincent 1982; Vincent 1981; Spaulding et al. 1994). The zonation relates to the tolerances of particular species to nutrient concentrations, temperature, and light.

Deep chlorophyll maxima (DCM) are often associated with the chemocline in meromictic lakes, creating sites of high phytoplankton biomass and productivity (Figure 3.13) (see also Section 3.5 and Figure 3.6) (Burch 1988; Vincent and Vincent 1982; Priscu et al. 1987). In the Dry Valleys lakes and Ace Lake these DCM are located at depths where only around 0.1%–1.0% of surface irradiation penetrates. Below the DCM dissolved oxygen is virtually absent and in this region photosynthetic bacterial chlorophyll is present. The chemocline is a region of high nutrient availability, where nutrients diffuse upwards from the nutrient-rich monomolimnion (see Table 3.3). Thus the DCM communities have access to nutrients but are confronted by extremely low levels of PAR and are adapted to undertaking photosynthesis under light-limited conditions. They have high photosynthetic efficiencies.

While PAR is of fundamental importance in driving photosynthesis, nutrient limitation, particularly of phosphorus, appears to be an important factor in controlling levels of primary production in the Dry Valleys lakes. Rates of carbon assimilation were

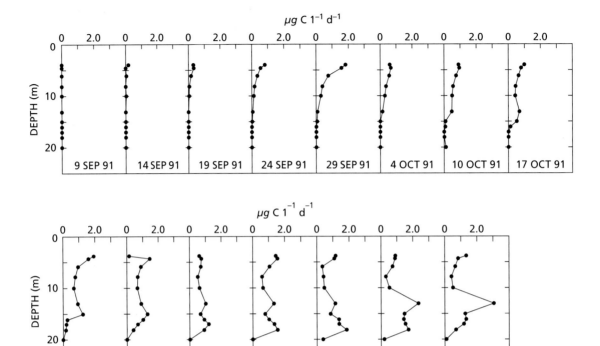

Figure 3.12 Rates of primary production in Lake Bonney during the spring/summer transition in 1991. From Lizotte et al. (1996). With the permission of Springer-Verlag.

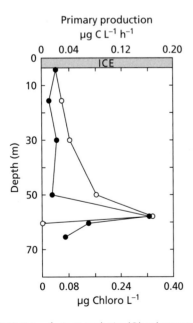

Figure 3.13 Rates of primary production (O) and concentrations of chlorophyll *a* (●) in relation to depth in Lake Vanda, Wright Valley, McMurdo Dry Valleys. Redrawn from Priscu et al. (1987).

enhanced in both lobes of Lakes Bonney and Frxyell when additional phosphorus was added to incubations. The only exception was in the DCM layer of the east lobe of Lake Bonney. The addition of nitrogen alone had no effect, however, the addition of nitrogen and phosphorus together had almost the same impact as the addition of phosphorus. Freshwater Lake Hoare differed in that the addition of phosphorus alone had no impact on carbon fixation (see Section 2.6.1). The saline lakes receive phosphorus through intermittent stream flow in the summer, but this study was conducted before stream flow occurred, indicating that phosphorus was probably mainly supplied by upwards diffusion across the chemocline (Priscu 1995 Dore and Priscu 2001). In contrast neither the meromictic lakes nor unstratified saline lakes of the Vestfold Hills appear to be nutrient limited. Both soluble reactive phosphorus and species of nitrogen were measurable in the water columns during the summer and over annual cycles. However, the upper water column of Ace Lake can become depleted in both nitrogen and orthophosphate, particularly in summer (Bell 1998; Bell and Laybourn-Parry 1999a; Laybourn-Parry et al. 2002).

3.7.2 Bacterial production

Data for bacterial production in saline lakes are limited to the Dry Valleys and Vestfold Hills lakes (Table 3.2). Rates of bacterial growth are relatively low in the Dry Valleys lakes compared with the coastal Vestfold Hills, reflecting lower rates of primary production. Where a lake receives allochthonous inputs, as is the case for Rookery Lake (lake 17, Table 3.1), very high rates of bacteria production are achieved, up to 7848 µg l^{-1} day^{-1} despite low temperatures (Table 3.2). In Antarctic lakes most of the DOC is derived from primary production within the system, as allochthonous sources are limited. Thus bacterial production is sustained by autochthonous carbon. In the case of Rookery Lake, high levels of bacteria growth are sustained by both autochthonous and significant allochthonous carbon and nutrients inputs from a nearby Adelie penguin rookery. These rates of bacterial production far exceed those for freshwater Heywood Lake, which is subject to allochthonous inputs from a fur seal wallow (see Table 2.2).

The DOC pool of Taylor Valley Dry Valleys lakes is derived from a number of sources: the exudation of labile DOC from the phytoplankton and the benthic algal mats, inputs from glacial streams and upward diffusion across sediments, and the chemocline in meromictic lakes. The bulk of DOC is derived from autotrophic production relative to stream input and upward diffusion. The calculation of carbon budgets for these lakes is dealt with in detail in Section 2.6.2; see also Table 2.8 for partial carbon budgets for Lakes Fryxell and Bonney. Bacterial production was two to three times higher than the DOC supplied from these identified sources in Lake Fryxell and 1.6 times greater in Lake Bonney (Takacs et al. 2001). The 'missing' DOC may be accounted or from the lysis of cells by grazers, but a major input may be from the recycling of carbon through viral lysis of bacterial and eukaryote cells. Studies in freshwater Antarctic lakes indicate that over an annual cycle between 20 and 60% of the DOC pool is attributable to viral lysis of bacteria (Säwström et al. 2008). Phosphorus availability limits phytoplankton production in Lake Bonney, (see Section 3.7.1). The bacterioplankton has also been shown to be phosphorus limited. The addition of 10 µM phosphorus increased thymidine incorporation in the mixolimnion (Ward et al. 2003).

These types of budget have not been developed for the lakes of the Vestfold Hills. Unlike the Dry Valleys these lakes are largely fed by snow-melt, which can vary considerably from year to year. Benthic algal mats are present but they are not as well developed or extensive as those seen in Dry Valleys lakes, and are unlikely to contribute high levels of photosynthate to the DOC pool. The majority of the DOC pool is probably derived from autotrophic production in the plankton, and in meromictic lakes, upward diffusion across the chemocline from biological processes in the monomolimnion. Viral abundance is high in saline lakes (see Section 3.6.3), although we do not have information on rates of infection or burst sizes. As is the case in freshwater lakes viral lysis of host cells may be important in recycling carbon in these systems. In Highway Lake, Ace Lake, Pendant Lake, and Lake Williams there was a decline in DOC availability over the summer. In most cases highest bacterial production occurred in early summer 1999/2000, while highest primary production was attained in December. However, there are inter-annual variations because, during a year-long study between January 2003 and January 2004, bacterial production peaked in January in Pendant and Ace Lakes, while primary production peaked in November and December in Pendant and Ace Lakes respectively. Among the lakes Lake Williams, the most saline of these lakes, had the highest summer levels of both primary and bacterial production (Tables 3.1 and 3.2) (Laybourn-Parry et al. 2002; 2007).

3.7.3 Heterotrophic grazing and carbon cycling

Creating models of carbon cycling for the complex communities of lower latitude lakes is a major challenge. Antarctic lakes have relatively simple microbial communities that appear to lend themselves more readily to the construction of models. A model of carbon flow in the mixolimnion plankton community, based on data from two summers (1996/1997 and 1997/1998) has been compiled for Lake Fryxell (McKenna et al. 2005). The model

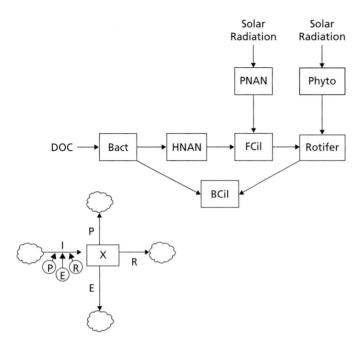

Figure 3.14 Planktonic food web for Lake Fryxell, Taylor Valley, McMurdo Dry Valleys. Phyto = phytoplankton, PNAN = phototrophic nanoflagellates, Bact = bacteria, FCil = flagellate feeding ciliates, BCil = bacterial feeding ciliates, HNAN = heterotrophic nanoflagellates and rotifers. Inset box defines fluxes of carbon for each trophic group (X); I = ingestion, P = production, E = egestion, R = respiration. From McKenna et al. (2005) with permission of Elsevier.

includes seven trophic groups: heterotrophic bacteria, photosynthetic nanoflagellates (PNAN), heterotrophic nanoflagellates (HNAN), ciliates (bacterial and flagellate feeding), phytoplankton, and rotifers (the top predator in this system) (Figure 3.14). Autotrophs and bacteria contributed around 94% of total community biomass, while grazers represented <6% of the total biomass. Table 3.8 shows a simplified version of the McKenna et al. (2005) model. Daily production to biomass ratios decreased with

Table 3.8 A simplified model of carbon flow through the mixolimnion planktonic food web of Lake Fryxell during the summers of 1996–1997 and 1997–1998 (from McKenna et al. 2005). Mean values are shown for each year in the waters column from immediately under the ice (4.5 m) down to 8 m. All biomass figures in $\mu g\ C\ l^{-1}$, respiration, production, and ingestion in $\mu g\ C\ l^{-1}\ day^{-1}$. Heterotrophic grazers include heterotrophic flagellates, bacterial and flagellate-feeding ciliates and rotifers.

Group and year	Biomass	Respiration	Production	Ingestion
autotrophs				
1996–1997	58.78 ± 71.1	16.11 ± 14.7	3.25 ± 3.2	–
1997–1998	192.80 ± 110.7	43.57 ± 23.4	5.91 ± 4.2	–
Heterotrophic bacteria				
1996–1997	14.39 ± 4.6	26.78 ± 8.5	4.23 ± 3.3	–
1997–1998	16.33 ± 8.0	30.64 ± 13.2	4.94 ± 3.0	–
Heterotrophic grazers				
1996–1997	4.37 ± 3.5	4.30 ± 3.0	0.47 ± 0.8	7.95 ± 5.9
1997–1998	8.32 ± 4.9	6.43 ± 3.3	0.20 ± 0.3	11.05 ± 5.8

depth in the mixolimnion and ranged between 0.318 ± 0.24 (1996/1997) and 0.350 ± 0.03 (1997/1998) for bacteria, 0.247 ± 0.44 (1996/1997) and 0.032 ± 0.01 (1997/1998) for autotrophs, and 0.097 ± 0.12 (1996/1997) and 0.018 ± 0.02 (1997/1998) for heterotrophic grazers. Flagellate-feeding ciliates, specifically *Plagiocampa* feeding on cryptophytes, constituted >80% of ciliate biomass and had production to biomass ratios of 0.182 ± 0.34 (1996/1997) and 0.020 ± 0.03 (199719/98). Their ingestion rates range from 3.12 to 4.56 cryptophytes cell day^{-1}. What is immediately evident from the mean values in Table 3.8 is that the heterotrophic grazers ingest the equivalent of bacterial and autotrophic production during the summer, although there were variations during the summer with highest grazing in January and lowest in October, while highest primary production occurred in December and bacterial production increased markedly in October to January. Another obvious feature is the inter-annual variation in both biomass of groups and heterotrophic grazer impact. There are very considerable differences in grazing rates from year to year (see Table 3.7).

While the structure of these systems is apparently simple, they are complicated by the fact that a significant portion of the phototrophic community is mixotrophic (see Section 3.6.4 *Flagellated Protozoa*) and effectively span trophic levels, being both autotrophic and heterotrophic. Indeed in Lake Fryxell there are times in the summer when cryptophyte grazing impact exceeds that of the heterotrophic flagellates (Roberts and Laybourn-Parry 1999). This is also true for mixotrophic PNAN in Lake Bonney during the March/April transition to winter (Thurman et al. 2012). In addition some mixotrophs and HNAN in Antarctic saline lakes can exploit the DOC pool as a carbon source (Laybourn-Parry et al. 2005). In Lake Bonney neither PNAN nor HNAN took up fluorescently labelled dextrans, indicating that they are not ingesting DOC. The ability of the Lake Fryxell flagellate communities to exploit DOC has not been investigated. A further complication that has not been considered in the model is viral induced mortality of the bacteria community. Limited evidence suggests this may be a significant factor in Antarctic lakes. Recycling of carbon by viral lysis short circuits carbon flow, returning DOC to the pool before it can be consumed by heterotrophic grazers. Thus the model should be viewed with some caution.

The top predators in these systems are rotifers, which contribute only a small proportion of heterotrophic biomass. They average between 1 l^{-1} to around 30 l^{-1} and they contributed only around 2% of community carbon flow. McKenna et al. (2005) suggest that major fluctuations in total community biomass patterns generally mirror seasonal levels of PAR. An in depth statistical analysis of physico-chemical parameters that control biomass abundance and primary production in freshwater Lake Hoare came to a similar conclusion. Here PAR and the availability of soluble reactive phosphorus drive autotrophic growth (Herbei et al. 2010). Thus these microbially dominated systems are controlled by bottom-up forces.

Winter heterotrophic dynamics in the Dry Valleys lakes are still largely a matter for conjecture. As indicated in Chapter 1, the logistics of conducting sophisticated research in the Dry Valleys in winter are extremely challenging. However, an extension of the research season was achieved in 2008. During the summer/winter transition in Lake Bonney the grazing impact of mixotrophic PNAN was considerable. On most occasions they removed more than 100% of daily bacterial production, at a time when primary production was declining as the light climate deteriorated. Here PNAN significantly outnumbered the HNAN and while HNAN had higher grazing rates (see Table 3.7), the mixotrophic PNAN imposed a much greater grazing impact, removing between 93 and 859 µg C l^{-1} day^{-1} compared with 6.5 and 48.6 µg C l^{-1} day^{-1} by the HNAN (Thurman et al. 2012).

A schematic of carbon flow in the water column of Ace Lake is shown in Figure 3.15. While we have data on the biomass and contribution to carbon flow of some of the groups in the mixolimnion, we lack data for the community in the monomolimnion. The mixolimnion phytoplankton includes the autotrophic ciliate *Mesodinium rubrum*, which during summer and spring contributes significantly to phytoplankton biomass (Figure 3.15). Rates of primary production are shown in Table 2.2. Heterotrophic ciliates usually contribute a small component of the heterotrophic community. The

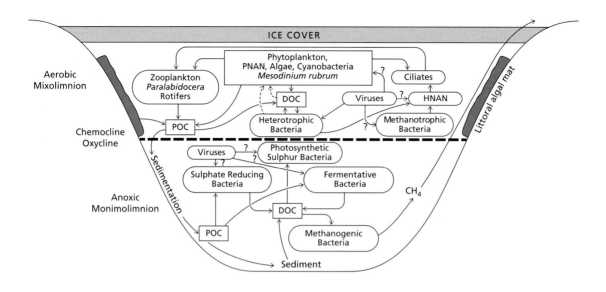

Season	Phytoplankton		Heterotrophic Bacteria	HNAN	Ciliates	Zooplankton
Winter	M. rubrum	0.06	27.25	11.08	0.01	0.59
	PNAN	0.23				
	Diatoms	0				
	Dinoflagellates	0.04				
	Picocyanobacteria	0.79				
	TOTAL	**1.12**				
Spring	M. rubrum	1.31	10.48	3.32	0.02	1.43
	PNAN	2.44				
	Diatoms	0.03				
	Dinoflagellates	1.19				
	Picocyanobacteria	0.21				
	TOTAL	**5.18**				
Summer	M. rubrum	22.82	12.20	9.58	5.88	10.78
	PNAN	4.68				
	Diatoms	0.07				
	Dinoflagellates	3.45				
	Picocyanobacteria	0.68				
	TOTAL	**31.7**				
Autumn	M. rubrum	0.41	15.5	10.69	0.12	1.32
	PNAN	0.32				
	Diatoms	0				
	Dinoflagellates	0.11				
	Picocyanobacteria	0.13				
	TOTAL	**0.97**				

Figure 3.15 Food web for meromictic Ace Lake, Vestfold Hills in the mixolimnion and monimolimnion, modified from Rankin et al. (1999). Dotted lines in mixolimnion indicate mixotrophic phototrophic nanoflagellates. Table below shows biomass for different trophic groups in the mixolimnion, from Bell (1998). All figures in μg carbon l^{-1}.

major grazers of the bacterioplankton are the HNAN and mixotrophic PNAN. HNAN remove between 13% and 92% of bacterial production per day, while PNAN remove between 0% and 45% of bacterial production per day over an annual cycle (Laybourn-Parry et al. 2005). Rotifers are sparse, the major zooplankter is the calanoid copepod *Paralabidocera antarctica*. Its feeding habits have not been elucidated, but it is likely to feed on elements of the phytoplankton and HNAN. Calanoid copepods feed raptorially or filter feed at low rates and can tackle particles in the range 5 to 100 μm (Moss 1980). Grazing rates derived from incubations with ^{14}C-methylamine hydrochloride suggest that the summer population of *Paralabidocera* can filter up to 3% of the water column per day (Swadling and Gibson 2000). These authors suggest that *Paralabidocera* may exert top-down control, however their grazing rate assumes no feeding selectivity and exploitation of all components of the phytoplankton. Ace Lake is a meromictic system where the mixolimnion waters may become periodically nutrient limited. Concentrations of chlorophyll *a* are low compared with unstratified neighbouring lakes (Table 2.2). The light climate (PAR) under annual ice is less severe than in the Dry Valleys lakes, but nonetheless is an important factor limiting carbon fixation, particularly in years when the ice holds in or only partially breaks out. Phytoplankton production is likely to be controlled by bottom-up factors rather than top-down grazing.

3.8 The biota of saline lake ice covers

Until fairly recently lake ice covers were not considered a potential habitat for life. The perennial ice covers of the Dry Valleys lakes can be very thick, up to 6 m, and contain considerable sediment blown onto the lakes from the surrounding ice-free unvegetated valley floors (Figure 1.33a). This dark material is heated by the sun and sinks down into the ice, where it forms layers and patches of aggregates and liquid water. Liquid water is present in the ice for about 150 days during the summer and up to 40% of the total ice cover volume may be liquid water (Priscu et al. 1998). Cyanobacteria are ubiquitous in glacial environments and are a common element in the ice covers of Dry Valleys lakes where they dominate the biomass (Priscu et al. 2005a). Aggregates of particles contain complex bacterial communities. Solitary and colonial cyanobacterial taxa are present, including *Phormidium*, *Oscillatoria*, *Nostoc*, *Scytonema*, and *Synechoccocus*. Very few eukaryotes colonize the ice.

The cyanobacterial assemblages have extremely low rates of photosynthesis, despite some high concentrations of chlorophyll *a* (range 0.05–58 μg l^{-1}). The maximum concentrations corresponded with layers of aggregate material. More than 98% of the autotrophic biomass in terms of chlorophyll *a* was associated with sections of ice core that contained sediment (Fritsen and Priscu 1998). Photosynthetic irradiance parameters vary between lakes (Table 3.9). These assimilation numbers and photosynthetic efficiencies were 10–100 fold lower than those recorded for the low light-adapted phytoplankton in the underlying water columns and sea algae adapted to low irradiances (Fritsen and Priscu 1998).

As indicated in Section 1.8.1, some cyanobacteria are capable of fixing atmospheric nitrogen. In Lake Bonney rates of nitrogen fixation ranged between 9.4 and 91.3 μmol N_2 mg^{-1} chlorophyll *a* h^{-1} (Priscu et al. 1998). The complex microbial consortia within the ice are enriched microzones where photosynthesis and bacterial production cycle carbon in what

Table 3.9 Photosynthetic efficiencies and assimilation numbers for lake ice communities in the Dry Valleys lakes. Numbers in brackets correspond with lake numbers in Table 3.1. Data taken from Fritsen and Priscu (1998).

Lake	Assimilation number (μg (C (μg Chl *a*)$^{-1}$ h^{-1})	Photosynthetic efficiency (μg C (μg Chl a)$^{-1}$ h^{-1} μmol photon m^{-2} s^{-1})
Lake Fryxell (1)	14.2 ± 0.97	17.0 ± 3.31
Lake Bonney East (6)	4.3 ± 0.23	4.9 ± 1.55
Lake Bonney West (4)	5.9 ± 0.46	4.5 ± 1.60

is effectively an extreme environment, comparable with glacial environments where microbial processes are now recognized as extremely important.

Freshwater Dry Valleys lakes such as Lake Miers and Lake Hoare also contain these types of microbial communities. Coastal oasis saline lakes in Antarctica have clear annual ice covers. The only organic matter are portions of benthic algal mats that have detached and become embedded in the ice. These are likely to provide microzones of activity, but at the time of writing they have not been investigated.

3.9 The benthic community

As outlined in Section 2.7.1, well-developed algal mats cover the benthic regions of the Dry Valleys lakes. Details of UV protective mechanisms and photosynthetic pigments are given in Section 2.7.1. Filamentous cyanobacteria dominate the biomass, particularly species of *Leptolyngbya* and *Oscillatoria*. A variety of diatoms live within this cyanobacterial matrix. The mats persist from year to year and possess areal pigment concentrations that exceed those of the phytoplankton. Mats are photosynthetically active at considerable depths where less than 1% of surface incident radiation penetrates. The depth to which light penetrates depends on the transparency of the ice cover and the water column. Lake Vanda in the Wright Valley has a relatively transparent ice cover compared with the lakes of the Taylor Valley allowing 18%–20% transmission of surface irradiation, compared with around 1%–4% in Lake Hoare (Hawes and Schwarz 2000; Hawes et al. 2001). Benthic cyanobacterial mats in the lakes of the Vestfold Hills, Rauer Islands, and Larsemann Hills contain *Leptolyngbya* and *Oscillatoria* and other genera, including *Nostoc, Phormidium*, and *Aphanocapsa* (Taton et al. 2006a; Hodgson et al. 2001b).

In Lake Joyce the cyanobacterial mats at 10 m and above are dominated by *Leptolyngbya* c.f. *antarctica*, *Oscillatoria* c.f. *sancta*, and *Phormidium* c.f. *autumnale*, in that order of abundance. A small number of pennate diatoms were associated with the mats. Below 10 m the cyanobacterial sheaths were largely empty. One of the interesting features of the mats in Lake Joyce is that they provide a picture of lake level change. Calcite microbialites are formed when environmental conditions are no longer suitable for sustaining photosynthesis and growth. These are effectively fossil benthic cyanobacterial and algal communities. The lake underwent a 7 m rise between 1973 and 2009. These mat structures were photosynthetically active between 1986 and 1997, but were barely photosynthetically competent in 2009 as a result of changes in their depth, and were transforming into microbialites (Hawes et al. 2011a).

Diatoms are common in the benthic communities of saline lakes and ponds of the Vestfold and Larsemann Hills, the Windmill Islands, and the Rauer Islands (Roberts and McMinn 1996; Hodgson et al. 2001b; Sabbe et al. 2004). They represent the second most abundant group after the cyanobacteria. There appear to be differences in the diatom assemblages in the various locations. For example, the hypersaline lakes of the Rauer Islands have a significantly different diatom flora to lakes of comparable salinity in the Vestfold Hills, despite their close proximity. The hyposaline lakes have diatom assemblages dominated by *Pinnularia microstauron* and *Luticola* cf. *muticopsis*. Many of the diatom assemblages found in the hyposaline lakes are similar to those found in the Maritime Antarctic (see Section 2.7.1). Hypersaline Vestfold Hills lakes are dominated by *Navicula directa*, and *Navicula* spp.

Shallow saline ponds are found inland of ice shelves, for example in the Pensacola Mountains. Forlidas Pond (82°27′S, 51°16′W) is a shallow 1.83 m deep water body that is now around 90 m in diameter. It is the remains of an extensive proglacial lake that had mid-Holocene water levels up to 17.7 m, as indicated by a series of lake terraces above the present water line. The pond freezes down to its bottom with a highly saline layer of slush at its base. The bottom and littoral zone were covered by red/brown cyanobacterial mats (*Leptolyngbya*) that were actively photosynthesizing, as indicated gas bubbles trapped against the lower ice surface. Diatoms were absent (Hodgson et al. 2010).

Heterotrophic bacterial diversity in mats has not been as well researched as in the plankton. Littoral algal mats collected from the moat of Lake Fryxell contained the anaerobic bacterium *Clostridium estertheticum* and some other psychroloterant strains, and among the aerobic bacteria *Flavobacterium hibernum*, *Janthinobacterium lividum*, and *Arthrobacter flavus* were the most common among the many strains isolated. Culture-independent 16S rDNA analysis of the prokaryote community produced sequences belonging to the class Proteobacteria, order Verrucomicrobiales, the class Actinobacteria, the Clostridium/Bacullus subphylum of Gram-positives and the Cytophaga-Flavobacterium-Bacteroidetes phylum. Archaea were represented by two phylotaxa, the majority being distantly related to *Methanoculleus palmolei* (89%–92% similarity) (Brambilla et al. 2001). The littoral mats of saline lakes in the Vestfold Hills and Lake Reid in the Larsemann Hills support a high diversity of bacteria based on isolated cultured strains. Strains belonged to the α-, β-, and λ-Proteobacteria, the high and low percent G + C Gram-positives, and the Cytophaga-Flavobacterium-Bacteriodetes groups (Van Trappen et al. 2002). Eighty-two isolates from Forlidas Pond (Pennsicola Mountains) belonged to the α-Proteobacteria, the β-Proteobacteria, Actinobacteria, Bacteriodetes, and Firmicutes. In some cases the phylotypes had no significant similarity to published sequences. However, the majority were found to have a cosmopolitan distribution (Peeters et al. 2011).

A comparison of the uncultured diversity from Lake Fryxell and Vestfold Hills mats with samples collected from sediments in the anoxic basins of meromictic Vestfold Hills lakes shows very little similarity, as might be expected given that the littoral samples were derived from an aerobic environment. In a range of Vestfold Hills lakes anoxic sediments, 16S rDNA clone library analysis revealed that 31% of the clones formed a novel deep branch within the G + C Gram-positive division. Others belonged to the δ-Proteobacteria (*Desulfosarcina* group, *Syntrophus* and *Geobacter/Pelobacter/Desulphuromonas* group). *Prochlorococcus* cyanobacteria constituted another abundant phylotype (Bowman et al. 2000).

Microbial mats provide a rich environment for Protozoa, as demonstrated in a study of Lake Fryxell (Cathey et al. 1981). Phototrophic flagellates included *Gymnodinium* spp., cryptophytes, *Ochromonas*, and *Chlamydomonas* spp. Heterotrophic flagellates included *Bodo* and *Desmarella*. Ciliates showed considerable diversity with 29 species recorded. Sarcodines which are common in environments providing surfaces on which to feed, were represented by several species of *Amoeba*. Rotifers belonging to the genus *Philodina* were found in Lakes Bonney, Fryxell, and Vanda, and tardigrades (*Hypsibius antarcticus*) occurred in Lakes Fryxell and Bonney. Nematodes were noted in Lake Bonney. They are a common component of benthic communities so they are likely to be present in other saline Dry Valleys lakes. The benthic fauna of the Vestfold Hills appears more diverse. Hyposaline lakes support playhelminthes, nematodes, tardigrades, and a range of rotifers species including *Encentrum spatitium, E. brevifulcrum, E. salinum, Notholca, Epiphanes senta, Lepadella patella, Adineta grandis*, and an unidentified bdelloid. Some of these species are also found in the freshwater lakes (Dartnall 2000). Crustacea are only reported from Ace Lake, where a small benthic harpacticoid (*Idomene scotti*) occurs in the mats above the anoxic monomolimnion (Rankin et al. 1999). Ace Lake is one of the most studied saline lakes. It is highly likely that benthic crustaceans occur in other hyposaline lakes; to date no one has looked for them.

3.10 Carbon cycling in the benthos

Cyanobacterial mats possess a range of photosynthetic pigments adapted to harvesting low levels of incident PAR. In Lake Vanda up to 6% of incident radiation reaches 30 m. This compares with only 1% at 10 m in freshwater Lake Hoare. As previously indicated, Lake Vanda has a relatively transparent ice cover and transparent waters. Chlorophyll *a* increases with depth in the mats from 7.4 ± 0.4 µg cm^{-2} at 8 m to 15.1 ± 0.5 µg cm^{-2} at 30 m. Phycoerthrin, which absorbs PAR between 500–600 nm, ranged from 2.4 ± 0.2 µg cm^{-2} at 10 m to 46.6 ± 3.1 µg cm^{-2} at 30 m. This pigment is probably very important at achieving efficient light utilization under conditions of dim blue-green light. Its high concentration at depth supports this view. Phycocyanin, which like phycoerythrin is another member of the light-harvesting phycobiliproteins, ranged from 5.6 ± 1.0 µg cm^{-2} at 10 m to 15.1 ± 1.5 µg cm^{-2} at 30 m. The high concentrations of phycobilins produced in response to low PAR may provide a competitive advantage in these types of lake. Rates of photosynthesis in the Lake Vanda mats ranged from 88.8 ± 4.56 µg C cm^{-2} h^{-1} to 84.72 ± 6.24 µg C cm^{-2} h^{-1}, with no obvious relationship to depth (Hawes and Schwarz 2000). These rates of photosynthesis are significantly higher than those reported for Lake Hoare (see Table 2.10). Cyanobacterial mats in Watts Lake in the Vestfold Hills had a mean rate of 74 µg C cm^{-2} day^{-1}. Like Lake Vanda, Watts Lake is covered by transparent ice, in this case annual ice, allowing good PAR transmission to the benthic zone. However, in Watts Lake phytoplankton production exceeded benthic production, whereas in Lake Hoare the reverse was the case. It is likely that in Lake Vanda benthic production exceeds plankton primary production as it does in Lake Hoare (see Section 2.7.2).

Unlike the plankton, mats accumulate biomass year on year. Maximum daily rates of carbon fixation of between 15–20 µg C cm^{-2} at 10 m are calculated, which equates to a vertical accumulation of material of approximately 0.9 mm at 4 m, 1.6 mm at 20 m, and 0.1 mm at 30 m each year. The carbon content of mats ranges from 5 to 8 mg C cm^{-2}, with highest levels between 8 and 20 m decreasing above and below this zone (Hawes et al. 2001).

Information on the individual productivity and biomass of the diatom communities of mats and the heterotrophic communities that exploit what must be a relatively resource rich environment are lacking. This is an area that begs investigation. Benthic mats are clearly important in contributing to the DOC pool in the water column and benthos, although the balance of production between the benthos and plankton can vary (see this section and Section 2.7.2). The DOC pool drives heterotrophic microbial production. In order to fully understand carbon dynamics and develop accurate models, we need data on the degree of photosynthate exudation from mats and from the plankton. In the majority of Antarctic lakes the DOC pool is largely derived from these autochthonous sources. At the moment we are largely extrapolating from lower latitude systems.

3.11 A unique Antarctic lake—Lake Vida

No book on Antarctic lakes would be complete without a mention of Lake Vida. In effect it is a small-scale subglacial system and could equally well be included in Chapter 6 on subglacial lakes. Lake Vida is a large lake situated in the Victoria Valley (see Figure 2.7). It is 3.5 km long and 1 km wide and has a very thick ice cover around 19 m in depth. It was originally thought that the lake was frozen to its base but ground-penetrating radar (GPR) revealed an underlying brine layer about 5 m deep with a calculated temperature of –10 °C to –13 °C. Two cores, one 14 m long and the other 15.8 m long showed that the ice contained sediments and microbial mat layers. The sediment in the upper 7 m of the ice indicates significant summer flooding, which has been more frequent in the recent history of the lake. Summer flooding from glacial melt cannot flow beneath the ice, as it does in many other Dry Valleys lakes, where moats develop around the edges in summer. In Lake Vida the meltwater contributes to the surface ice (Doran et al. 2003; Murray et al. 2012).

The base of the 15.8 m core was composed of wet saline ice with a temperature of –11.5 °C. The brine trapped in the ice was predominately NaCl with a salinity of 245‰, the equivalent of ×7 seawater. The ratio of salts was similar to ratios seen in other Dry Valleys lakes, suggesting that Lake Vida has a similar history. ^{14}C analysis of microbial mats layers frozen into the ice indicate that the lakes history extends back at least 2800 ^{14}C years. Measurable photosynthesis and bacterial production occurred within water melted from the ice core where organic layers were present. The rates were very low but nonetheless suggest that the microbial communities trapped in the ice for hundreds of years are able to become metabolically active when liquid water is available (Doran et al. 2003).

Detailed analysis of the underlying anoxic brine, using small subunit rRNA analysis, revealed a diverse bacterial community including representatives from eight bacterial phyla, including some unique sequences. The assemblage was most similar to that living in the subglacial brine of Blood Falls and is distinct from the communities seen in the monmolimnia of other Dry Valleys saline lakes (Murray et al. 2012). The communities within the overlying ice associated with organic matter also show a high degree of diversity and include Bacteria, Cyanobacteria, and Algae (Mosier et al. 2007). The bacteria in the brine layer live under very high levels of reduced metals, ammonia, hydrogen, and DOC. There are also high concentrations of oxidized species of nitrogen, super saturated nitrous oxide and nitrate, and sulphate. Bacterial numbers ranged between 0.1 and 0.6×10^9 l^{-1}, which falls within the range seen in other Antarctic saline and freshwater lakes (see Chapter 2 and this chapter). Interestingly, very small bacteria or ultramicrobacteria of <2.0 µm diameter were more abundant, having concentrations between 49 and 60×10^9 l^{-1}. Ultramicrobacteria also predominate in glacial ice (Miteva et al. 2004; Miteva and Benchley 2005) and in the Lake Vida Ice cover (Mosier et al. 2007). Very low rates of bacterial growth were sustained in the brine at –12 °C to –13 °C under anoxic conditions. The lower brine layer of Lake Vida is therefore a sustainable, encapsulated system with high microbial diversity but very low rates of metabolism (Murray et al. 2012).

CHAPTER 4

Epishelf lakes

4.1 Introduction

As indicated in Section 1.5, epishelf lakes are unusual lacustrine systems that are either freshwater directly overlying seawater, or freshwater systems with a direct connection to the sea through a conduit underneath an ice shelf (Figure 4.1). They are effectively tidal freshwater systems, and are often fringed by rafted ice (see Figure 1.10a). These types of lake are almost unique to Antarctica. They occur in the Arctic, but are rapidly disappearing due to the collapse of ice shelves, resulting from climate warming. Climate warming is three times greater in the Arctic than the global average (Vincent et al. 2001; Veillette et al. 2008). A continuous ice shelf fringed Ellesmere Island northern coast (Canadian High Arctic) in 1906, as documented by the Peary expedition. Detailed analysis of maps from this time suggests that there may have been 17 epishelf lakes retained by these ice shelves. By 2008 the Milne Fjord was the sole remaining deep epishelf lake in the Arctic (Veillette et al. 2008).

A well-documented example of the demise of an Arctic epishelf lake is that of the Disraeli Fjord on Ellesmere Island, Canada (82°N–83°N) (Mueller et al. 2003). RADARSAT imagery of the Ward Hunt Ice Shelf taken in 1998 and 1999 revealed that it resembled occasional images taken since 1954. There was no obvious fracturing other than peripheral tide cracks. In 2001, a helicopter survey showed a clear north–south fracture extending from the southern margin of the Ward Hunt Ice Shelf to Disraeli Fjord, effectively cleaving the ice shelf into two. This destroyed the integrity of the ice shelf and resulted in the draining of Disraeli Fjord. Historical data showed changes in the stratification and temperature profiles of the epishelf lake as its freshwater progressively drained (Figure 4.2). The decline in the freshwater layer was probably due to catastrophic drainage through fractures in the Ward Hunt Ice Shelf.

Some of the larger epishelf lakes are listed in Table 2.1. They are among some of the largest and deepest lakes in Antarctica, among which Beaver Lake in McRobertson Land, adjacent to the Amery Ice Shelf (lake 1, Table 2.1, see Figure 2.5), ranks as the most extensive and deepest. The freshwater of Beaver Lake sits on saline water, as shown in Figure 4.1a. The Bunger Hills is an ice-free area inland of the Shackleton Ice Shelf and contains five epishelf lakes: White Smoke Lake, Lake Pol'anskogo, Southern Lake, Transkriptsii Gulf, and Northern Lake (Gibson and Anderson 2002) (Figure 2.3). Most of these lakes fit the model in Figure 4.1b. The Schirmacher Oasis lies inland of the Novolazarevskaya Ice Shelf. It has a number of epishelf lakes including Lakes Zigzag, Predgornoye, Ozhidaniya, Karovoye, Prival'noye, and Kholodnoye (Richter and Bormann 1995). Like the Bunger Hills these epishelf lakes appear to be connected to the sea by a conduit under the ice shelf and do not appear to have haloclines, that is, they are freshwater to their bases. Lake Moutonée and Ablation Lake are impounded at Ablation Point, behind the George VI Ice Shelf on Alexander Island on the Antarctic Peninsula (Heywood 1977; Smith et al. 2006) (Figure 2.9). These are the documented epishelf lakes of Antarctica. There maybe other epishelf lakes in Antarctica yet to be described.

In spite of their relative abundance in Antarctica, epishelf lakes are poorly researched compared with other freshwater and saline lakes. The paucity of data relates mainly to the fact that they are not located close to permanent research stations, making

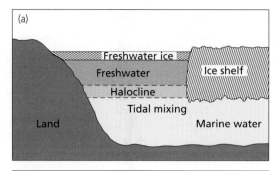

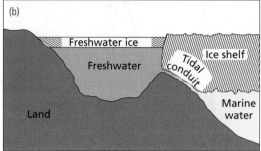

Figure 4.1 Structure of epishelf lakes: A type A, where freshwater sits directly on seawater, e.g. Beaver Lake in the Amery Oasis. B type B, where the freshwater body is connected by a conduit beneath a glacier or ice shelf to the sea, e.g. Schirmacher Oasis epishelf lakes.

the logistics for researching them challenging. Special expeditions have to be mounted to study them, or relatively long distance daily expeditions from permanent stations by fixed-wing aircraft or large helicopters are required. These types of research project are very much at the mercy of weather conditions and usually generate short-term data sets.

4.2 Formation and physico/chemical characteristics of epishelf lakes

4.2.1 Geomorphology

Epishelf lakes vary in age from around 3000 yrs B.P. (White Smoke Lake, Bunger Hills) to basins that were formed during the Pliocene and probably filled after the last glacial maximum (LGM) (e.g. Beaver Lake, Amery Oasis, MacRoberston Land). White Smoke Lake lies on the southern edge of the Bunger Hills and is bounded by the Apfel Glacier, under which there is a connection to the sea. The lake is relatively small with an area of 0.83 km^2 and a depth of 90 m. Its ice cover ranges between 2.3 and 3.7 m and develops a moat in summer where it connects with the land. The moat is poorly developed where it abuts the glacier (Doran et al. 2000; Gibson and Andersen 2002). It has two basins, of which the eastern basin contains a well-developed sediment that is carbon dated to >3000 yrs B.P. In contrast, the western basin has little sediment and is probably only around 100 years old, and was probably formed by the retreat of snowfields and to a lesser extent the retreat of the Apfel Glacier (Doran et al. 2000).

Beaver Lake has an area of around 800 km^2 and a depth of 435 m, with the freshwater/marine interface between 220 and 260 m (Bardin et al. 1990). The floating tongue of the Charybdis Glacier covers the outer reaches of the lake and is the major source of freshwater input (see Figure 2.5). At its southwestern edge the Pagadroma Gorge connects Beaver Lake with the Radok Lake basin. Periodically, Lake Radok drains into Beaver Lake (Wand et al. 2011). The lake has a clear permanent ice cover that ranges in thickness from around 3.5 m 1 km offshore to 4 m 7 km offshore (Laybourn-Parry et al. 2002, 2006).

The Beaver Lake basin was excavated by the Nemesis Glacier and its ancestral equivalents flowing from the north, probably well before the Pliocene (5300–2588 million yrs B.P.). Then, the Amery Oasis was less well uplifted than it is now, when sea level relative to the land was higher and when glaciers were less incised. Based on a range of geomorphological evidence, Adamson et al. (1997) suggest that the Beaver Lake basin was partially emptied in the last Pleistocene glacial period and refilled in the Holocene. More recently, 100 m sediment cores taken from a depth of 54 m in the lake have provided more detail (Wagner et al. 2007). Carbon dating of the sediment was not possible because of particles of coal that would have produced erroneous bulk carbon ages. The mineralogy of the lowest section of the core (100–91 cm) suggests sedimentation in a glacial or subaerial environment. The next section, between 91 and 81 cm, was laid down under glacial conditions, probably during the advance of the glacier into the Beaver basin. Overlying this, the sediments between 81 and 31 cm were either emplaced by mass movements from the slopes above the basin, or created by in situ subaerial weathering

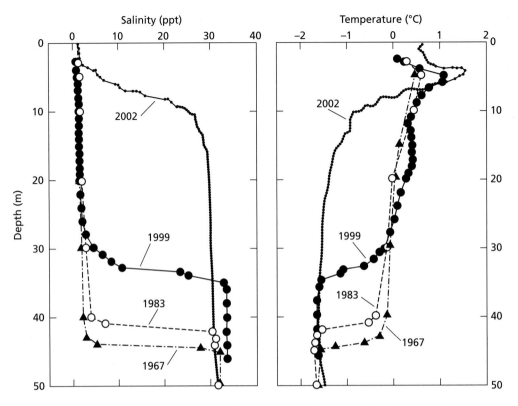

Figure 4.2 Profiles of salinity and temperature between 1967 and 2002 in the Disraeli Fjord, Canadian High Arctic, showing changes as the lake was lost due to the disintegration of the impounding ice shelf. From Mueller et al. (2003) with the permission of the American Geophysical Union (Wiley and Sons).

of older glacial and colluvial sediments. The latter is the most likely scenario. The upper 31 cm of the core is glacio-marine in origin and was laid down under the conditions that prevail today.

The historical interpretation of this sediment profile is that, following the initial glacial or subaerial setting, the south-western part of the basin was overrun by continental ice (paleo Nemesis Glacier). During the last glacial cycle, the Nemesis Glacier was at least 150 m higher than at present, intruding the Lake Terrasovoje Basin in the northern part of the Amery Oasis (see Figure 2.5), and the south-western part of the Beaver basin at the core site. Based on the timing of the deglaciation of Lake Terrasovoje and Lake Radok, it is likely that glacial retreat in the Beaver basin occurred at around 12,500 yrs B.P. This was followed by deposition in subaerial conditions (core section 81–31 cm). Exactly when the basin underwent a marine transgression is extremely difficult to pinpoint without carbon dating, but it likely to have occurred at any time in the Holocene, which confirms the interpretation of Adamson et al. (1997). Following marine transgression, freshwater inflow from Lake Radok and glaciers would have come to overly marine waters.

The present day George VI Ice Shelf sits between the Antarctic Peninsula and Alexander Island (Figure 2.9). On the east coast of the island the ice shelf impounds two epishelf lakes, Lake Mountonée and Ablation Lake. While the present day George VI Ice Shelf has survived the impact of warming it may be close to the limit of stability. If it were to collapse, the two lakes would cease to be lacustrine environments and would transform into marine embayments. There is evidence that this has happened in the past. Analysis of sediment cores from Lake Mountonée show five units. The lowest (unit 1: 537–522 cm) and the top unit (unit 5: 236–0 cm) are typical of sedimentary conditions in modern day epishelf lakes. Units 2 (522–490 cm) and 4 (302–326 cm) are separated by unit 3, a layer of clast-rich unsorted ice-rafted debris that was probably laid down rapidly from the melting of icebergs. Units 2 and 4 contain marine diatoms and foraminiferans with a clear marine signature in carbon isotopes and were laid down under marine conditions. The Lake Mountonée core shows that the George VI Ice Shelf disappeared and reformed during the Holocene. It collapsed at around 9595 yrs B.P. and had reformed by 7945 yrs B.P. When it was absent, marine sediments were laid down (units 2 and 4). The breakup of the ice shelf would have been accompanied by widespread iceberg formation. Their melting caused the rapid deposition of unit 3 (Bentley et al. 2005).

4.2.2 Physico/chemical characteristics

There are considerable variations in the physical characteristics of epishelf lakes. While most of those that have been studied have consistently cold-water columns, temperature profiles vary (Figure 4.3). Where temperatures are taken over a summer, as was the case in Beaver Lake (Figure 4.3), the upper waters retained a consistent pattern that was the same in 2000, while the lower waters below 60 m showed a 1 °C variation. In the other example shown in Figure 4.3 (Transkriptsii Gulf), there was quite a lot of variation in the upper waters between years. In this lake and other epishelf lakes in the Bunger Hills, there was a significant thickening of ice covers between 1992 and 2000. For example in Transkriptsii Gulf from 2.6 to 3.9 m and in White

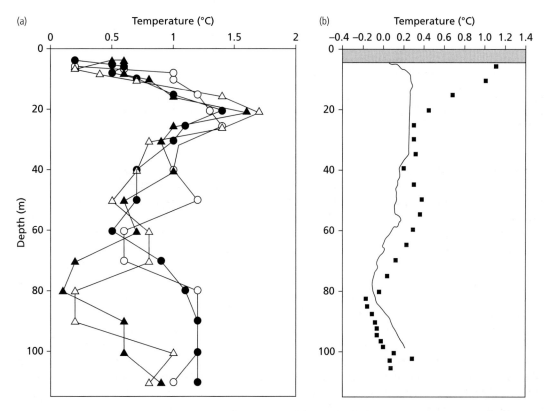

Figure 4.3 Temperature profiles for **A** Beaver Lake, Amery Oasis over the summer of December 2002 to January 2003: ○ 3rd Dec, ● 14th Dec, △ 23rd Dec, and ▲ 4th Jan. From Laybourn-Parry et al. (2006) with the permission of Blackwell Publishing Ltd. **B** Two temperature profiles for Transkriptsii Gulf, Bunger Hills; solid line a single profile in summer 2000 and squares a profile in summer 1992. Redrawn from Gibson and Andersen (2002).

Smoke Lake from 2.3 to 3.7 m (Gibson and Andersen 2002).

Epishelf lakes are hydrologically complex. They are fed by direct glacial melt, or by glacial and snow melt via another lake in summer; for example, Lake Radok into Beaver Lake and Lake Glubokoye into epishelf Lake Prival'noye in the Schirmacher Oasis (Wand et al. 2011; Loopmann and Klokov 1988). They are tidal because they are either freshwater directly overlying marine water, or they have a connecting conduit to the sea under an ice shelf or glacier. In the latter case there may or may not be a halocline where marine water has become trapped in basins. Most appear to have perennial ice covers from which water is lost by ablation. The major source of salt in the lakes is the underlying marine waters. Salt water entering the basins occurs as a result of maintaining hydrostatic equilibrium. During periods of negative water balance at the lake surface, when inputs of water from terrestrial sources (glaciers, other lakes) is less than that lost by ablation from the ice cover, the boundary between the freshwater and the saline marine waters moves closer to the surface of the lake. The morphometry of the lake basins varies and in some cases there may be a number of basins separated by sills. Saline water may flow over sills into basins that are normally isolated from the marine connection. When the surface water balance returns to positive, the boundary between freshwater and marine water will lower. Where isolated basins have been filled during negative water balance, the inflowed marine water will be trapped and remain saline (Gibson and Andersen 2002). Because of the tidal nature of these lakes there will be some degree of turbulence at the halocline, particularly in lakes with a constricted conduit. This will result in marine waters being diluted by freshwaters, representing another source of freshwater loss from these systems.

The tidal range in epishelf lakes can be quite large, for example White Smoke Lake has a range of around 1 m (Doran et al. 2000), Transkriptsii Gulf 1.3 m and Pol'anskogo 1.3 m (Gibson and Andersen 2002). The tidal characteristics of Beaver Lake have been particularly well studied (Galton-Fenzi et al. 2012). There is a regular tidal signal that is lagged and attenuated from tides beneath the adjacent Amery Ice Shelf. The rocky shoreline of the lake provides a stable environment for vertically referencing sea level. The tides in the lake lag behind the tides underneath the Amery Ice Shelf by around an hour, and there is also an amplitude attenuation. Tidal amplitude ranged from 0.308 m in 1990/1991 to 0.292 m in 2002. Over an eleven-year period both the lag phase and the amplitude attenuation have increased. The changes are probably due to a thickening of the Amery Ice Shelf. Synthetic Aperture Radar (SAR) interferograms reveal that the inlet to the lake has narrowed due to a decrease in the net basal melt of the ice shelf.

Levels of Photosynthetically Active Radiation (PAR) in the water column of lakes is a function of the transparency of the ice and the clarity of the water column. The ice cover of Beaver Lake is clear, containing no obvious sediment load, allowing good transmission of surface irradiation (Figure 4.4). Between 8.25% and 14.8% of PAR received at the ice surface penetrates to the water immediately under the ice. The data shown in Figure 4.4 relate to the top 5–30 m of the water column (the length

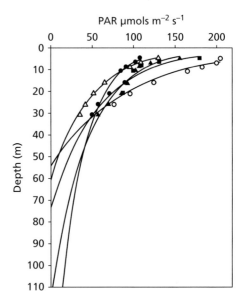

Figure 4.4 Photosynthetically active radiation (PAR) profiles in Beaver Lake December 2002 and January 2003. Sensor cable length was 30 m; below this depth the lines are extrapolated. ○ 3rd Dec, ● 14th Dec, △ 23rd Dec, ▲ 4th Jan, and ■ 28th Jan. From Laybourn-Parry et al. (2006) with the permission of Blackwell Publications Ltd.

of the light sensor), but extrapolation of these data suggests that light penetrates to the lake bottom on occasion (Laybourn-Parry et al. 2006). In Ablation Lake, between 15% and 20% of surface PAR penetrated the ice in January and February (Heywood 1977). The perennially ice-covered lakes of the Bunger Hills, including epishelf lakes, contain little sediment (Doran et al. 2000), and like Beaver Lake and Ablation Lake probably allow good PAR penetration.

There are few data on the nutrient chemistry of epishelf lake water columns (Table 4.1). The limited information indicates that these systems are particularly deficient in orthophosphate, but there are times and strata in the water column where species of nitrogen may also limit bacterial and primary production. Concentrations of chlorophyll a are correspondingly low. A single sampling of Ablation Lake and Lake Moutonnée in February and December respectively revealed concentrations of around 0.5 µg l^{-1} (Table 4.1) and in the Bunger Hills, Transkriptsii Gulf had very low levels in the water column in February (Klokov et al. 1990). The data for Beaver Lake cover five samplings over December and January and reveal seasonal changes and an increase with depth in the water column (Figure 4.5). The sampled water column had a depth of 105 m to the sediment and was thus well above the halocline at 220–260 m. Surprisingly, highest concentrations of chlorophyll a were recorded in the deeper waters.

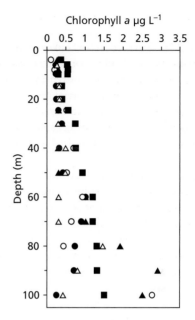

Figure 4.5 Chlorophyll a profiles in the water column of Beaver Lake, Amery Oasis during December 2002 and January 2003. ○ 3rd Dec, ● 14th Dec, Δ 23rd Dec, ▲ 4th Jan, and ■ 28th Jan. From Laybourn-Parry et al. (2006) with the permission of Blackwell Publications Ltd.

This may be related to nutrient release from the sediments. As indicated in Figure 4.4, PAR penetrated the entire water column.

Dissolved organic carbon data are only available for the water column of Beaver Lake (Table 4.1).

Table 4.1 Inorganic nutrients, dissolved organic carbon (DOC), and chlorophyll a in the upper freshwater layer of epishelf lakes.

Lake	NO$_3$-N µM l^{-1}	NH$_4$-N µM l^{-1}	PO$_4$-P µM l^{-1}	DOC µg l^{-1}	Chl a µg l^{-1}
Lake Moutonnée[1,2] 70°51'S, 68°20'W 5–35 m in water column	0–0.78	0–1.54	0–0.013	–	< 0.5*
Ablation Lake[1] 70°49'S, 68°27'W 5–55 m in water column	0–0.46	0–36.78	0–0.021	–	0.5–0.65*
Beaver Lake[3] 70°48'S, 68°15'E 4–100 m in water column	0.05–3.52	0.90–10.3	0–0.20	20–380	Mean < 1.0
Transkriptsii Gulf[4] 66°15'S, 100°35'E 3–93 m in water column	0.04–0.21*	–	0.06–0.08*	–	0.12–0.27*

1: Smith et al. 2006; 2: Heywood 1977; 3: Laybourn-Parry et al. 2006; 4: Klokov et al. 1990.
*based on a single sampling.

The concentrations are low, the lower values being close to the limits of detection. Large freshwater lakes such as Crooked Lake, Lake Druzhby, and Lake Hoare possess higher concentrations of DOC (Table 2.3). At times the low concentrations of DOC in Beaver Lake may limit bacterial growth (see Section 4.4). Particulate organic carbon concentrations are also low, so that the total organic carbon pool was usually below 600 µg C l^{-1} (Laybourn-Parry et al. 2006).

Concentrations of dissolved inorganic carbon are relatively high in White Smoke Lake, 520–770 µg C l^{-1}, so that water at all depths is saturated with respect to CO_2. Saturation with CO_2 is atypical for perennially ice-covered lakes, as outlined in Section 3.5, so that primary production may be limited by a lack of inorganic carbon (Doran et al. 2000).

4.3 The planktonic biota of epishelf lakes

Data on the biota of epishelf lakes are sparse. Bacteria concentrations have only been enumerated in Beaver Lake (Laybourn-Parry et al. 2001b, 2006). Two samplings in February 2000 revealed a mean maximum concentration of $7.60 \times 10^7\ l^{-1}$ at two sites, one inshore at 50 m and the other 7 km offshore with a depth of 110 m. More detailed sampling between December 2002 and February 2003 at a site 7 km offshore showed a range from 9.3 to $14.0 \times 10^7\ l^{-1}$. Concentrations did not show any great variation within the water column, but there were some temporal variations (Figure 4.6) with lowest numbers in early December and highest in late December. These concentrations of bacteria are similar to the minimum values recorded in large ultra-oligotrophic freshwater lakes such as Crooked Lake and Lake Druzhby (see Table 2.4).

Viruses were present in the water column of Beaver Lake, ranging from 0.014 to 3.02×1–$2\ 10^9\ l^{-1}$, with highest concentrations below 30 m. Their numbers increased progressively over the summer, differing from the pattern of bacterial abundance, so that virus to bacterial ratio (VBR) achieved highest values in late January (3.32–7.33) (Figure 4.7). The lysogenic cycle was present, on occasion 87% of the lysogenic bacteria were present in the top 30 m of the water column (Laybourn-Parry et al. 2013). Virus concentrations and VBR in Beaver Lake are within the same order of magnitude as those recorded in the waters of Crooked Lake, Lake Druzhby, and other continental freshwater lakes (see Table 2.5). In Beaver Lake there was a significant correlation between viral abundance and chlorophyll a, DOC, ammonium, and nitrate concentrations, but not between bacterial abundance, temperature, or orthophosphate. Nutrients indirectly affect viruses through direct effects on their hosts. In polar lakes, bacterial growth may be limited by the availability of phosphorus (Dore and Priscu 2001; Säwström et al. 2007c). In Antarctic lakes, significant correlations between orthophosphate and virus abundance have been noted (Lisle and Priscu 2004; Madan et al. 2005). The lack of any impact from orthophosphate limitation in Beaver Lake is difficult to explain. Highest viral concentrations occurred at a time when orthophosphate was not detectable in the water column.

The major predators of bacteria, the heterotrophic nanoflagellates (HNAN), showed considerable variation in abundance over December and January in Beaver Lake (Figure 4.6). Highest abundance was recorded in early December. The same was true of phototrophic nanoflagellates (PNAN) (Figure 4.6). Those found in the lower waters had intense autofluorescence, indicating active chlorophyll. Euglenids were common in the upper water being replaced by small colonial chlorophytes in the lower waters. Cryptophytes occurred throughout the water column. Prasinophytes also occurred and, based on electron microscopic analysis, the dominant species was *Mantoniella* sp. HPLC analysis of photosynthetic pigments confirmed the microscopic analysis. Alloxanthin indicated that cryptophytes dominated throughout the water column, as did chlorophytes (chlorophyll b, lutein, neoxanthin, and vioaxanthin) (Laybourn-Parry et al. 2006). Maximum concentrations of HNAN and PNAN in Beaver Lake during the summers of 2000 and 2002/2003 exceed those reported for other large freshwater continental systems (see Table 2.4). This is somewhat surprising given that bacterial concentrations are similar. This begs the question of what sustains these larger HNAN populations.

Diatoms are apparently absent from the phytoplankton of Lake Moutonée and Ablation Lake in the Antarctic Peninsula (Smith et al. 2006), and

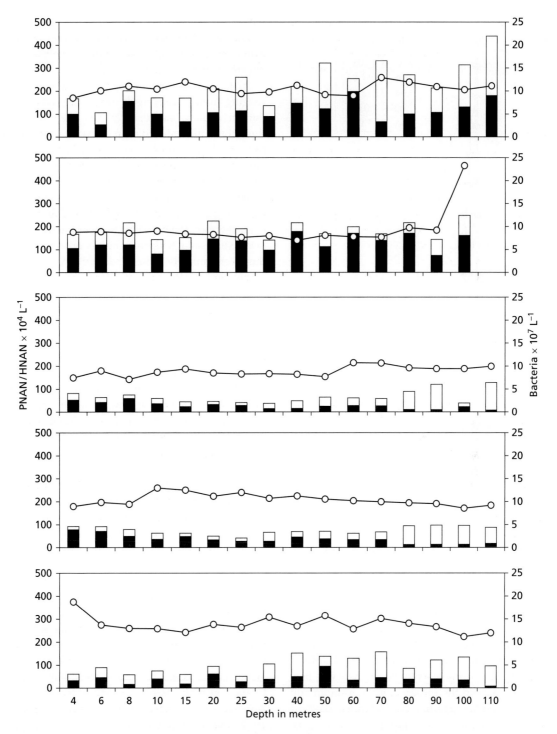

Figure 4.6 Concentrations of heterotrophic nanoflagellates (solid sections of columns) and phototrophic nanoflagellates (open sections of columns) and bacteria (open circles plot) in relation to depth in Beaver Lake, Amery Oasis during December 2002 and January 2003. In sequence from top panel down: 3rd Dec, 14th Dec, 23rd Dec, 4th Jan, and 28th Jan. From Laybourn-Parry et al. (2006) with the permission of Blackwell Publications Ltd.

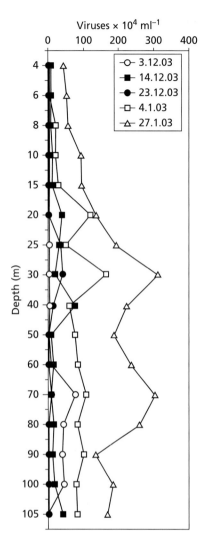

Figure 4.7 Concentrations of viruses in Beaver Lake, Amery Oasis during the summer of December 2002 and January 2003. Key shown on graph. From Laybourn-Parry et al. (2013), with the permission of Blackwell Scientific Publications.

from Beaver Lake in the Amery Oasis (Laybourn-Parry et al. 2001, 2006; Cremer et al. 2004). HPLC derived data on photosynthetic pigments in Beaver Lake showed the presence of fucoxanthin and diadinoxanthin, suggesting the presence of diatoms and/or chrysophytes or synurophytes. However, since microscopic analysis did not reveal diatoms it is likely that the pigments were indicators of the latter two groups. Diatoms are not common in the plankton of freshwater continental Antarctic lakes (see Chapter 2, Section 2.5.4), so their absence in the extremely oligotrophic waters of epishelf lakes is not surprising.

Ciliated protozoans are sparse in Beaver Lake, ranging from a few cells l^{-1} up to 120 cells l^{-1}, although usually less than 50 l^{-1}. They occurred throughout the water column and were represented by a small number of species. Scuticociliates, *Askenasia* sp., and *Monodinium* sp. predominated with the occasional appearance of a plastidic *Strombidium* sp. and *Mesodinium* sp. These ciliates were probably exploiting bacteria and HNAN and PNAN as food. On a few occasions heliozoans were noted. However, ciliate and heliozoan numbers were so low they were unlikely to have exerted any significant grazing impact. Rotifers were very rare, on only one occasion was a single specimen of *Notholca* seen, from seven sampling profiles of the water column (Laybourn-Parry et al. 2001b, 2006).

Beaver Lake supports a population of the calanoid copepod *Boeckella poppei* (Figure 1.31a). The individuals are dwarf forms. They were most common lower in the water column and on any one sampling occasion the maximum number collected from 80 litres of water was 10 individuals (adults, copepodites, and nauplii). One of the females was gravid with only four eggs in her egg sac, thus fecundity appears low (Laybourn-Parry et al. 2001, 2006). A net haul through the top 3 m of the water column retained 10 adult females, one adult male and 23 copepodite stages (Bayly and Burton 1993). The average length of the females was 1.05 mm, whereas the average length of females from other locations in Antarctica is 2.2–3.3 mm. *Boeckella poppei* also occurs in Lake Radok and Lake Terrasovoje in the Amery Oasis, so it is clearly a species endemic to this area. *Boeckella poppei* has been recorded in Ablation Lake on the Antarctic Peninsula, where its density was estimated at 0.5 individuals l^{-1} (Heywood 1977). This species occurs in western Antarctica (see Chapter 2, Section 2.5.5), but the only location in which it has been recorded in eastern Antarctica is the Amery Oasis (Bayly and Burton 1993; Bayly et al. 2003). Another calanoid copepod, *Gladioferens antarcticus*, occurs in low numbers in epishelf lakes in the Bunger Hills (Bayly et al. 2003). Dwarfism and low fecundity are clearly adaptations to an extreme cold and depauperate environment. It is likely that growth is extremely slow and the life cycles span several years. It is probable that, like *Daphniopsis studeri* in the freshwater lakes of the Vestfold Hills, *Boeckella* remains active throughout the year and does not exploit resting eggs or diapause.

In Beaver Lake *Boeckella* faecal pellets were conspicuous. These were well colonized by bacteria and flagellates and provide sites for enhanced bacteria production. Attached bacteria have higher rates of production that those of a free-floating bacteria (Grossart and Simon 1993). It is likely that the copepods exploit these faecal pellets as one of their food sources (Laybourn-Parry et al. 2006).

4.4 Carbon cycling in the plankton of epishelf lakes

Primary production determinations in epishelf lakes are limited to single data sets for Ablation Lake and Lake Moutonée on the Antarctic Peninsula and Lakes Prival'noye and Karovoye in the Schirmacher Oasis (Table 4.2). The most detailed data sets relate to Beaver Lake, where determinations were undertaken on five occasions during December and January in 2002/2003 and in February 2000 (Table 4.2). The rates of carbon fixation achieved in epishelf lakes are similar to rates reported for large freshwater continental lakes (Table 2.2). Highest photosynthetic rates in Beaver Lake corresponded to highest concentrations of chlorophyll a (see Figure 4.5) which occurred between 90 and 100 m. Here, highest primary production was achieved in late December and early January, 10.92 µg C l^{-1} day^{-1} and 13.90 µg C l^{-1} day^{-1} respectively. In 2000, rates recorded in February were higher. Given that inter-annual variation appears to be a feature of Antarctic water bodies, it may be the case that the summer of 1999/2000 was a more productive year. All of these freshwater bodies have continuous low water temperatures and suffer from nutrient limitation, factors that inhibit photosynthesis. As indicated in Figure 4.4, PAR penetrated to 100 m in the Beaver Lake water column just above the lake sediment, which may have supplied nutrients to drive the enhanced rates of photosynthesis at this great depth.

Chlorophyll a specific rates of photosynthesis or the assimilation number (µg C fixed (µg chl a)$^{-1}$ l^{-1} h^{-1}) ranged from 0.03 to 0.58 throughout the water column, while photosynthetic efficiency (µg C fixed (µg chl a)$^{-1}$ h^{-1} µmoles m^{-2} s^{-1}) ranged between 0.003 and 0.014 in the top 25 m of the water column for which PAR data were available (see Figure 4.4 where values below 30 m were extrapolated). These values are lower than those reported for Crooked Lake and Lake Druzhby, where chlorophyll a specific rates of photosynthesis ranged between 0.10 and 44.95 and 0.06 and 12.92 respectively and photosynthetic efficiency ranged between 0.005 and 5.19 and 0.03 and 1.23 respectively (Bayliss et al. 1997; Henshaw and Laybourn-Parry 2002). In a broader polar context, the maximum

Table 4.2 Primary and bacterial production in epishelf lakes in the Antarctic Peninsula, the Amery Oasis, and the Schirmacher Oasis.

Lake	Primary production µg C l^{-1} day^{-1}	Bacterial production µg C l^{-1} day^{-1}
Ablation Lake[1]	Max. 6.0 at 3 m February*	–
Lake Moutonée[1]	Max. 12.0 at 5 m December*	–
Beaver Lake[2,3] In 6–100 m water column during Dec to Jan 2002/2003 In top 4–19 m of water column on 15th Feb 2000	Min. 2.14 ± 0.05 on 23rd Dec Max. 13.90 ± 3.74 on 28th Jan Min 19.7 ± 18.0* on 15th Feb Max 25.5 ± 15.6* on 15th Feb	0.050–0.288 0.32–1.15
Prival'noye[4]	0.4*	–
Karovoye[4]	5.0*	–

*indicates a single determination. Data for Beaver Lake cover five determinations between 3rd December and 28th January 2002/2003 and February 2000.
1: Heywood 1977; 2: Laybourn-Parry et al. 2006; 3: Laybourn-Parry et al. 2001b; 4: Kaup 1995.

assimilation numbers for Beaver Lake are comparable to those reported for High Arctic Canadian Lake Char and Meretta Lake (0.16–0.40 and 0.62–0.77 respectively) (Markager et al. 1999). Overall photosynthetic parameters in Beaver Lake indicate a system that is constrained by extreme physical and chemical conditions.

At the time of writing, bacterial production has only been measured in Beaver Lake (Table 4.2). The rates are low: in the summer of 2002/2003 bacterial production was on average less than 10% of primary production. These rates of bacterial production are low in comparison with those reported for large continental freshwater lakes (see Table 2.2). In Crooked Lake and Lake Druzhby, bacterial production represented 30% and 22% of primary production respectively. Concentrations of DOC are very low in Beaver Lake (Table 4.1) and this, together with nutrient limitation and constant low temperatures, undoubtedly limits bacterial growth. The DOC pool is derived from autochthonous sources, mainly from the phytoplankton. Exudation of photosynthate (DOC) by the phytoplankton is probably very low.

The HNAN community of Beaver Lake have clearance rates of 1.1 nL h^{-1} $cell^{-1}$, which gives ingestion rates ranging from 2.7 bacteria $indiv^{-1}$ day^{-1} to 3.6 bacteria $cell^{-1}$ day^{-1}. The ingestion rates of the HNAN are significantly lower than those reported for other freshwater lakes (see Table 2.9). It is likely that they are also taking up DOC, as these grazing rates are unlikely to meet their carbon requirements. In spite of low grazing rates, HNAN removed between 22.5% and 26.7% of bacterial production. Highest grazing rates occurred in early to mid December, falling to lower rates in late December and January. Some of the cryptophytes were mixotrophic, exploiting bacteria as a food source, but there were insufficient data to allow estimates of grazing impact (Laybourn-Parry et al. 2006).

It is obvious that our knowledge of carbon flow in epishelf lakes is severely limited. These unusual lake systems beg to be more thoroughly researched, particularly since they are vulnerable to the effects of climate change and disintegration, as has been well documented in the Arctic (see Section 4.1).

4.5 The benthic communities of epishelf lakes

There is limited information on the benthos of epishelf lakes. The benthic region can be freshwater where the connection to the sea is via a conduit, or it may be saline (see Figure 4.1). A benthic fish was caught in a net haul from the water column of Beaver Lake. It was tentatively identified as *Trematomus scotti*, a marine benthic feeder (Cromer et al. 2005b). Four specimens of *Tematomus bernacchii*, also a marine benthic species, were caught in a trap at 70 m in Ablation Lake in water with a salinity above 30‰ (Heywood 1977). The specimens appeared to be a smaller growth form of *T. bernacchii* found in the open sea off Terre Adelie and in McMurdo Sound. The species are not adapted to freshwater and Heywood (1977) noted that the fish were suffering from acute distress when brought to the surface through freshwater. These are the only records of fish in Antarctic lakes. They are not part of the freshwater community of epishelf lakes, but live in the saline underlying waters, where there must be a sufficient benthic food source to support them. Heywood (1977) collected an unidentified cyclopoid copepod from the lower saline waters of Ablation Lake.

While diatoms are absent from the plankton of epishelf lakes they do occur in the benthos. Surface sediment transects of Ablation Lake and Lake Moutonée which covered the freshwater into the saline benthic zones of each lake showed distinct zones of diatoms. Diatoms clustered into four zones in Lake Moutonée (Figure 4.8). Zone 1 had relative high abundances of freshwater diatoms, all above the halocline. As the depth increased freshwater brackish species occurred, followed by brackish and then marine species. In Ablation Lake, the diatoms clustered into three zones, freshwater, brackish, and marine. The freshwater species *Achnanthes muelleri* dominated the freshwater zone. The brackish zone (zone 2) was characterized by marine taxa such as *Diploneis* sp. and the marine region (zone 3), by the presence of *Diploneis* sp. Amphora copulate and *Achnanthes breviceps*. These diatom assemblages have a limited species diversity compared with other Antarctic lakes (Smith et al. 2006). Interestingly, a 30 cm sediment core taken from Beaver Lake

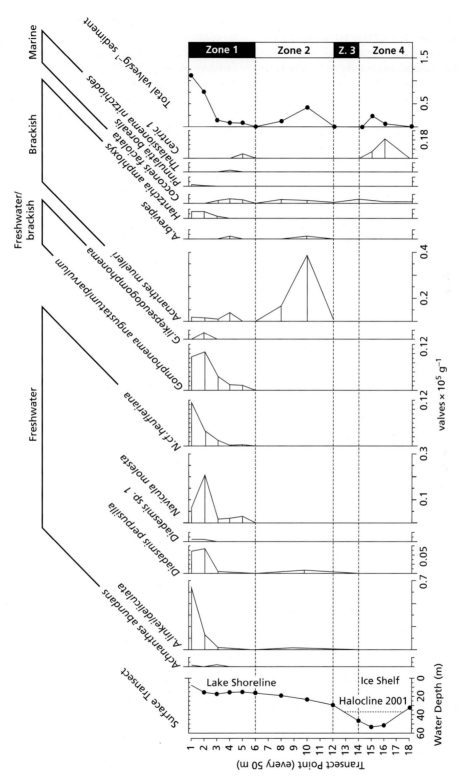

Figure 4.8 Benthic diatoms in a transect of surface sediments in Lake Moutonée, Antarctic Peninsula. The diatom data are absolute abundance (valves g⁻¹ sediment). The left-hand panel shows a water depth profile and the position of the halocline. From Smith et al. (2006), with the permission of Springer.

contained no diatoms, while neighbouring Lake Radok and nearby Lake Terrasovoje possessed a benthic diatom community (Cremer et al. 2004). The lack of a benthic diatom community in Beaver Lake is particularly intriguing.

To date there is no information on the productivity of epishelf lake benthic communities. In particular, we lack information on processes involved in carbon cycling for both the plankton and benthos of these unusual lakes.

CHAPTER 5

Lakes and ponds on glaciers and ice shelves

5.1 Introduction

Lakes and ponds form on glacial surfaces and on ice shelves (see Figures 1.6 and 1.7). These lakes are relatively short-lived compared with lakes that have evolved in rock basins and fluvial hollows (see Chapters 2–4). Indeed some of them are very ephemeral. Our knowledge of cryolakes or supraglacial lakes is limited and is mostly related to their physical and chemical characteristics. They are apparently 'extremely low in biota' (Leppäranta et al. 2012). We do, however, have detailed biological data on cryoconite holes on glaciers from both the Arctic and Antarctic. These are effectively straight-sided, small diameter, shallow (up to 50 cm) mini-lakes that are common on glacier surfaces. Glacial drainage encompasses cryoconite holes, thus the biological activity associated with the sediment or cryoconite will be transported into cryolakes. Biological activity in supraglacial lakes is likely to be similar to cryoconite holes. Ponds and small lakes on the ablation zone of ice shelves are also shallow systems and display a wide range of chemical conditions, ranging from freshwater to hypersaline. All of these systems freeze to their bases in the winter.

Supraglacial lakes and englacial lakes serve as important temporary storage sites for surface meltwater on glaciers. Englacial lakes are usually small pockets of water that are the closed-off remains of crevasses or meltwater tunnels. In contrast with temperate latitude glaciers, where supraglacial lakes develop and drain during the ablation season, in cold polar locations, supraglacial lakes can persist throughout the ablation season. Where there is a large amount of supraglacial debris differential melting causes the development of extensive water-filled hollows and sink-holes (Benn and Evans 2010). Drainage from supraglacial lakes plays an important role in establishing hydrologic connections between surface and sub-glacial hydrological systems. The connections are called moulins (Figure 5.1), and can act as fast flowing conduits from the glacier surface to the subglacial environment. Supraglacial lakes are shallow being of the order of a few metres deep, but can have a large surface area of several square kilometres (Leppäranta et al. 2012).

The McMurdo Ice Shelf carries the most extensive area of ablation zone ponds in Antarctica. The water bodies of this area have been particularly well researched because they are in close proximity to McMurdo Station (USA) and Scott Base (New Zealand). Such water bodies have also been described from other Antarctic ice shelves, for example the George VI Ice Shelf which lies adjacent to the Antarctic Peninsula. Here they vary in size from a few m^2 to several km^2 in area and from a few cm to metres deep. The lakes usually have a ribbon-like structure (Reynolds 1981). Such ribbon-shaped lakes and ponds are also found on ice shelves in the high Arctic, for example the Ward Hunt Ice Shelf (Mueller et al. 2005). These ablation zones on ice shelves allow sediment that has been incorporated into or deposited on the ice to migrate and accumulate towards the surface, thereby promoting the development of surface meltwaters that form aquatic ecosystems (Hawes et al. 2008b).

Antarctic Lakes. Johanna Laybourn-Parry and Jemma L. Wadham.
© Johanna Laybourn-Parry and Jemma L. Wadham 2014. Published 2014 by Oxford University Press.

Figure 5.1 A moulin through which supraglacial lakes drain. Photo courtesy of D. Chandler.

5.2 Supraglacial lakes

5.2.1 Types of cryolakes

The Greenland Ice Sheet is undergoing accelerated melting as a result of climate warming. Unsurprisingly there has been a good deal of focus on supraglacial lakes and their drainage to the subglacial environment in this region, as glaciologists attempt to predict the future consequences of warming (see Figure 7.3). Thus, much detail on these types of lakes is available from the Arctic. For example a large number of supraglacial lakes in West Greenland were observed over 10 consecutive melt seasons using satellite derived data over a study area of approximately 16,500 km^2 (Liang et al. 2012). In warmer years, lakes experienced more extreme drainage and filling events and shorter mean lake duration (i.e. lakes disappeared earlier). The annual maximum lake area was used as a proxy for lake volume. It showed no correlation with annual melt intensity, leading to the conclusion that the lakes accommodate increased surface meltwater flow by draining more frequently and earlier in the melt season. Moreover, in warmer years more lakes appeared at higher elevations. What this study illustrates extremely well is that cryolakes are delicate dynamic systems that respond rapidly to climatic change, and as such are important sentinels in the polar regions, where climate warming is enhanced relative to lower latitudes.

Information on Antarctic cryolakes is sparse in comparison with the Arctic. However, the limited data suggest that there is a range of types. Coastal glaciers often develop temporary lakes, which quickly drain through moulins. These are so short lived that they are unlikely to develop any significant biological activity. Studies on inland glaciers, like those of the McMurdo Dry Valleys, indicate that here cryolakes are relatively long lived, possibly around 2,000–4,000 years. These lakes are usually formed at the base of an ice cliff from the coalescence of cryoconite holes that scavenge debris and water from surrounding melt features (Bagshaw et al. 2010b; Bagshaw et al. submitted). They grow larger over time as they move down what is effectively an ice valley on the glacier, characterized by a lake and riffle structure (Figure 5.2). Eventually they disintegrate at the glacier snout, where many of them feed into valley floor lakes. Cryolakes are shallow, around 2 m of water maximum, and

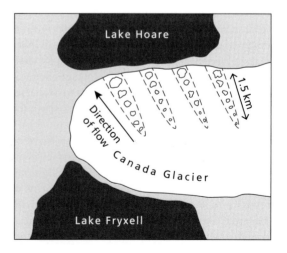

Figure 5.2 Schematic of the development and drainage system of cryolakes on the Canada Glacier (Taylor Valley).

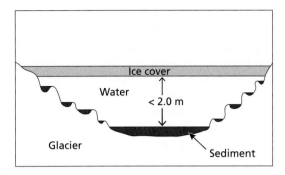

Figure 5.3 Section of a cryolake, note riffle effect caused by water flow through the lake.

are permanently ice-covered with ice that thins to around 16 cm in January (Bagshaw et al. submitted). As the lakes mature they develop a substantial sediment load up to around 0.5 m. In section they have a terraced structure with a sediment layer in each 'terrace' (Figure 5.3). The terraced structure is the result of water flowing around the edges of the lakes in the ice valley drainage system.

Supraglacial drainage has been investigated in detail on the Canada Glacier (Taylor Valley, McMurdo Dry Valleys). Here, flow largely occurs under a layer of up to 0.5 m of surface ice. The system is made up of a series of interconnecting channels, riffles, and cryoconite holes. Around 50% of the cryoconite holes are in hydrological contact, while the rest are isolated. In winter the drainage system freezes solid. It was assumed that cryolakes melted progressively during the ablation season and that lake drainage was a gradual process. However, observations on the Canada Glacier during the summer of 2008/2009 revealed that episodic filling and drainage of the surface lakes occurs. Thus these lakes are hydrologically dynamic with short water residence times. The fluxes of water in one lake were monitored before, during, and after it was subject to flooding. During flooding the lake volume increased by ~300% (from 180 to 630 m^3). This lake would have flooded in 2.3 hours if the maximum discharge had been maintained and outflow was negligible. It would have drained in 88 hours if the outlet stream flowed at the usual rate and it received no further water inputs. In fact, the lake drained by ~260 m^3 after 5 days as inflows declined and outflows increased as a result of flow melting and enlargement of the outflow channel (Bagshaw et al. 2010). Short water residence times suggest that planktonic activity is probably not significant and that the sediment in these systems is where biological activity is concentrated. Cryolakes freeze to their bases in winter, so their biota has to be able to survive in resting stages.

A third type of cryolake has been described from a blue ice region in Dronning Maud Land (see Figure 1.1). These lakes were located at the sides of nunataks close to the Finish Station, located at 73°2.5'S, 13°24.4'W. They appear to be relatively long-lived as the first investigation occurred in 2004/2005 and the lakes were still in existence in 2010/2011 (Leppäranta et al. 2012). The summer development of Lake Suvivesi (7 km^2) was studied in detail. There appear to be inter-annual differences in the timing of the development of a lake profile (Figure 5.4). The summer growth of the lakes is forced by solar radiation, which is primarily dependent on cloudiness. In the summer of 2004/2005 the first profiling took place on 15th December and revealed a thin layer of liquid water beneath an ice cover. The water was underlain by a slush layer, which in turn sat on hard ice which can be regarded as the lake bottom (Figure 5.4). During the course of the summer the liquid water layer expanded, the ice cover thinned, and a further slush layer developed, sandwiched by the hard ice. Interestingly, a sediment-rich pocket was found below the hard ice lake bottom within the glacial ice. In 2010/2011 the first profile undertaken on 10th December revealed only glacial ice, but by 17th December a slush layer had developed and the lake's development followed a similar pattern to that observed in the earlier study (Figure 5.4). Similar patterns of lake development were observed in the other lakes nearby.

Spatial sampling showed that Lake Suvivesi had a patchy structure rather than being a well-defined three-dimensional body. These variations were attributed to variations in albedo and the light attenuation coefficient. The lake bottom was uneven, with deeper hollows created by sinking stones and sand that absorbed solar heat and sank into the ice. The lake water was extremely clear with a very low conductivity of 1–10 μS cm^{-1}, indicating low levels of dissolved salts. The waters were well oxygenated, with dissolved oxygen concentrations of

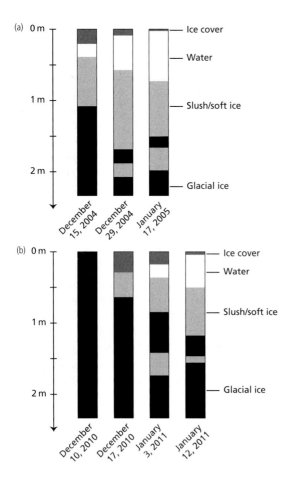

Figure 5.4 Profiles of Lake Suvivesi in the blue ice area of Dronning Maud Land, A summer 2004/2005, B summer 2010/2011. From Lepparanta et al. (2012) with the permission of Cambridge University Press.

water pressure became sufficiently high, it might cause cracks and subsequent draining of the water through a moulin (Leppäranta et al. 2012). Water is likely to be lost from the ice surface by sublimation. Thus the residence time of the liquid water during the summer is probably long and this allows a build-up of biological activity, which will persist from year to year, surviving the winter in resting stages or cryopreservation.

5.2.2 The physical/chemical environment and biology of cryolakes

At the time of writing, our knowledge of the chemistry and biology of cryolakes is very limited. There is an indication of the level of biological activity from a study of three supraglacial ponds on the lower Darwin Glacier (Victoria Land) conducted in December 2007 and January 2009 (Webster-Brown et al. 2010). The conductivity of their waters was similar to the surrounding ice (1.3–6.6 µS cm^{-1}), as were the concentrations of DOC and nutrients. Nitrate (NO_3-N) concentrations ranged between 0.00016 and 0.0177 µM l^{-1} (mean 0.0009), ammonium (NH_4-N) between 0.055 and 0.776 µM l^{-1} (mean 0.229), soluble reactive phosphorus (PO_4-P) between 0 and 0.02 µM l^{-1} (mean 0.006), and DOC between 0 and 7.89 mg l^{-1} (mean 0.66). The mean nutrient and DOC values are similar to those recorded for large ultra-oligotrophic, continental freshwater lakes (see Table 2.3). Concentrations of chlorophyll a ranged from below the limit of detection to 2.46 µg l^{-1} (mean 0.21). These values are within the range reported from ultra-oligotrophic freshwater systems (see Table 2.2). Unlike other ponds in the vicinity, the cryoponds lacked any benthic mats. Their waters contained a range of species in low concentrations, including a range of cyanobacterial species. The species diversity of cyanobacteria was relatively high and included *Chroococcus* sp, Oscillatoriales, *Cyanothece* sp., *Phormidium* spp., *Planktolyngbya* spp., *Pseudoanabaena*. spp., and unidentified picocyanobacteria. Diatoms and a small desmid were also collected from the plankton. Mean heterotrophic bacteria concentrations were 1.5×10^7 l^{-1}, an order of magnitude lower than concentrations reported for ultra-oligotrophic freshwater lakes (See Table 2.4).

10–12 mg l^{-1}. Intriguingly, we are told that low levels of phytoplankton were observed in the summer of 2010/2011 (Keskitalo et al. unpublished, quoted in Leppäranta et al. 2012). The pockets of sediment found in the hard ice are interesting and may be vestiges of warmer summers when the lake had a deeper profile. While the waters may support some biological activity, the sediment pockets and layers are also likely to be biologically active. The rocks and sediment originate from the nearby Nunataks.

Unlike the more dynamic lakes on the Canada Glacier, the lakes in the blue ice region in Dronning Maud Land do not appear to have inflows or outflows. There is the possibility that if liquid

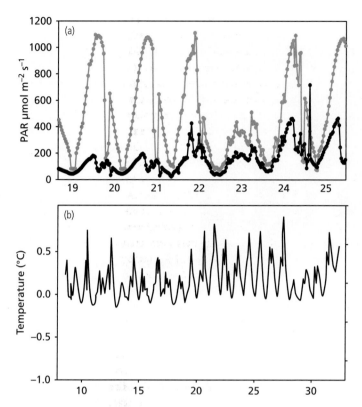

Figure 5.5 (a) Photosynthetically active radiation (PAR) in a cryolake on Joyce Glacier (Garwood Valley) during January 2010. Grey line surface PAR, black line PAR under ice in the cryolake. (b) Temperature profile for a cryolake on the Joyce Glacier in January 2010. Data courtesy of E. Bagshaw et al.

Cryolakes on the Joyce Glacier (Garwood Valley, Dry Valleys) were investigated for 25 days during January 2010. PAR levels attenuated significantly below the ice surface and 99% had attenuated before it reached the surface sediment, indicating that primary producers on the sediment are adapted to low light levels, as are the phytoplankton in the freshwater and saline lakes of the Dry Valleys floors (see Chapter 3) (Figure 5.5a). Water temperatures were close to freezing throughout January (Figure 5.5b). At the time of writing, we do not have any in situ process-related data, such as photosynthesis or bacterial production, for cryolakes. However, cryoconite holes on glacier surfaces have been subject to detailed study in Antarctica and to a greater extent in the Arctic (Figure 5.6). They are effectively mini-lakes that are part of the surface drainage system of glaciers and as such drain into supraglacial lakes. While they are not directly analogous to large water bodies on ice sheets and glaciers, they do provide us with an indication of the potential productivity of glacial aquatic environments. Concentrations of bacteria in cryoconite hole water are of the same order of magnitude as found in cryoponds on the Darwin Glacier (Webster-Brown et al. 2010) (Table 5.1). Bacterial production in cryoconite hole waters is within the ranges reported from continental freshwater lakes (see Table 2.2). However, it is clear from Table 5.1 that the sediment layer or cryoconite is the site of greatest productivity. Primary production data are only available for the sediment. On the Canada Glacier (McMurdo Dry Valleys) the rate of carbon fixation was between 0.4 and 1.4 µg C g^{-1} sediment day^{-1}, while further north on the Sørsdal Glacier (Vestfold Hills) it ranged between 0.21 and 4.82 µg C g^{-1} sediment day^{-1} (mean 2.10 ± 1.5) (Bagshaw et al. 2011; Hodson et al. 2013). The cryoconite sediment of cryoconite holes is oxic, whereas the deeper sediment of cryolakes is likely to have a profile that ranges from oxic to anoxic. In the deeper layers where oxygen becomes deleted, sulphate reduction followed by methanogenesis is probable.

The view that the icy cryosphere is another biome in the biosphere is now gaining wide acceptance

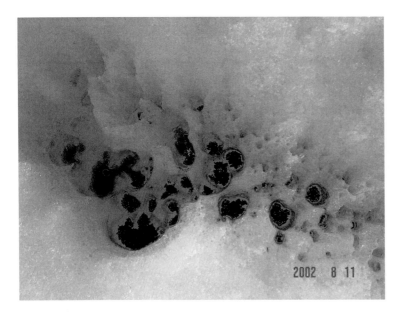

Figure 5.6 Cryoconite holes on an Arctic glacier. Note the sediment (cryoconite) at the bottom of the hole. Photo J. Laybourn-Parry.

(Anesio and Laybourn-Parry 2012). As climate warms, the amount of melt during the ablation season will increase, leading to increased biological activity in supraglacial lakes, ponds, streams, and cryononite holes. This in turn has an effect on the albedo which impacts on the rate of melt. Much more research effort is needed in this area of polar limnology if we are to gain detailed information on carbon cycling on glaciers and its potential responses to long-term climate change.

5.3 Ice shelf ponds and lakes

One of the most extensive regions of ponds is found on the ablation zone of the McMurdo Ice Shelf. The area is an interconnected meltwater system with

Table 5.1 Bacterial concentrations, bacterial production in cryoconite holes in Antarctica and the Arctic for comparison.

Glacier and location	Cryoconite (sediment)		Overlying water	
ANTARCTICA	Bacteria × 10^7 l^{-1}	Bac Prod. µg C g^{-1} day^{-1}	Bacteria × 10^7 l^{-1}	Bac Prod. µg C l^{-1} day^{-1}
Canada McMurdo DV[1]	20.0	3.0–3.4	7.9	<1.0
Hughes McMurdo DV[1,2]	4.5	0.55	1.3	0.096
Commonwealth McMurdo DV[1,2]	11.5	0.55	5.2	0.096
Sorsdal Vestfold Hills[3]		0.036		–
ARCTIC				
Midtre Lovénbreen Svalbard[2]	3.9×10^9 g^{-1}	0.95	5.6	0.13
Austre Brøggerbreen Svalbard[2]	9.9×10^8 g^{-1}	0.21	7.0	0.07

1: Foreman et al. (2007); 2: Anesio et al. (2011); 3: Hodson et al. (2013). DV = Dry Valleys

two distinct types of topography (see Figure 1.7); undulating ice, covered with 10–20 cm of moraine material and pinnacle ice. The ponds are found in an area of undulating ice and range considerably in size from around 1m^2 up to 30,000 m^2. However, they are shallow with depths <2.0 m. The ponds account for about 30% of the ablation zone. Pinnacle ice is much less stable and supports temporary ponds and streams, and suffers continuous ice erosion during the summer. The ponds situated on the undulating ice have a wide range of salinity from freshwater to hypersaline, while those on the pinnacle ice range from freshwater to hyposaline (Vincent and James 1996). Although the ice shelf is moving slowly and the surface is in a state of constant change, some of the ponds have life spans of decades (Howard-Williams et al. 1989). During winter the ponds usually freeze to their bases.

The ponds on the undulating ice have well developed benthic cyanobacterial mats (Howard-Williams et al. 1989, 1990; Vincent et al. 1993; Fernández-Valiente et al. 2001; Jungblut et al. 2005). Common species forming the mats were *Phormidium autumnale*, *P. fragile*, *P. frigidum*, *P. laminosum*, *Oscillatoria deflexa*, *O. limosa*, *Lyngbya* spp., and *Leptolyngbya* spp. The hypersaline ponds also contained *Oscillatoria priestleyi*. Diatoms are common in the cyanobacterial mat matrix, including *Pinnularia cymatopleura* and *Nitzschia antarctica*, while in the hypersaline ponds *Tropodineis laevissima* occurs. The heterocystous genera *Anabaena*, *Nodularia*, and *Nostoc* were present, but were rare to subdominant. Similar cyanobacterial assemblages form mats in ice shelf ponds in the High Arctic (Mueller et al. 2005). Rotifers inhabit the benthic mats with up to seven species being recorded. Their distribution is negatively correlated with conductivity, the most saline ponds containing only *Philodina* species (Suren 1990).

The three-dimensional structure, determined by electron and confocal microcopy, shows vertical stratification of groups of cyanobacteria and mineral sediments with a high content of extra-cellular polymeric substances and large void spaces that are occupied by water (de los Ríos et al. 2004). Cyanobacterial mats dominated by *Phormidium* and *Oscillatoria* spp. have a high degree of vertical colour zonation. The surface is red or orange, underlain by a blue-green layer where most of the chlorophyll *a* is located. The surface layer is enriched by carotenoids and other components that attenuate UVB radiation. In addition, trichome motility related to light intensity is a characteristic of *Oscillatoria* dominated mats. The trichomes migrate upwards in dimmer light (Vincent et al. 1993).

The sediments beneath the accumulations of cyanobacterial biomass are anaerobic and the domain of anaerobic bacteria (see Chapter 3, Section 3.6.1). Methanogenic and sulphate-reducing bacteria in this environment are subject to the effects of freeze/thaw cycles. When frozen, methanogenesis and sulphate reduction decreased, but when thawed and incubated at 4 °C in low sulphate (or chloride) sediment, total carbon and electron flow were faciliated by acetate-driven sulphate reduction and H_2-driven methanogenesis. In contrast, in high sulphate sediments, sulphate reduction was the major process mediating carbon and electron flow under both thawed and frozen conditions. (Mountford et al. 2003).

The water column overlying the mats supports a planktonic community. A summer study of twenty ponds, with salinities ranging from freshwater to hypersaline, revealed heterotrophic bacterial concentrations ranging from 0.46 to 1.64×10^9 l^{-1} and chlorophyll *a* concentrations from 1.1 to 78.2 μg L^{-1} (James et al. 1995). Chlorophyll *a* concentrations in the plankton are two orders of magnitude lower than in the benthic mats (Howard-Williams et al. 1989). The picocyanobacterium *Synechococcus* was particularly common in Salt Pond, the most saline of the study ponds. Filamentous cyanobacteria also occurred in the plankton, particularly *Oscillatoria* and *Anabaena* spp. The phytoplankton was dominated by cryptophytes, *Ochromonas* and *Chroomonas*. Cryptophytes are a conspicuous component of the phytoplankton of hyposaline, saline, and hypersaline lakes elsewhere in Antarctica (see Chapter 3, Section 3.6.4). The ponds supported a relatively diverse ciliate plankton of up to 22 species. Many of the species have also been recorded in the McMurdo Dry Valleys lakes (see Section 3.6.4, Table 3.6). However, the genus *Vorticella*, which is a very efficient grazer of bacteria, is very common in the ponds, while it is not common in the lakes. Rotifers are rare in the plankton, the only species seen was *Philodina*, which had probably migrated upward from the benthic mats. Water conductivity or salinity is regarded as the main factor determining community structure in ice shelf melt ponds (James et al. 1995; Sutherland 2009).

Like shallow ponds on land, the winter sees the water column freezing to its base, or close to it. In summer the shallow ice shelf ponds are usually stratified with a layer of warmer, more saline water overlain by a layer of mixed, less saline, cooler water. Water temperatures reach close to 10 °C and there is a wide variation over the year, as is shown in Figure 5.7 for Fresh Pond. Here, the point of freezing is shown by the marked temperature decline. Conductivity changes during the freezing process. This is because the volume of water diminishes as the ice grows in thickness. For example, in Brack Pond conductivity increased on ice formation to 1.5 times the summer levels. By April, three ponds (Legin, Orange, and Egg) had between 8% and 15% of the their original water volume. The stratification seen during the summer ice-free phase usually disappears as the upper water gains salts excluded from ice formation. As the ice forms it is clear and highly transparent, but as it thickens it becomes opaque due to the inclusion of air bubbles. Light transmission decreases so that by the end of March it can be as low as 5% of surface PAR. This is not only the result of changes in the nature of the ice cover but also the onset of the polar night (Hawes et al. 1999; Safi et al. 2012).

The annual freeze/thaw pattern imposes considerable challenges on the biological community. Most studies have occurred in summer during the thaw phase, while the freezing phase is poorly investigated. Ephemeral ice develops on the ponds during January and is persistent by February, at which point its thickness increases steadily at a rate of 1.5–1.6 cm day^{-1}. By early April, ponds may have frozen to their bases or retain a layer of liquid water that is hypersaline. During the freeze-up process the salts in the water becomes progressively concentrated, from around 2 to 5‰ to hypersaline (Hawes et al. 2011b). During the freeze-up photosynthesis continues and increases dissolved oxygen; the majority of the photosynthesis being achieved by the benthic mats (Hawes et al. 2011c).

Changes in conductivity have a very marked impact on the physiology of benthic mat communities (Hawes et al. 1999). Incubations under a range of conductivity conditions showed a significant drop in gross photosynthesis relative to controls (Figure 5.8). Photosynthesis declined above 10 mS cm^{-1}, whereas respiration showed little change below conductivities of 80 mS cm^{-1}. When mats were returned to pond water they recovered their photosynthetic capabilities within 48 hours. Moreover, sub-zero temperatures did not appear to have any impact on photosynthesis or respiration.

As freezing progresses, the plankton community undergoes major changes as it faces physico/chemical challenges (Safi et al. 2012). For the first month after ice formation, chlorophyll a increased in both the upper mixed layer and on the transition boundary or chemocline in Legin, Orange, and Egg Ponds to between 24 and 30 μg l^{-1} by early March. Heterotrophic bacteria were most abundant on the chemocline, increasing in numbers during ice formation and attaining a maximum in March (across the three ponds between 6 and 23 × 10^9 l^{-1}). Their production declined by 37% between February and early March and by 74% between March and

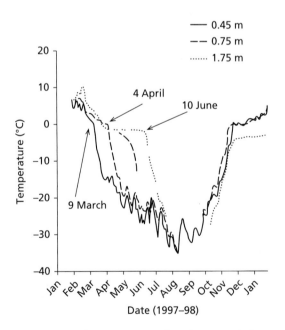

Figure 5.7 Annual temperatures in Fresh Pond, Amery Ice Shelf pond. From Hawes et al. (1999), with the permission of Cambridge University Press.

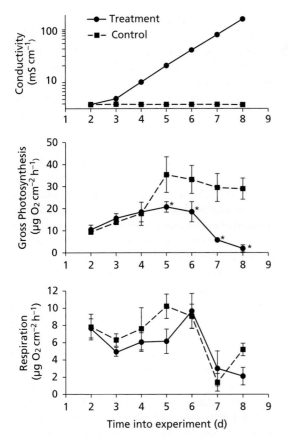

Figure 5.8 Net photosynthesis and respiration of isolated cores of benthic mats under increasing conductivity, compared with controls incubated in pond water. Significant difference between treated samples and controls is indicated by an asterisk. From Hawes et al. (1999), with the permission of Cambridge University Press.

and light climate deteriorated. During this transition grazing pressure on heterotrophic bacteria and pico-phytoplankton increased, contributing to their decline. Around 34% of bacterial production was grazed in February, increasing to 15 times production in March. As the final freeze-up approached, all components of the community declined. During the summer months, in a range of ponds, grazing impact by ciliates was minimal, usually less than 5% and heterotrophic flagellate impact was undetectable (James et al. 1995). Thus during the annual freeze/thaw cycle through spring, summer, and autumn there are very significant changes in trophodynamics and carbon cycling in the plankton.

During summer, mat communities can achieve carbon fixation rates of 24–96 µg C cm^{-2} day^{-1}, which lies within the range reported for benthic mats in lakes (see Table 2.10). Nutrient limitation is not a constraint as both nitrogen and phosphorus are abundant in the interstitial waters. All of the mat communities achieved saturation at 15% of ambient midday irradiance, and in short-term incubations there was no evidence of photo-inhibition, even under full sunlight (Vincent et al. 1993). Since nitrogen is apparently not limiting to production, it is hardly surprising that nitrogen-fixing heterocystous cyanobacteria are not a common component of pond benthic mats. Nonetheless significant nitrogen fixation does occur, mediated by *Anabaena*, *Nodularia*, and *Nostoc*. Based on acetylene reducing assay (ARA), nitrogen fixation contributes an estimated 1 g m^{-2} $year^{-1}$, considered to be the highest nitrogen input to these systems (Fernández-Valiente et al. 2001). The estimated nitrogen requirement of the benthic communities is around 3 g m^{-2} $year^{-1}$, of which nitrogen derived from precipitation contributes around 36 mg N m^{-2} $year^{-1}$; the remainder coming from nitrogen fixation and internal biogeochemical cycling (Hawes et al. 1993).

Ice shelf ponds represent extreme aquatic environments, experiencing radical changes in physico/chemical conditions over the annual freeze/thaw cycle. Such conditions confront their extremophile communities with major physiological challenges.

early April. Picophytoplankton was most dense on the chemocline but declined after February. Phytoflagellates also declined as freezing progressed and they encysted. Heterotrophic flagellates declined less rapidly and gradually came to represent the larger portion of the flagellate community. A slower decline was observed in the ciliated protozoan community. They came to dominate the remaining biomass by late March and early April. In effect the system moved from an autotrophic system to a net heterotrophic system as freezing

CHAPTER 6

Subglacial lakes

6.1 Introduction

Subglacial lakes are large bodies of water that exist between the ice sheet sole and the underlying bedrock or sediments. Melting conditions exist over approximately 55% of the Antarctic Ice Sheet bed (Pattyn 2010) (Figure 6.1), and within these areas >380 subglacial lakes have been now been identified (Wright and Siegert 2011) (Figure 1.11), the largest of which is Subglacial Lake Vostok. The study of these lakes has featured highly on the International Science Agenda over the last decade (Priscu et al. 2005b) for three primary reasons. Firstly, these lakes present conditions that may be favourable for microbial life. Secondly, sediments deposited at the bed of lakes may reveal clues to the past glaciation and climate history of Antarctica, and thirdly, the presence of lakes and the movement of water between them have the potential to influence ice flow rates.

Subglacial lakes were first identified beneath the ice sheet via the analysis of airborne radio-echo sounding data in the late 1960s (Robin et al. 1970). Lakes showed up as strong and relatively constant reflections compared with the ice–bedrock interface which were very flat in relation to the surrounding topography (Figure 6.2). The original work identified just 17 lakes, with later inventories recording 77 lake locations (Siegert et al. 1996). This was later updated to 145 lakes by Siegert (2005), which has since been revised several times, with the most recent inventory including over 380 lakes beneath the ice sheet (Wright and Siegert 2011). Areas of flat surface topography in satellite images such as ERS-1 and the Moderate Resolution Imaging Spectroradiometer (MODIS) (Bell et al. 2006) have also been linked to the presence of lakes beneath the ice sheet. More recently, satellite laser altimeter data have provided a new tool with which to identify and examine the behaviour of subglacial lakes (Fricker et al. 2007; Wingham 3–93 m in water column. 2006; Smith et al. 2009; Gray et al. 2005). These data sets show changes in ice surface elevation over time, which reflect the dynamic filling and draining of lakes. About a third of subglacial lakes at one time have been found to be what we may term as 'active lakes' (Smith et al. 2009) and which are thought to be linked to a subglacial water system that ultimately discharges subglacial meltwater at the coast.

Of the vast array of lakes in Antarctica, three subglacial lakes have been identified for direct sampling via national drilling campaigns. These are Subglacial Lakes Vostok, Ellsworth, and Whillans (Table 6.1). In addition, a former subglacial lake, Hodgson Lake (Hodgson et al. 2009b; Hodgson et al. 2009a) has also been studied (Table 6.1). The environmental constraints of entering and sampling Antarctic subglacial lakes are challenging and all lake access campaigns have been accompanied by carefully planned environmental protection protocols, involving clean and sterile ways of sampling lake waters (Siegert et al. 2012; Priscu et al. 2013a). Lake Vostok has perhaps, historically, generated the most scientific interest of all of the lakes because of its large size and fortuitous sampling of frozen lake waters via the accretion ice (see Section 6.3.1). As data emerge from these direct drilling campaigns, it is very likely that our understanding of subglacial lake ecosystems and their hydrological behaviour will be dramatically transformed and this will influence the direction of the next decade of research into Antarctic subglacial aquatic ecosystems.

Antarctic Lakes. Johanna Laybourn-Parry and Jemma L. Wadham.
© Johanna Laybourn-Parry and Jemma L. Wadham 2014. Published 2014 by Oxford University Press.

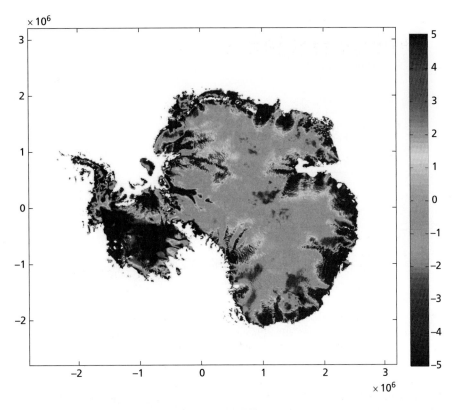

Figure 6.1 Estimated Antarctic basal melting and freezing rates in mm yr^{-1}. Melting, shown in warm colours, is positive and freezing is negative and shown in cold colours. Redrawn by the authors of: Tulaczyk and Hossainzadeh (2011). Values that are higher or lower than ±5 mm yr^{-1} have been truncated. Calculations were performed in a similar manner to (Pattyn 2010). (See Plate 13)

6.2 Distribution and physiographic characteristics of subglacial lakes in Antarctica

The majority of subglacial lakes in Antarctica are less than 10 km in length, with a modal size of 5 km (Wright and Siegert 2011). Various estimates of lake volume have been made, giving minimum estimates of 4000–12,000 km^3 (Dowdeswell and Siegert 1999), and higher estimates with the inclusion of more recently identified lakes, 9000–16,000 km^3 (Wright and Siegert 2011). These are significant volumes, and account for 10%–20% of the annual volume of meltwater produced by basal melting beneath the entire Antarctic Ice Sheet (65 km^3) (Pattyn 2010). The lakes that have been identified to date are not evenly distributed around Antarctica, which may reflect either the bias of identification techniques or underlying physical controls. There is a tendency for lakes identified by radio-echo sounding to increase in abundance towards the ice divide and under thicker ice (Wright and Siegert 2011). Lakes identified from changes in surface elevation (i.e. active lakes) are typically observed under thinner ice and with no clear pattern of location relative to the ice divide (Wright and Siegert 2011). This may be a real trend or may reflect the less effective operation of RES in warm areas of high crevassing near the margins, and the worsening coverage of ICESat towards the pole. Districts with particularly high numbers of lakes exist, including the Ridge B/Vostok area, Dome C area, Hercules Dome, and South Pole (Dowdeswell and Siegert 1999).

Subglacial lakes display a wide range of physiographic settings and evolutionary histories. Their distribution is not random and lakes tend to be

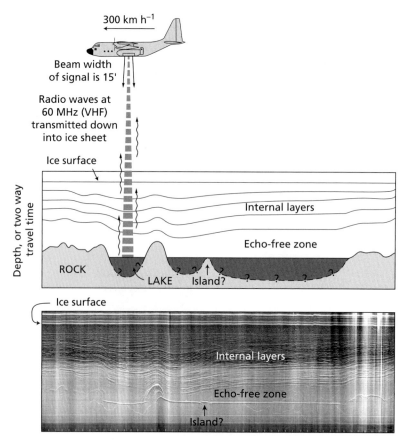

Figure 6.2 Diagrammatic explanation of the application of radio-echo sounding to the identification of Subglacial Lake Vostok and other lakes. Subglacial lakes can be easily identified in these data by their uniformly strong and flat appearance. Bed-rock reflections show as hyperbolae in these data (Siegert et al. 2001). With the permission of Nature Publishing Group.

Table 6.1 A summary of the major features of the three Antarctic subglacial lakes which have been targeted for entry and sampling.

Name/location	Area/dimensions (km²)	Topographic setting	Sediment (water depth)	Mode of drainage	Drilling technology
Subglacial Lake Vostok, East Antarctica[1,2]	~16,000	Subglacial Basins/ Rift Valley	Several 100 m (up to 1000 m depth, mean of several 100 m)	No episodic drainage	Thermo-mechanical
Subglacial Lake Whillans, Whillans Ice Stream, West Antarctica[3]	60	Dynamic ice stream	Yes (2 m)	Episodic drainage	Hot water
Subglacial Lake Ellsworth, Ellsworth Mountains, West Antarctica[4,5]	18	Mountainous lake	Yes (150 m)	Unknown	Hot water
Lake Hodgson, Alexander Island, Antarctic Pensinsula[6,7]	3	Former subglacial lake	Yes (90 m)	Unknown	Hot water

1: Siegert et al. (2001); 2: Thoma et al. (2008), 3: Fricker et al. (2007), 4: Woodward et al. (2010); 5: Ross et al. (2011), 6: Hodgson et al. (2009a); 7: Hodgson et al. (2009b).

associated with one of several lake-forming environments. Within the ice sheet interior, lakes are often found in both the large hypothesized subglacial basins and subglacial highlands located between basins (Dowdeswell and Siegert 2003; Tabacco et al. 2006). Within the interior of subglacial basins, the topography can be very flat. Hence, although some very large lakes are reported in these areas (e.g. Lake Concordia and Vincennes Lake) (Tabacco et al. 2006), lakes are likely to be shallow and may be easily confused with water-saturated sediments (Carter et al. 2007). Further lakes in the ice sheet interior may be found, either associated with high subglacial mountain ranges or perched on the flanks of these mountain ranges (Wright and Siegert 2011). The former category includes Subglacial Lake Vostok, which occupies a rift valley (Studinger et al. 2003) and Subglacial Lake Ellsworth within the submerged Ellsworth Mountains (Woodward et al. 2010). Around the ice sheet margins are two different categories of lakes, those that tend to be associated with the region where fast-flowing ice starts to dominate closer to the margin, or those located beneath the trunks of ice streams. Examples of the latter are the lakes beneath the Siple Coast Ice Streams, such as Subglacial Lakes Whillans, Engelhardt, and Mercer (Fricker et al. 2007). The latter group tend to be smaller in size and exhibit dynamic filling and draining behaviour, many having been identified via satellite altimetry data (Smith et al. 2009).

6.3 Detailed studies of subglacial lakes

6.3.1 Lake Vostok

Lake Vostok has captured the imaginations of scientists worldwide due to its vast size (~250 km long and 50 km wide), which is comparable with that of Lake Ontario, one of the Great Lakes of North America. The lake comprises two basins, a small, shallow north embayment and a deeper, larger southern basin and is thought to act as a hydrologically closed system (Figure 6.3). A number of ice cores have been recovered from Lake Vostok since the 1960s. The upper 3310 m of these cores have been used for palaeoclimate analysis (Jouzel et al. 1987; Petit et al. 1997). The 1998 core halted some 120 m above the lake, but revealed a lowermost section of refrozen water, identified to be formed by freezing of lake waters onto the underside of the ice sheet. This accreted ice section is 210 m thick, and has been analysed in detail for its chemical and biological composition (Priscu et al. 1999a; Karl et al. 1999; Siegert et al. 2003; Christner et al. 2001; Bulat et al. 2009). This accretion ice is chemically distinct from the overlying ice core, displaying a very low total solute content, large ice crystals (Jouzel et al. 1999), and visible debris inclusions (Simoes et al. 2002; Souchez et al. 2003; Royston-Bishop et al. 2005). The lake is believed to predate glaciation in East Antarctica. The temperature regime within the lake exhibits spatial heterogeneity because of the differences in the lake water surface elevation. In general, there is freezing in the south part of the lake, where accretion ice is formed, and melting in the north. Approximately 300–400 m of sediments have been estimated to exist at the lake floor, some of which are calculated to be preglacial (Filina et al. 2008) (Figure 6.3). Information about the chemistry and biology of lake waters is likely to increase dramatically over the next few years, since the lake was entered directly in January 2013 by a Russian drilling team, during which time lake water rose up and froze in the borehole with plans for subsequent analysis.

6.3.2 Lake Ellsworth

This is a small (18 km^2) long lake located close to the ice divide at the head of Pine Island catchment in West Antarctica (Siegert et al. 2004). It was first revealed by radio-echo sounding work in West Antarctica, which showed the presence of a 10 km long lake beneath some 3.4 km of ice near the Ellsworth mountains (Siegert et al. 2004). The lake itself is located within a narrow, steep-sided subglacial valley (Woodward et al. 2010) (Figure 6.4) but it is unknown whether it is connected to any wider hydrological system. While theoretical work has indicated the potential for the formation of accretion ice within the northwest part of the lake, geophysical survey work has not indicated any accretion ice to date. A UK team attempted to penetrate the lake unsuccessfully in 2012, but new plans are being devised to repeat the experiment. The drilling and environmental stewardship protocols to do this have been published (Siegert et al. 2012).

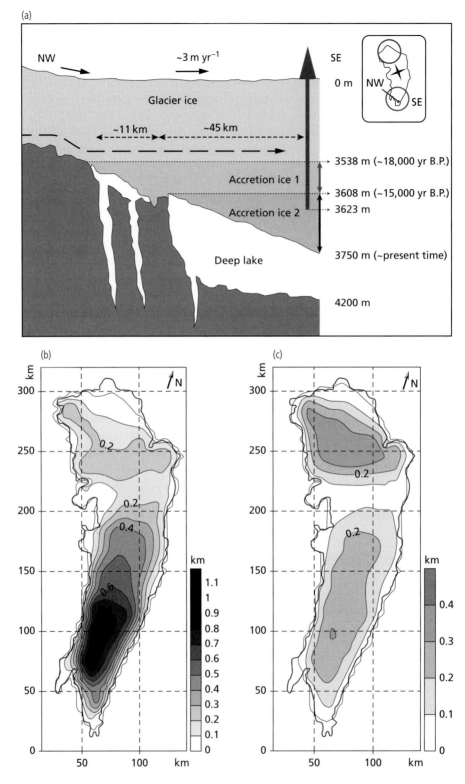

Figure 6.3 (a) Conceptual model of Subglacial Lake Vostok, showing the lake cross-section along a flow line (ice flow is NE to SW) (De Angelis et al. 2004). The two main areas of accretion ice are shown. Type 1 contains debris and Type 2 is clean, the thick vertical arrow denotes the site where the accretion ice has been cored. (b) Predicted water depths (resolution = 0.1 km). (c) Sediment thicknesses (contour interval = 0.1 km) in Subglacial Lake Vostok as inferred from airborne gravity data (by inversion), where the red line indicates the lake coast line from radar data (Filina et al. 2008). With the permission of Elsevier. (See Plate 14)

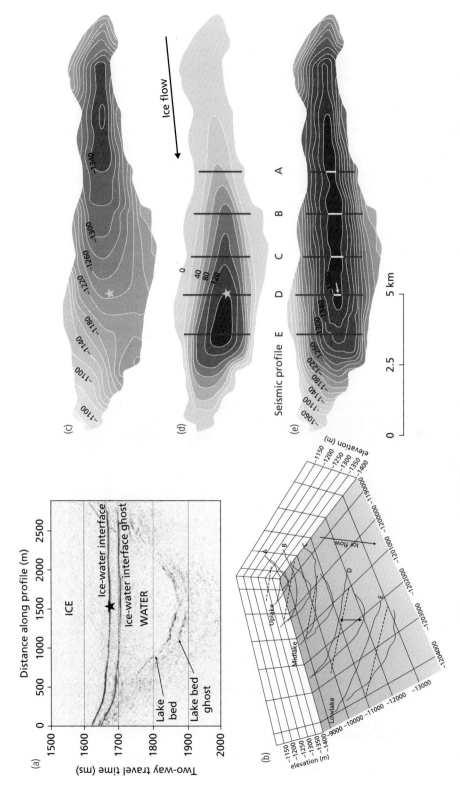

Figure 6.4 Seismic reflection data from Subglacial Lake Ellsworth (adapted from (Woodward et al. 2010). (a) An example of a seismic reflection data profile, with the main reflectors and ghosts identified (profile D). (b) An up-lake (into ice flow) 3D representation of the lake surface (red lines) and bed (blue lines) identified from five seismic profiles, with the drill site indicated by a black line in profile D. Black dashed lines indicate the critical boundary pressure for each seismic line (ice thickness = 3170 m). (c) Ice-water surface. (d) Water column thickness (relative to WRS-84 ellipsoid). (e) Lake bed topography (relative to WRS-84 ellipsoid), where yellow stars indicate drill site locations. Red lines in (d) and (e) indicate the measured positions of the lake bed (and water column thickness) from the seismic data and the parts highlighted in white are parts of the lake where the bed elevation is −1380 m. With the permission of the American Geophysical Union. (See Plate 15)

6.3.3 Lake Whillans

Subglacial Lake Whillans is a small (60 km²), shallow, downstream lake located some tens of kilometres from the grounding line of the Whillans Ice Stream (Fricker et al. 2007). Lake Whillans was penetrated in January 2013 via the WISSARD: Whillans Ice Stream Subglacial Access and Research Drilling Project, when several metres of water were found within the lake system at the time of drilling through 800 m of ice. Stringent sterility and cleanliness protocols were employed in order to minimize forward contamination of lake waters (Priscu et al. 2013a). Subglacial Lake Whillans (Figure 6.5) contrasts with all other lakes targeted for drilling access in that it is shallow, highly dynamic, and fills and drains on decadal timescales. The lake is located beneath the Whillans Ice Stream approximately 120 km from the grounding line, and was first documented in Fricker et al. (2007), based upon vertical movement of the ice surface, which indicated filling and draining of the lake. The lake coincides with a zone of low subglacial pressure, which would drive water flow into the lake. A cluster of similar such lakes are present beneath the Whillans and Mercer Ice Streams, including Subglacial Lakes Engelhardt, Mercer, and Conway. The predicted flow path for meltwaters, calculated on the basis of hydropotential gradients, do not show any likely connections between these lakes and Lake Whillans. Rather, waters from Lake Whillans are predicted to take a flow path to the Ross Ice Shelf, where they are discharged into marine waters. Without drainage events, such as that exhibited by Subglacial Lake Whillans, numerical modelling work predicts that very little water would emerge (Figure 6.5) (Carter and Fricker 2012).

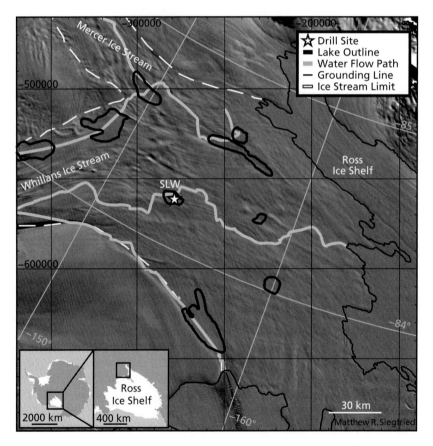

Figure 6.5 Map of Subglacial Lake Whillans (SLW) (Priscu et al, 2013a), re-drawn by M.Siegfried and the surrounding areas, showing the location of the lake (star) and the predicted flow path for water to the Ross Ice Shelf (grey line). With the permission of Cambridge University Press.

6.3.4 Hodgson Lake

Hodgson Lake (72°00.549'S, 068°27.7080'W) is a small former subglacial lake located beneath perennial ice on Alexander Island, west of the Antarctic Peninsula. The lake was submerged beneath the glacier (297–465 m ice thickness) during the last glacial, but has become exposed at the ice margins during the Holocene as a result of glacial retreat since c. 13.5 kyr B.P. (Hodgson et al. 2009b) (Figure 6.6). Pre-Holocene subglacial lake sediments are present on the lake floor and have been cored. The lake has remained isolated from the atmosphere since its emergence due to the development of thick ice cover, and is fed by subglacial waters. Hence, it provides a good analogue site for true subglacial lakes and for the development and refinement of protocols for lake entry, sampling, and sample processing on site.

6.4 Formation of subglacial lakes and hydrological conditions

In order for a subglacial lake to form, the boundary between the ice and the underlying substratum must be at the melting point for ice. This is a few degrees below 0 °C beneath Antarctica, because of the depression of the freezing point of water due to the pressure of the overlying ice column. In Antarctica, lake water derives from the gradual melting of the underside of the ice sheet due to geothermal heating at rates which may be just a few mm per year (Fahnestock et al. 2001). However, the ice sheet margins tend to exhibit higher basal melt rates of 5–6 mm yr^{-1} (Joughin et al. 2004), which reflects the frictional and strain heating of ice as it flows. Water will always flow from areas of high hydraulic potential to low hydraulic potential (or pressure), and lakes form in regions of low subglacial pressure or low hydraulic potential gradient (Fricker et al. 2007). For lakes that episodically drain, this may be true during the fill phase but is reversed during the drain phase, where timescales for lake filling and draining are of the order of months to years (Smith et al. 2009) (Table 6.1).

Lakes also display contrasting hydrological behaviours, ranging from lakes that are hydrologically closed systems to highly dynamic active lakes which regularly exchange water with the surrounding basal environment via lake drainage events. The dynamism of subglacial lakes certainly seems to increase towards the ice sheet margins (Smith et al. 2009), although this may partly reflect biases in the measurement technique. This has led to a view that the interior lakes (e.g. Lake Vostok) may be more hydrologically stable compared with the 'cascading' lakes located around the ice sheet margins, which drive water export towards the grounding line (Bell 2008) (Figure 6.7). Because of the sheer size of East Antarctica and the diverse range of physical environments beneath the ice sheet, the East Antarctic Ice Sheet includes the greater number of subglacial lakes and includes members from each lake environment type. The arising differences in conditions and hydrological connectivity are likely to have an important bearing upon the local physical and geochemical conditions of the lake habitat. It is certainly true to say that many lakes are part of a connected subglacial hydrological system supplied by melting ice and episodic discharge of lake water.

The connectivity of subglacial lakes is likely to be important for life since it provides a means of flushing the habitat and removing reaction products that may have reached saturation, and of supplying new nutrients and organic carbon to sustain microbial life. The discharge of meltwaters from lakes during drainage events can be facilitated by very large subglacial flood events, where the water fluxes may reach some 100 m^3 s^{-1}. This is of a similar order of magnitude to run-off rates observed during Jökulhlaups (Walder and Fowler 1994), and also in rivers draining Greenland Ice Sheet catchments during the peak melt season (Bartholomew et al. 2011). Recent work shows a very large water discharge from Lake Cook, where the volume of water lost during drainage was 2.7 km^3, resulting in almost 50 m of ice surface lowering (McMillan et al. 2013). There is mounting evidence that subglacial meltwater is exported at the ice sheet margin. Land-based and sub-marine geomorphological data display paleo-channels and other hydrological features indicative of the occurrence of outburst floods (Sugden et al. 2006; Lowe and Anderson 2003); also supported by direct observations of flood events at the ice margin (Goodwin 1988). Traces of channels incised into the base of a large number of ice shelves around

164 ANTARCTIC LAKES

(a) Pre-LGM subglacial Hodgson Lake occupies a cavity under the overriding ice sheet.

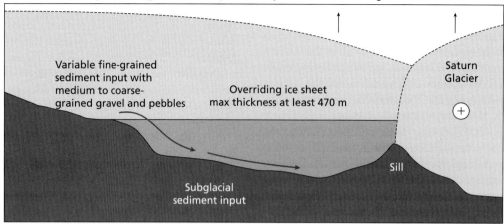

(b) LGM-Early Holocene onset of deglaciation in the catchment.

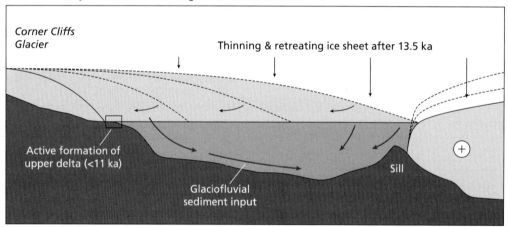

(c) Mid-late Holocene: Lake level falls, elevated deltas become isolated, glaciers reach present configuration.

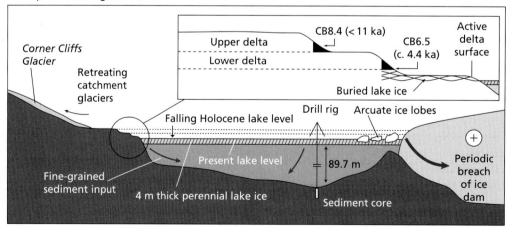

Figure 6.6 Conceptual model of the late Pleistocene history and development of Hodgson Lake (Hodgson et al. 2009a). With the permission of Elsevier.

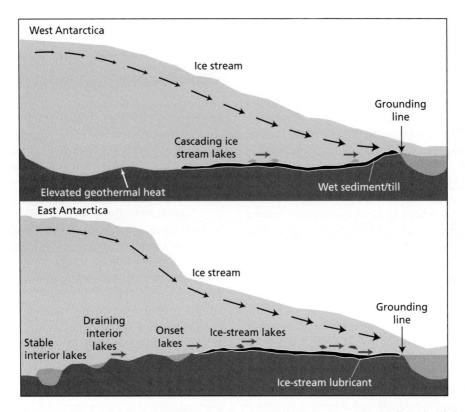

Figure 6.7 Conceptual model of the distribution of liquid water beneath the West and East Antarctic Ice Sheets (Bell 2008). With the permission of Nature Publishing Group.

the continent have also recently been observed in satellite images of the ice shelf surface and suggest meltwater export by large subglacial channels (Le Brocq et al. 2013). The presence of sand and clay deposits in ice-marginal core records also indicates that sediment-laden meltwater plumes originate beneath the ice sheet (Lowe and Anderson 2002). Submarine groundwater discharge has also recently been measured from the East Antarctic coast, at rates that were two orders of magnitude higher than rates reported for mid-latitude sites (Uemura et al. 2011). The total annual amount of water produced beneath the ice sheet is estimated at 65 km^3 (Pattyn 2010), which is an order of magnitude lower than run-off estimates from the Greenland Ice Sheet (Bamber et al. 2012). However, the prolonged residence times for subglacial meltwater beneath the Antarctic Ice Sheet, due to slow water flow rates and long flow paths, is likely to result in much higher concentrations of solute being acquired by meltwaters (Wadham et al. 2010b; Wadham et al. 2013). The solute export from sub-Antarctic meltwaters is estimated to be of a similar order of magnitude to some of the planet's largest rivers (Wadham et al. 2010b).

Some lakes exhibit zones of net freezing of lake water onto the underside of the ice sheet, forming refrozen lake ice known as 'accretion ice'. The distribution of melting and freezing is determined by the surface slope of the ice sheet, which affects the ice thickness above a lake and hence, the degree to which the freezing point of water is depressed below 0 °C in different areas. Typically, where the ice is thickest, melting conditions are more likely to occur, and where it is thinnest zones of freezing may prevail. This is particularly true of Lake Vostok, where melting occurs in the north of the lake where ice is thicker, and freezing occurs in the south of the lake (Siegert et al. 2000). This, together with the associated spatial distribution of heat, and hence water density in the lake, is thought to drive a weak but continuous water circulation which may be important for redistribution of solute and biota. It is also important for understanding the chemical records held within the accretion ice. The accretion ice of Lake Vostok represents the earliest data set available on subglacial lake chemistry and biology (Priscu et al. 1999a; Karl et al. 1999; Jouzel et al. 1999), but its interpretation is challenging because the lake water chemical and biological composition is altered during freezing. A more detailed discussion on accretion ice can be found in Section 6.4.

A strong motivation for the search for life in subglacial lakes relates to the hypothesized isolated nature of these sub-surface ecosystems, which may have a strong influence on the evolution of lake geochemical conditions and also of lake biota. The age of lakes beneath the ice sheet is unknown and is probably highly variable. Certainly, interior lakes which occupy subglacial troughs or rift valleys are thought to have existed for a similar length of time as the ice sheet itself, which in East Antarctica is of the order of at least 20 million years (Huybers and Denton 2008). Clearly, subglacial lakes located in more marginal locations and in West Antarctica, may have a much shorter lifetime due to the dynamics of ice sheet growth and decay on millennial timescales (Pollard and DeConto 2009) (see Section 1.4, Chapter 1). Although some lakes may be very old, the concept that many form part of an interconnected drainage system beneath the ice sheet suggests that microbial life within the lakes technically has not been isolated over such long timescales.

6.5 Geochemical conditions in subglacial lakes

Very little is known about the geochemistry of meltwaters in Antarctic subglacial lakes, but geochemical conditions are critical in influencing what levels of nutrients and energy sources might be available to sustain microbial life (Section 6.6), and therefore which types of biota might be supported. As for any subglacial aquatic environment, meltwater chemistry is affected by features of the hydrological environment, including water residence times, rock:water ratios, bedrock mineralogy, hydrological connectivity, and the prevalence of melting and freezing processes within a lake. These are likely to vary widely between different Antarctic lakes, giving rise to heterogeneity in lake chemistry across Antarctica.

The chemical composition of meltwaters within subglacial lakes derives from a balance between solute inputs and outputs. Solute inputs to lakes are sourced from the melting of the basal layers of the ice sheet, in addition to inputs from the chemical dissolution of minerals within sediments in and around the lake. It is probable that vertical gradients in solute composition will be apparent in lakes, as sediments at the lake bed geochemically weather and add solute and dilute ice melts into surface lake waters. A summary of primary geochemical processes identified within subglacial environments worldwide and within Antarctic subglacial lakes can be found in Table 6.2. Losses of solute from lakes may occur via biological activity, particularly of nutrients, and via precipitation of minerals such as calcium carbonate, once the saturation index is attained. In addition, biological activity and redox processes may result in the transformation of solute from one form to another. In hydrologically active lakes, additional solute inputs and outputs may occur via the gain or loss of water from the lake during fill and drain periods. Where there is melting and re-freezing within different sectors of the lake, solute may become concentrated in lake waters due to solute rejection from water as it freezes to form ice (Siegert et al. 2003). This is particularly likely to be the case in lakes which do not appear hydrologically connected, where continual melting and re-freezing processes may serve to increase solute concentrations in the lake over time.

In common with many surface freshwater Antarctic lakes (Chapter 2), subglacial lakes are thought to display ultra-oligotrophic conditions, being depleted in nutrients required for life. Concentrations of solute and nutrient levels present in Lake Vostok (Karl et al. 1999) and Hodgson Lake (Hodgson et al. 2009b) are low, and similar to concentrations reported from global subglacial waters (Table 6.3). Dissolved organic carbon concentrations, inferred from Lake Vostok accretion ice, are also very low and of the order of 100 µM. Since these are environments that are essentially out of contact with the atmosphere, it is also likely that

Table 6.2 Summary of the dominant geochemical weathering processes found in subglacial environments and evidence for their occurrence in Antarctic subglacial lakes.

Reaction	Observations worldwide	Subglacial lakes
Proton supply mechanisms		
1. Carbonation $CO_2 + H_2O \leftrightarrow HCO_3^- + H^+$	Widespread	Likely
2. Microbial Respiration $C_{org} + O_2 \leftrightarrow CO_2$	Present on glaciers overriding organic-rich sediments[1,2]	Inferred in Lake Vostok[3,4] and Hodgson Lake[5]
3. Sulphide oxidation (oxic) $FeS_2 + 3.5O_2 + H_2O \leftrightarrow 2SO_4^{2-} + Fe^{2+} + 2H^+$	Widespread	Inferred in Lake Vostok[6] and Hodgson Lake[5] from the presence of sulphate
4. Sulphide oxidation (anoxic) $FeS_2 + 14Fe^{3+} + 8H_2O \leftrightarrow 2SO_4^{2-} + 15Fe^{2+} + 16H^+$	Widely identified[7,8,9]	No direct evidence
Rock chemical dissolution reactions		
5. Carbonate hydrolysis $CaCO_3 + H_2O \leftrightarrow Ca^{2+} + HCO_3^- + OH^-$	Widespread[10]	Carbonate hydrolysis/dissolution inferred from calcium in Lake Vostok accretion ice[4,6]
6. Carbonate dissolution (fuelled by CO_2) $CaCO_3 + CO_2 + H_2O \leftrightarrow Ca^{2+} + 2HCO_3^-$	As for 5 above	As for 5 above
7. Silicate dissolution (fuelled by CO_2) Anorthite (calcium feldspar) $CaAl_2Si_2O_8 + 2CO_2 + 2H_2O \leftrightarrow Ca^{2+} + 2HCO_3^- + H_2Al_2Si_2O_8$	Widespread, enhanced in long residence time subglacial waters[2]	Silicate dissolution inferred from the presence of crustal sodium in Lake Vostok accretion ice[4,6]
8. Carbonate dissolution fuelled by sulphide oxidation $4FeS_2 + 16CaCO_3 + 15O_2 + 14H_2O \leftrightarrow 16Ca^{2+} + 16HCO_3^- + 8SO_4^{2-} + 4Fe(OH)_3$	As for 4 and 5	As for 4 and 5
9. Silicate dissolution fuelled by sulphide oxidation Alkali feldspar $4FeS_2 + 16Na_{1-x}K_xAlSi_3O_8 + 15O_2 + 86H_2O \leftrightarrow 16(1-x)Na^+ + 16xK^+ + 8SO_4^{2-} + 4Al_4Si_4O_{10}(OH)_8 + 32H_4SiO_4 + 4Fe(OH)_3$	Widespread, enhanced in long residence time subglacial waters[2]	Silicate dissolution inferred from the presence of crustal sodium in Lake Vostok accretion ice[4,6]
Redox reactions		
10. Denitrification $1.25CH_2O + NO_3^- + H^+ \rightarrow 0.5N_2 + 1.25CO_2 + 1.75H_2O$	Inferred from depletion of nitrate in valley glacier run-off[11]	None identified
11. Dissimilatory iron reduction $CH_3COO^- + 8Fe^{3+} + 3H_2O \rightarrow 8Fe^{2+} + HCO_3^- + CO_2 + 8H^+$	Iron reduction coupled to sulphur cycling reported in Blood Falls, Antarctica[12]	None identified
12. Sulphate reduction $SO_4^{2-} + 2CH_2O \rightarrow H_2S + 2HCO_3^-$	Isotopic evidence from a large polythermal glacier[8].	None identified to date
13. Methanogenesis $CH_3COOH \rightarrow CH_4 + CO_2$ $CO_2 + 4H_2 \rightarrow CH_4 + 2H_2O$	Experimental evidence from subglacial sediments[13,14]	Likely[14], particularly in lake sediments[6]. No methane clathrates in Lake Vostok accretion ice. Methane found in Lake Whillans[15]

1: Wadham et al. (2010a); 2: Wadham et al. (2010b); 3: Karl et al. (1999); 4: Christner et al. (2006); 5: Hodgson et al. (2009b); 6: Siegert et al. (2003); 7: Bottrell and Tranter (2002); 8: Wadham et al. (2004); 9: Wynn et al. (2006); 10: Anderson (2007); 11: Hodson et al. (2005); 12: Mikucki et al. (2009); 13: Boyd et al. (2010); 14: Wadham et al. (2012);15: Priscu et al. (2013b).

Table 6.3 Comparison of Subglacial Lake Vostok predicted chemistry with that from other lakes and glaciers worldwide.

	DOC (µM)	Bacteria (cells mL^{-1})	Ion Concentration (µeq L^{-1})							Σmajor ions (mM)	notes
			Na$^+$	K$^+$	Mg^{2+}	Ca^{2+}	Cl$^-$	SO$_4^{2-}$	HCO$_3^-$		
Average glacial ice	16	120	2.4	0.40	0.72	2.2	2.8	3.6	–	0.012	From (De Angelis et al., 2004), meteoric ice, Vostok Ice Core
Accretion Ice – Type 1	65	260	22	0.32	12	14	17	18	–	0.083	From (Christner et al. 2006)
Accretion Ice – Type 2	35	83	0.92	0.13	0.3	2.0	0.94	0.3	–	0.005	From (Christner et al. 2006)
PREDICTED LAKE WATER CHEMISTRY											
Embayment Water (Christner et al. 2006)	160	460	10000	14	5400	5200	7300	22000	–	49.9	Predicted from average Type 1 and 2 accretion ice respectively, Partitioning coefficients from McMurdo Dry Valleys (Lake Bonney) (Christner et al. 2006)
Main Lake Water (Christner et al. 2006)	86	150	430	5.9	138	740	400	360	–	2.07	
Embayment waters (Priscu et al. 1999)	–	–	200	–	275	115	54	1150	n/a	1.79	(Priscu et al. 1999) Contains a surplus of negatively charged ions, partition coefficients from perennial lake ice, Lake Hoare, McMurdo Dry Valleys
Embayment waters (Siegert et al. 2003)	–	–	700	–	350	270	461	444	289	2.51	From filtered type 1 sample, partition coefficients (0.0018) after (Souchez et al. 2000)
SUBGLACIAL WATERS WORLDWIDE											
Hodgson Lake, Antarctica	47.2	–	440	14	92	67	496	66	53	1.23	Mean of 3 lake samples (Hodgson et al. 2009b)
Casey Station outburst	–	–	6600	200	200	1000	2200	600	3500	14.3	(Goodwin 1988)
Kamb Ice Stream	–	–	20000	600	6200	12800	1100	34000	4100	78.8	(Skidmore et al. 2010)
Bindschadler Ice Stream	–	–	35000	700	17200	18000	2000	62000	7500	142	(Skidmore et al. 2010)
Typical glacial runoff	–	–	100	20	200	400	100	200	400	1.42	(Skidmore et al. 2010)

sub-oxic or anoxic conditions will develop. Certainly, the recent sampling of Lake Whillans has indicated an oxygen-depleted environment (Priscu et al. 2013b). Anoxia is likely to be most strongly developed within lake floor sediments, where microbial respiratory processes and oxygen-consuming geochemical weathering reactions such as sulphide oxidation (Table 6.2) are likely to remove dissolved oxygen from porewaters faster than they can be replenished by diffusion (Siegert et al. 2003). Similar enhancements in anoxia have been reported beneath valley glaciers (Bottrell and Tranter 2002; Wadham et al. 2004), where anoxic processes such as denitrification, iron reduction, sulphate reduction, and ultimately methanogenesis have been reported (Table 6.2). In contrast, lake surface waters may exhibit higher dissolved oxygen concentrations as small amounts of oxygen are supplied from the release of air bubbles during the melting of basal ice layers (Siegert et al. 2003). These processes, which add and consume oxygen, may drive the development of chemical gradients within lakes, enabling a diverse range of anaerobic and aerobic microbial communities to evolve. The gas content of lake waters is poorly constrained, although it may aid with identifying geochemical processes such as sulphate reduction and methanogenesis, which produce gas. Trace quantities of clathrate (crystal lattices formed by water molecules which form a cage around gas molecules under conditions of low temperatures and high pressures) (Lee and Holder 2000) have been found in Lake Vostok accretion ice (Siegert et al. 2001). The clathrate present in Lake Vostok is thought to mainly comprise nitrogen and oxygen arising from the progressive expulsion of air from accretion ice during freezing, and its concentration in the lake assuming a closed system (McKay et al. 2003). Less gas build-up is likely within lakes that act as open hydrological systems (Siegert et al. 2012).

Most of what is known regarding biogeochemical processes within subglacial lakes arises from the analysis of the accretion ice of Lake Vostok and theoretical work, supplemented more recently with direct samples recovered from former Hodgson Lake (Hodgson et al. 2009b) and Subglacial Lake Whillans (Priscu et al, 2013). Subglacial Lake Vostok is the most intensively studied subglacial lake within Antarctica, and the chemical composition of the accretion ice (see Section 6.3.1, Figure 6.3) has been used as a proxy for the lake water chemistry. This is based upon the premise that concentrations of solute species and particles within the accretion ice reflect those in lake water prior to freezing, where differences in the concentration are proportional to the partition coefficients which apply during the freezing of water (Priscu et al. 1999a; Christner et al. 2006; Siegert et al. 2001). Two types of accretion ice have been reported, with type 2 (3610–3623 m) located closer to the lake surface and type 1 above it (3539–3609 m). The vertical variations in accretion ice chemistry reflect horizontal gradients in lake chemistry, where type 1 accretion ice is thought to form from freezing of embayment waters and type 2 via freezing of the main lake waters (De Angelis et al. 2004) (Figure 6.3). The composition of meltwaters entering Lake Vostok via basal melting may be inferred from the chemical composition of the meteoric component of the Vostok ice core (Legrand et al. 1988), which displays very low concentrations of solutes (Table 6.3). Predictions of the chemical composition of lake waters based on the chemistry of the accretion ice have led to vastly different estimates (Priscu et al. 1999a; Souchez et al. 2000; Siegert et al. 2003). This arises from the uncertainty in the selection of partition coefficients, and the use of different accretion ice samples or processing methods. Use of unfiltered samples, which display amplified solute concentrations due to chemical weathering of particulate material, produces maximum estimates of lake solute concentrations (Souchez et al. 2000). Conversely, by assuming that pockets of lake water may become trapped within the freezing ice crystal matrix during freezing and resulting in more concentrated accretion ice, gives rise to minimum estimates of lake water solute concentrations (Siegert et al. 2003). Table 6.3 compares predictions of lake water chemistry based upon filtered accretion ice samples and standard partition coefficients. The envelope in predicted lake solute concentrations still spans several orders of magnitude. The single study that has predicted solute concentrations for both embayment and main lake waters shows substantial differences between the two, with embayment waters being more concentrated and similar to sub-ice stream porewater concentrations (Christner et al. 2006). All other estimates of lake water chemistry bear resemblance to the total solute content typical of glacial run-off worldwide, and are of the order of a few mM (Skidmore et al. 2010). Samples from Hodgson Lake show low concentrations of solute. These are similar to bulk glacial run-off worldwide, indicating a subglacial environment where much lower rock:water interaction has taken place (Hodgson et al. 2009b).

The detailed composition of the accretion ice can also be employed to infer geochemical weathering reactions within the lake (Siegert et al. 2003; Christner et al. 2006), which may subsequently be compared with the chemistry of subglacial meltwaters around the globe (Wadham et al. 2010b). For example, a reaction that is typical of long residence time meltwaters beneath glaciers is coupled sulphide oxidation/carbonate dissolution (Tranter et al. 2002) (Table 6.2). Sulphides are continually regenerated by crushing of bedrock and are rapidly oxidized in the presence of dissolved oxygen or microbially under anaerobic conditions using Fe(III) (Bottrell and Tranter 2002; Wadham et al. 2004). The protons generated by this oxidation reaction can be used to dissolve rock minerals, most notably carbonate and silicate minerals. Once carbonate minerals in rock flour become exhausted or waters attain saturation with respect to carbonate minerals, silicate minerals may be preferentially dissolved by the protons generated during sulphide oxidation (Wadham et al. 2010b). The complete coupling of carbonate dissolution and sulphide oxidation is indicated by a gradient of 1 for a line of best fit when HCO_3^- is plotted against SO_4^{2-} for glacial meltwaters (Reaction 8, Table 6.2). This arises from the stoichiometry of the reaction, which generates 16 molar equivalents of SO_4^{2-} and the same of HCO_3^- (Table 6.2). A gradient of <1 indicates that a proportion of the protons generated by sulphide oxidation are being used to dissolved silicate minerals. A consistent pattern should be displayed in plots of (Ca^{2+} + Mg^{2+}) versus SO_4^{2-}, where coupling of sulphide oxidation and carbonate dissolution yields a gradient of 2 based upon the stoichiometry of Reaction 8 (Table 6.2). When the gradient of the line of best fit for the associations of HCO_3^- versus SO_4^{2-} is plotted against that for (Ca^{2+} + Mg^{2+}) versus SO_4^{2-} (Wadham et al. 2010b), the Lake Vostok data (De Angelis et al. 2004) displays very low gradients and plots in a similar space to other large ice sheet catchments (Figure 6.8). This suggests that enhanced silicate dissolution is apparent within Lake Vostok, indicative of prolonged residence times. In support of this are the low ratios of divalent cations (Ca^{2+} + Mg^{2+}), sourced from carbonate dissolution, relative to monovalent cations (Na^+ + K^+), sourced from silicate dissolution (Figure 6.8). Enhanced silicate dissolution may arise from either attainment of carbonate saturation in lake waters or exhaustion of carbonate minerals. The concordance of Lake Vostok ion ratios with those from other long residence time ice sheet systems suggests that similar processes are likely to prevail in subglacial lakes where prolonged rock water contact is apparent.

6.6 The biota of subglacial lakes

As previously indicated, many subglacial lakes are very old and have been isolated from the atmosphere for millions of years. Thus they may have a biota that includes new and novel species. Consequently, sampling of subglacial lakes requires stringent aseptic techniques, as outlined in Section 6.1. The recent successful entry of Lake Whillans using sterilized hot water drilling technology exemplifies this approach (Priscu et al. 2013a). Similar protocols were applied in the unsuccessful attempt to enter Lake Ellsworth (Siegert et al. 2006). This approach was developed to ensure that samples taken from the two lakes were not contaminated with microorganisms from the surface. There has been much debate about the drilling and eventual sampling of Lake Vostok, which used thermo/mechanical drilling procedures (Table 6.1) that did not have well developed cleanliness and sterility protocols.

At the time of writing, our knowledge of life in subglacial lakes is limited. While the Russians entered Lake Vostok in early 2013 and there are media reports of bacteria in the waters and dispute as to whether they are endemic or contaminants, there are as yet no reports in the scientific literature. At present we have data on accretion ice from Lake Vostok (see Section 6.4), which is formed from the lake waters and contains organisms from that water (Priscu et al. 2008), and samples taken from Lake

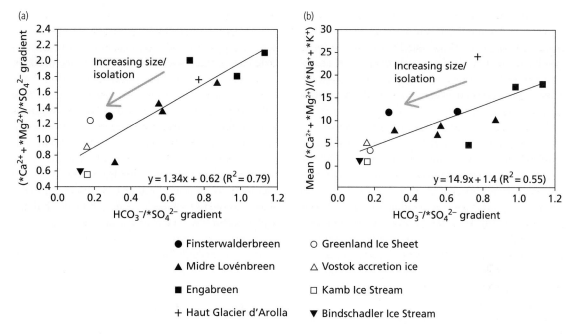

Figure 6.8 (a) associations between the gradients of the line of best fit in scatter plots of crustal ($Ca^{2+} + Mg^{2+}$) versus SO_4^{2-} and HCO_3^- versus SO_4^{2-} for global subglacial meltwaters and (b) association between the mean crustal ratio of ($Ca^{2+} + Mg^{2+}$: $Na^+ + K^+$) in global subglacial meltwaters and the gradients of the line of best fit in scatterplots of HCO_3^- versus SO_4^{2-}. (Wadham et al. 2010b). With the permission of the American Geophysical Union.

Whillans (Christner et al. 2014 submitted). As indicated in Section 6.3.3, Lake Whillans is dynamic and flushes regularly, whereas Lake Vostok is an ancient stable lake. The two systems are very different and this is likely to be reflected in their microbial communities.

Major ions and organic carbon all decrease with depth in Lake Vostok accretion ice. Depth in the ice represents a proxy for increasing distance from the shoreline, suggesting that there may be greater potential for biological activity in the shallow shore waters (Christner et al. 2006). In Table 6.3, accretion ice type 1 derives from a shallow embayment in the south-western area of the lake, while accretion ice type 2 is ice formed over deep water. The organic carbon originates from the lake sediments or the overlying ice sheet (Christner et al. 2006). Incubations of melted accretion ice with ^{14}C-labelled acetate and glucose resulted in the release of ^{14}C-CO_2, indicative of metabolically active cells, but at low levels of activity (Karl et al. 1999). Highest concentrations of bacterial cells in melted accretion ice reached 380 ± 53 cells ml^{-1}. This is two to seven times higher than in the overlying glacial ice (Christner et al. 2006). The bacterial cells exhibited a spectrum of cells sizes with small coccoid cells (0.1–0.4 µm) representing about half of the assemblage, while the remainder was made up of rods and vibrios (Karl et al. 1999). Molecular analysis revealed phylotypes that belonged to the Alpha-Proteobacteria, Beta-Proteobacteria, and Gamma-Proteobacteria, the Firmicutes, the Actinobacteria, and Bacteriodites lineages (Priscu et al. 1999a; Christner et al. 2001; Bulat et al. 2004; Christner et al. 2006). The closest relatives of some of the phylotypes provide clues to the functional dynamics of the lake plankton community. Among the Beta-Proteobacteria, clone sequences were most closely related (6%–97%) to aerobic methylotrophic species in the genera *Methylobacillus* and *Methylophilus*. These bacteria use C-1 compounds as a substrate (e.g. methanol, formate, carbon monoxide) and carbon assimilation is via a ribose monophosphate pathway, suggesting possible niches for methylotrophy in Lake Vostok

(Christner et al. 2006). The thermophilic facultative chemolithotrophic bacterium *Hydrogenophilus thermoluteolus*, a member of the Beta-Proteobacteria, has been identified from accretion ice. It is an inhabitant of hot springs and may use hydrogen for energy and carbon dioxide as a carbon source (Bulat et al. 2004).

Two ecological scenarios have been proposed for Lake Vostok (Christner et al. 2006). Firstly, organic carbon derived from the sediments and/or the overlying ice sheet supports an exclusively heterotrophic community. Bacterial cell concentration in the near-surface water is calculated to lie between 140 and 770 cells ml^{-1}. Based on the calculated rate of organic carbon input, the predicted cell numbers, a cell carbon content of 10 fg cell^{-1}, and assuming no abiotic sinks for organic carbon within the lake, a positive organic carbon flux would provide a heterotrophic bacterial community with 0.49–3.8 × 10^{-4} organic C (g cell C)$^{-1}$ h^{-1}. Even if all the organic carbon proves a suitable substrate, this would not meet the theoretical carbon demand of the community. It would be sufficient to support maintenance, but not growth. However, Christner et al. (2006) indicate that their calculations are conservative. What can be said with certainty is that this lake is likely to be extremely unproductive, even more so than the large surface freshwater lakes of Antarctica (see Chapter 2).

The second scenario is one of a supplemental microbial food-web-based chemolithotrophic primary production. It is suggested based on the presence of thermophilic bacteria and other evidence that a geothermal system may exist beneath Lake Vostok's waters. Thus a thermophilic chemolithotroph may exist in faults beneath the lake (Bulat et al. 2004). While this possibility has been considered by Christner et al. (2006), they suggest that a chemolithotrophic-based ecosystem is conceivable without geothermal activity. A range of reduced compounds may be present to support biogeochemical reactions. Oxidants (oxygen and nitrate) would be supplied by the ice sheet and by chemical weathering of bedrock and sediment (e.g. SO_4^{2-} from sulphide oxidation, see Table 6.2). Basal ice continually melts onto the lake, providing glacial debris containing sulphide and iron minerals as well as organic material. This input provides material for biological processes mediated by microorganisms. Such microbially mediated chemical weathering interactions have been described in hypoxic and anoxic glacial beds (see Table 6.2) (Tranter et al. 2002).

At the time of writing, the first data from Lake Whillans have been submitted (Christner et al. 2014 submitted). The data suggest that the lake supports a metabolically active, phylogenetically diverse microbial ecosystem that functions in the dark at sub-zero temperatures (-0.49 °C). Initial analysis of Lake Whillans water indicates a pH from 8.0 to 8.2 and a conductivity at around 720 μS cm^{-1} making it freshwater. δ^{18}O-H$_2$O values of –38‰ indicate that the water within the lake is of glacial origin. Inorganic nutrient concentrations are low, with ammonium (NH$_4^+$) at around 2.4 ± 0.6 μmol l^{-1}, orthophosphate (PO$_4^{3-}$) 3.1 ± 0.7 μmol l^{-1}, and nitrate (NO$_3^-$) 0.8 ± 0.5 μmol l^{-1}. Isotopic data point to microbially driven processes for nitrate production (Priscu et al. 2013b). Dissolved organic carbon concentration was 221 ± 55 μmol L^{-1}, which is similar to the Dry Valleys lakes and twice that inferred from the accretion ice for Lake Vostok water (Christner et al. 2006; Christner et al 2014 submitted) (Table 6.3).

Epifluorescent microscopy revealed a high morphological diversity among the bacterial community, as was the case for Lake Vostok accretion ice (see above). Cell concentrations in the water column showed a maximum of 1.3 × 10^5 cells ml^{-1} (Christner et al. 2014 submitted). These concentrations are many orders of magnitude higher than in Lake Vostok, based on an analysis of accretion ice. The uptake of tritiated thymidine and ^{14}C leucine indicated maximum rates of bacterial production between 2.9 ng C L^{-1} day^{-1} to 13.7 ng C L^{-1} day^{-1} respectively, an order of magnitude lower than rates seen in the waters of the Dry Valleys lakes (see Tables 2.2 and 3.2). Bacterial carbon demand was calculated at between 105 ng C L^{-1} day^{-1} to 23 ng C l^{-1} day^{-1}. Organic carbon production via chemoautotrophy achieved 32.9 ng C L^{-1} day^{-1} indicating that autotrophic production is able to meet a least part of the heterotrophic carbon demand.

SSU rRNA sequence analysis amplified from water column samples revealed OTUs that classified into 32 bacterial phyla and 2 archaeal phyla. Among the Bacteria 53% of the sequences affiliated with the Proteobacteria, mostly the beta and delta classes and 11% within the Actinobacteria. Most of the Archaea were classified as Thaumarchaeota with one OUT representing the fifth most abundant phylotype. Many of the abundant phylotypes among the Bacteria were closely related to chemolithoautotrophic species that use reduced nitrogen, iron or sulphur compounds as sources of energy.

The sediments of Lake Whillans are defined as a till, which is a weak and easily deformed sediment because of its high water content. The sediment column shifted from oxidized at the surface to reduced at depth. Methane was present with a flux from 40 cm to the surface, and was probably derived from beneath the ice sheet . Bacterial production measured in the sediments was lower than in the water column. Based on the incorporation rates of thymidine and leucine the production rates were 46.4 ng C day^{-1} g dry wt^{-1} and 0.9 ng C day^{-1} ng dry wt^{-1} respectively. As in the water column the majority of the OTUs from the surficial sediment classified within the Proteobacteria (mostly beta and gamma classes). The most abundant phylotypes were related to chemolithoautotrophs or species that exploit C-1 hydrocarbons as carbon and energy sources.

The surface sediment of former subglacial Hodgson Lake and successive layers down to 300 cm supported concentrations of bacteria up to 4.4×10^7 cells g^{-1}. Culture-independent molecular analysis of the Hodgson Lake sediment revealed a preponderance of Actinobacteria, Proteobacteria, and Planctomycetes, with lower percentages of Chloroflexi, Spirochaetes, Firmicutes, Chlamydiae, Nitrospirae, and Bacteroidetes (Pearce et al. 2013).

Within the next few years more information from Lake Whillans and Lake Vostok data will be published, as further sampling and analysis is completed, and will eventually be augmented by data from Lake Ellsworth and other subglacial lakes. We will then gain a more comparative picture of biodiversity, and importantly biological functioning, in different types of subglacial lakes under the Antarctic ice sheet.

CHAPTER 7

Conclusions and future directions

7.1 Antarctic lakes in a global context

As previous chapters have illustrated, Antarctica has a wide range of lake types. While freshwater and saline lakes are found worldwide, some lake types are almost unique to Antarctica. Epishelf lakes are largely confined to Antarctica; their Arctic equivalents are disappearing rapidly due to climate warming and ice shelf collapse (Veillette et al. 2008). Ice shelf lakes are restricted to Antarctica and the Arctic, while cryolakes are restricted to the polar regions and glaciers at lower latitudes. Subglacial lakes are widely distributed under the Antarctic continental ice sheet (Figure 1.11) (Wright and Siegert 2012) and are only now beginning to reveal their secrets following the entry into Lake Whillans and Lake Vostok in 2012/2013.

The majority of freshwater and saline lakes in Antarctica lie at the oligotrophic end of the trophic spectrum. The lakes differ from lower latitudes in having a plankton dominated by microorganisms (bacteria, algae, and protozoa) with sparse metazoan plankton. Among the eukaryotic microbial community, species diversity is usually low in comparison with temperate and tropical systems. As yet we cannot assess the comparative diversity of the Bacteria and Archaea, however, there is some indication of endemism. The benthic regions are frequently clothed with extensive and often spectacular cyanobacterial mats, and in some cases with moss beds. The benthic heterotrophic communities have low diversity and lack the complex arthropod, annelide, molluscan, and microfaunal constituents seen at lower latitudes (Rautio et al. 2008). Continuous low temperatures and low annual PAR impose considerable physiological constraints. Many of the organisms that have colonized Antarctic lakes have evolved adaptations that enable them to thrive in an extreme environment.

Rates of photosynthesis and bacterial production outlined in previous chapters are within the ranges reported for lower latitude oligotrophic lakes. While the phenomenon of mixotrophy occurs in lower latitude lakes, it is fully exploited by some groups of phytoflagellate in Antarctica, allowing growth under light-limited conditions and in winter when PAR is absent. Populations of cryptophytes and *Pyraminmonas* sp. maintain active winter populations, feeding on bacteria and DOC enabling them to 'hit the deck running' in the short austral summer. Unlike many (though not all) of their relatives in temperate lakes, planktonic crustaceans such as *Daphniopsis studeri* and *Paralabidocera antarctica* remain active throughout the winter in continental lakes, and do not appear to overwinter in resting stages or resting eggs. There would be insufficient time for a population to emerge from resting eggs, undergo their development, and reproduce during the short summer. They have low fecundity, but they have no predators so they maintain their sparse populations. Fish have not been recorded in Antarctic lakes. The only exception is the species *Tematomus*, a marine benthic fish that has been found in the underlying marine waters of a few epishelf lakes.

As one would expect, Antarctic lakes have most in common with their Arctic counterparts. Most Arctic lakes are freshwater. Lakes in the low Arctic are typified by a well-developed zooplankton and fish communities, for example Toolik Lake in Alaska, which is the site of an NSF Long Term Ecosystem Research Program, while high Arctic lakes have greatly reduced community complexity and are more comparable with Antarctic lakes. Low

Antarctic Lakes. Johanna Laybourn-Parry and Jemma L. Wadham.
© Johanna Laybourn-Parry and Jemma L. Wadham 2014. Published 2014 by Oxford University Press.

Arctic lakes have well-vegetated catchments and receive significant allochthonous carbon and nutrient inputs, while high Arctic polar desert lakes are similar to Antarctic lakes in being dependent on autochthonous carbon (Vincent et al. 2008).

In summary, when set in a global context Antarctic lakes are typified by truncated food webs, low species diversity, and low productivity.

7.2 Inter-annual variations and longer-term trends

Marked inter-annual variation is not a normal feature of lakes in the temperate latitudes. There are a number of lakes in these regions that have been subject to long-term study, for example Lake Windermere, Lake Constance, and Lake Mendota. These lakes have undergone progressive eutrophication or enrichment over decades, but the changes in their biota and productivity have been gradual. Generally variations in the abundance and species composition of the phytoplankton seen in a given lake over an annual cycle are often broadly repeated year after year (Reynolds 1984). A good example is England's largest natural lake, Lake Windermere in the English Lake District. The lake has been undergoing eutrophication or nutrient enrichment, as a result of human influence, since the 1800s, a situation made worse in 1850 with the coming of the railway to Windermere, opening it up to tourism. The process of eutrophication continued until the 1970s and 1980s. A major problem was the increase in phosphorus loading, which was ameliorated in 1992 with the advent of phosphorus stripping of effluent at the local wastewater treatment plant. Over this timescale there was a gradual increase in phytoplankton biomass, with acceleration after 1920, and substantial increases in cyanobacteria in the deep north basin. In the shallower south basin, production increased but there were abrupt shifts in community composition where filamentous Cyanobacteria, or siliceous algae, chlorophytes, and cryptophytes assumed greater importance (Elliott 2012; McGowan et al. 2012). These changes are largely the result of human impact. However, nutrient enrichment enhanced the lakes' response to climate change, although disentangling the effects of the two factors is very difficult.

While we do not have such long-term data sets for Antarctic systems, where there are data sets extending back over several decades, for example for the Vestfold Hills, Signy Island, and the Taylor Valley lakes, there is clear evidence of considerable inter-annual variation in production and/or species composition. Antarctic systems are not subject to direct human impact, so changes are likely to be driven by both local and large-scale climatic impacts. Where freshwater lakes are subject to human impact because they are used as a water supply for a research station, the impacts are seen very quickly (see Section 2.1).

Episodic warming events occur in Antarctica and can have a profound impact on primary production. One such event occurred in the Dry Valleys in the summer of 2001–2002, increasing glacial runoff, causing record stream discharges, increased lake water levels, and thinning of the perennial ice covers. Long-term databases, like those for the Dry Valleys, have allowed a detailed assessment of the impact of these hydrological events. Integrated primary production decreased by 23% in Lake Bonney West, correlating with a 95% decrease in PAR at 10 m as a result of stream-induced turbidity. In contrast, integrated primary production in Lake Fryxell was greater by fivefold than in previous years, when stream flow was relatively low. The increased production was mainly attributed to increased loading of nutrients from the stream inflow. Such events have impacts in the following years. In the summer of 2002–2003, increased primary production was observed in both lobes of Lake Bonney and directly under the ice of Lake Fryxell, reaching the highest recorded value for ten years. In Lake Fryxell, highest production was recorded early in the summer and then decreased, contrary to the usual pattern. It is likely that advected nutrients from the flood stimulated an early peak in primary production (Foreman et al. 2004).

Changes are also apparent when there is no major event to which they can be attributed. For example, a study of Lake Fryxell phytoplankton over the years between 1987 and 1991 revealed inter-annual differences in species diversity, cell abundance, and biomass. In the summers of 1987 and 1988, *Chroomonas lacustris* dominated the community, while between 1989 and 1991 *Cryptomonas* sp. dominated.

There were also considerable differences in the occurrence of other species from year to year (Spaulding et al. 1994). Another example of inter-annual variability in bacterial and phytoflagellate abundance in Lake Hoare is illustrated in Table 2.7 (Chapter 2). In the Vestfold Hills, Pendant Lake supported a bloom of *Stichococcus* in the summer of 1999/2000, with chlorophyll *a* concentrations up to 29.6 µg l^{-1} under an ice cover that held in all summer (Laybourn-Parry et al. 2002). Concentrations of chlorophyll *a* are usually between 3.7 and 7.5 µg l^{-1} in other years, when the community is dominated by cryptophytes. Studies of Ace Lake which covered seven summers between 1979 and 2002/2003 show considerable fluctuations in maximum chlorophyll *a* concentration, from <1.0 to 5.7 µg l^{-1}. In some of these years the ice broke out completely, in others only partially. There is no clearly identified explanation for these yearly variations, but it is likely that these extreme ecosystems are responding to short-term local climatic variations.

There are, however, clear longer-term trends. The McMurdo Dry Valleys showed a net cooling of 0.7 °C between 1966 and 2000, which elicited a rapid ecosystem response. Here, small changes in summer temperature and solar radiation strongly impact the local hydrology, melting glacier ice, and runoff to streams, lakes, and the soil. Lake levels rose from 1903 to around 1990 on average by 16 cm year^{-1}. Lake levels have receded in response to cooler summers since 1990. Cool and quiescent conditions in summer have reduced sublimation loss from the lakes, but not sufficiently to compensate for reduced stream flow (Doran et al. 2002b). The flooding event in the summer of 2000–2001, previously discussed, reversed the decline in lake levels and volumes. Warming trends are obvious on Signy Island in the Maritime Antarctic (see Figure 2.10). Here the ice cap has receded by 45% since 1951. During this time the summer open water phase on the lakes has increased. Mean winter water temperatures increased by 0.9 °C between 1980 and 1995 for nine of the island's lakes. This increase in temperature relates to the earlier break-out of ice, allowing the sediments and water to absorb more solar energy, creating a forcing effect. In winter, heat is released from the sediment to the water column where the overlying ice provides insulation against heat loss to the atmosphere. These changes in the physical environment have resulted in an increase in phytoplankton productivity (Quayle et al. 2002).

It is evident that Antarctic lakes can be regarded as sentinels of climate change. They appear to respond rapidly to short-term and longer-term climatic perturbations, unlike most lakes at lower latitudes. Since the lakes are removed from direct anthropogenic influence, the observed changes are undoubtedly attributable to local and larger-scale climate change.

7.3 The gaps in the data—the way forward

One of the problems with many Antarctic lake data sets is that they have limited data relating only to summer, and in some cases limited samples. In effect they represent a snapshot of what goes on during an annual cycle. It is rather like seeing a few minutes of a movie and having to guess the entire plot from that information. This is the nature of research in Antarctica. It is only where there are substantial resources and complex logistics that long-term and detailed data sets can be achieved. The USA, Australia, the United Kingdom, and New Zealand all support large and well-resourced limnological programmes with a focus on processes. It is clear from the preceding chapters that the majority of the detailed biological process-related data pertain to the Dry Valleys, the Vestfold Hills, and Signy Island. The McMurdo Dry Valleys LTER has provided large long-term detailed data sets that are allowing us determine how lake ecosystems function and their response to climate change. Lake research has largely ceased at Signy Island, but there is an excellent legacy of data, much of which covers annual cycles. The Vestfold Hills has also generated annual data sets over the last three decades, although because of the nature of UK and Australian science funding they are not continuous like those of the LTER.

There are a number of areas that beg to be addressed. The lack of detailed winter data for the Dry Valleys is a major logistic challenge, but in recent years a number of summer seasons have been extended to cover spring and autumn. While we have a lot of data on the mixolimnia of meromictic saline lakes, with only a few exceptions our knowledge of biogeochemical processes in the monimolimnia is limited, yet these regions of the lake water column undoubtedly contribute substantially to carbon and nutrient cycling in these lakes. Many articles on freshwater lakes are focused on basic limnology, with limited information on the processes that provide an insight into carbon cycling and ecosystem function. The growth of molecular biology has provided a valuable opportunity to characterize the biodiversity of lake communities, particularly the prokaryotes. However, in many cases important data on the physical and chemical environment is lacking. It is not enough to know what populates a given lake, we really need to understand the role of the various organisms in the ecosystem; an approach typified by Karr et al. (2003, 2005).

As indicated in Chapter 4, epishelf lakes are poorly researched. There is only limited information on biological processes in these unusual systems. This largely relates to the fact that these lakes are not near permanent research stations. Cryolakes are also poorly understood (Chapter 5). The growing interest in supraglacial biogeochemistry will undoubtedly redress this dearth of information in the future. Subglacial lakes are effectively the last great exploration challenge in Antarctica (see Chapter 6 and Figure 1.11). Drilling through up to 4 km of ice to reach the lakes and then ensuring that the collected samples are sterile and uncontaminated is no easy task and requires huge investment of logistics and funding. The failure of the UK-led Lake Ellsworth venture illustrates the challenge. Just like the early days of Antarctic exploration the road to success is paved with success and disappointment. Shortly after the disappointment of Lake Ellsworth, the WIZZARD programme successfully entered Lake Whillans to take both water and sediment samples.

7.4 Future directions

The application of in situ sensors to Antarctic lakes offers an opportunity to gain detailed data from lakes and glaciers over annual cycles. While these techniques are being applied to Antarctic limnology, there is considerable scope for wider application as the technology improves. Automated meteorological stations are now widely used, and often the data can be uploaded remotely by satellite. Automated remote samplers have been deployed in Lake Fryxell that have taken samples and fixed them in Lugol's iodine over the winter (McKnight et al. 2000). In some Dry Valleys lakes, PAR is measured continuously at the surface and at 10 m, and the data are downloaded on site (Foreman et al. 2004). A more sophisticated integrated platform, designed to measure water column temperatures and PAR at a range of depths down to 20 m, as well as ice thickness and surface meteorological conditions, was successfully deployed on the ice of Crooked Lake (Figure 7.1) (Palethorpe et al. 2004). Data were collected every 5 minutes and were uploaded at intervals via the Iridium satellite throughout the winter and spring (Figure 7.2). Figure 7.2 clearly shows variations in PAR on a daily

Figure 7.1 Photograph of a remote-sensing platform on Crooked Lake in winter. Solar panels charged a battery to drive the system. The platform had PAR sensors and temperature sensors suspended at different depths in the water column and a series of systems designed to measure ice thickness. Photo Malcom Foster.

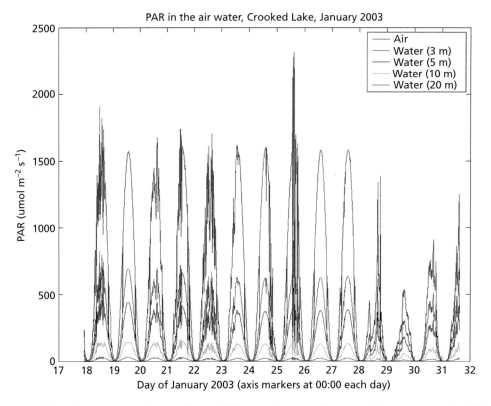

Figure 7.2 Daily PAR data from Crooked Lake for part of January 2003. Data collected every five minutes. The curves show marked differences in the patterns of PAR, for example the 19th, 26th, and 27th were cloudless sunny days, while the 18th, 20th, and other days were sunny with some cloud, resulting in a more jagged set of curves and the 29th was extremely overcast. (See Plate 16)

basis in the water column and marked differences between sunny days when PAR is high, and cloudy days when it is low. This type of remote sensing is described as a wireless network; these are systems designed to collect and transmit data back to the laboratory (Porter et al. 2005). Such platforms provide data at more frequent intervals throughout the months of the year, when fieldwork is particularly challenging. There is considerable scope for the deployment of more sophisticated sensors and sampling devices in the future. Temperature microstructure profiling systems (at millimetre intervals) are increasingly being used in lakes and the sea to capture short-term mixing dynamics. The inhospitable conditions in lakes make many parameters intractable using in-situ chemical sensors developed for temperate regions. Simple parameters such as dissolved oxygen have been measured continuously using in-situ fibre optic technology (Bagshaw et al. 2011). Optical sensors that measure nitrate, coloured dissolved organic matter (CDOM), and other chemicals, and microelectrodes that measure biological processes like respiration and photosynthesis in situ are now becoming available to be integrated into wireless networks. The remote sampling devices used in subglacial lake exploration exemplify this approach (see Chapter 6).

Satellite imagery is widely used in marine science and has potential application in limnology. The monitoring of the extent of ice cover and its breakout and formation times is of particular importance to polar limnology. For example, imagery from an Advanced Very High Resolution Radiometer (AVHRR) deployed on the National Oceanographic Atmosphere Administration (NOAA) satellites has been used to study how Canadian Arctic lake ice cover has changed over recent decades (Latifovie and Pouliot 2007). The mapping of Antarctica's subglacial lakes illustrates very well how advances in technology and analysis have provided much more accurate and detailed data. The first mapping of these extraordinary lakes was achieved by airborne radio-echo sounding (RES) between 1967 and 1979. This was achieved through collaboration between the Scott Polar Institute (UK), the US National Science Foundation, and the Technical University of Denmark (SPRI/NSF/TUD). Some twenty years later, developments in satellite radar and laser altimetry have permitted the time-dependent analysis of ice surface elevation changes identifying basal water bodies, and allowing the gains and losses of water in those water masses to be detected (Wright and Siegert 2012). As a consequence over 380 subglacial lakes have now been identified. The discovery via these remote-sensing studies that many sub-Antarctic lakes are hydrologically connected and play a role in moving large volumes of meltwater from the ice sheet interior to its margins is of relevance to marine ecosystems that act as a sink for this water. Not only is a large proportion of the total subglacial meltwater production likely to pass through lakes, but the dynamic fill–drain behaviour of lakes imparts unique discharge dynamics to water export at the ice margin. The total amount of water exported at the ice margins in Antarctica may be of the order of 65 km^3 (Pattyn 2010), which is an order of magnitude lower than runoff estimates from the Greenland Ice Sheet (Bamber et al. 2012), and it is thought to be rich in nutrients (e.g. iron). A topic of importance for the future study of subglacial lakes will be determination of the wider role of the export of this nutrient-rich meltwater in the productivity of marine ecosystems (Wadham et al. 2013).

The MODIS (Moderate Resolution Imaging Spectroradiometer) is a key instrument on the Terra and Aqua satellites (NASA). Data from MODIS has been used to study the development cryolakes or supraglacial lakes on the Greenland Ice Sheet (Sundal et al. 2009; Liang et al. 2012). Figure 7.3 shows

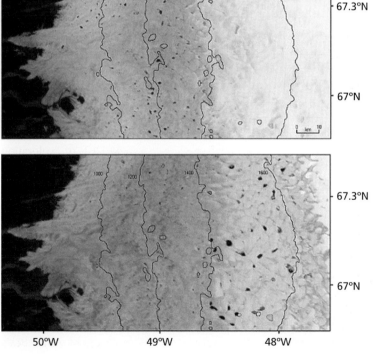

Figure 7.3 MODIS true colour images of cryolakes on the Greenland Ice Sheet. The top image was taken on the 11th of June while the lower image was taken on the 16th of July. The black lines show the altitude of the ice surface. From Sundal et al. (2009), reproduced with the permission of Elsevier.

satellite images that track the evolution of supraglacial lakes over the summer. This type of approach has yet to be applied to Antarctic supraglacial lakes. Unmanned aerial vehicles (UAVs) are increasingly being used to map glacier surfaces to assess the extent of cryoconite holes (e.g. Hodson et al. 2007) and have the potential for detailed investigations of supraglacial lakes. Technology is continually being developed and improved, and will undoubtedly provide new tools for the future investigation of lakes in one of the most difficult environments for limnological research on the planet.

Glossary

Algae a diverse group of photosynthetic organisms that range from unicellular to multicellular (e.g. seaweeds). Unicellular members include the diatoms and desmids.

algal mat mats that usually form on the bottom or benthic regions of lakes and streams. They are mainly composed of filamentous Cyanobacteria that provide a matrix for diatoms, bacteria, and a range of protozoans and small metazoans.

allochthonous derived from outside the system; in the case of lakes, nutrients and carbon inputs from the catchment.

amictic lakes that have no pattern of thermal water stratification during an annual cycle where the water column is always mixed.

amoebae members of the protozoan phylum Sarcomastigophora, subphylum Sarcodina. Includes naked, testate amoebae, heliozoans, and foraminiferans.

annual ice cover ice that forms and melts out each year. Typically found on lakes in coastal oases. Usually up to around 2 m thick and highly transparent.

anoxic lacking in oxygen.

ARA acetylene reducing assay is based on the nitrogenase enzyme responsible for N_2-fixation, also reducing C_2H_2 (acetylene) to C_2H_4 (ethylene), thereby providing a useful assay for the quantification of the N_2-fixation process.

Archaea a group of prokaryotic microorganisms previously grouped with the Bacteria as the Archaebacteria, now regarded as a distinct domain in the living world. The other two domains are Bacteria and Eukaryota.

assimilation number chlorophyll *a* specific rate of photosynthesis expressed as µg C µg chl a^{-1} h^{-1}, or as other units of weight.

autochthonous derived from within the system, for example carbon fixed by photosynthesis within a lake is described as autochthonous carbon.

autotroph means a 'self feeder', organisms that use light energy in the process of photosynthesis, e.g. cyanobacteria and algae (phototrophs), or organisms that derive energy from inorganic chemical reactions (chemotrophs).

Bacillariophyceae the diatoms, a major class within the algae. They have a distinctive siliceous cell wall and are common in the plankton of temperate lakes and in the benthos of lakes worldwide.

Bacteria prokaryotic microorganisms that form a major domain in the living world. The other two domains are the Archaea and the Eukaryota.

bacterial production bacterial growth, usually expressed as a unit weight of carbon produced per unit time per unit volume of water. Measured by following the uptake of radiolabelled thymidine or leucine in incubations.

BAS British Antarctic Survey.

benthic living on the bottom of a lake, stream, river, or the sea.

benthic mosses mosses that form dense mats on the bottom of some Antarctic lakes.

benthos the community of organisms that live on the bottom of a lake or other aquatic environment.

biomass total volume of a population or community per unit volume or area in aquatic environments, or weight or area in terrestrial environments. Usually expressed as units of carbon.

blue ice area exposed areas of the Antarctic ice sheet where there is no net addition or subtraction of snow. Any snowfall is equalled by sublimation.

B.P. before present.

calanoid a type of microcrustacean belonging to the order Calanoida in the sub-class Copepoda (phylum Arthropoda). They are common in the plankton and benthos of lakes.

carotenoids photosynthetic pigments forming two broad groups, the xanthophylls and the carotenes. They serve two functions: they absorb light energy to drive photosynthesis and they protect the cell's photo system from photo damage.

Cenozoic the most recent geological era, which began 65 million years ago.

chemocline the transitional boundary between the two water layers in meromictic lakes over which there are major gradient changes in dissolved oxygen, salinity, and nutrients.

chemotrophic bacteria bacteria that obtain energy by the oxidation of electron donors in their environments. These molecules can be organic (chemoorganotrophs) or inorganic (chemolithotrophs).

chemotrophic redox transformations oxidation/reduction chemical transformations by which some microorganisms obtain energy.

chironomid belongs to a family of insects called the Chironomidae which include the midges. The larval stages are usually aquatic and benthic.

chlorophyll *a* the most widespread plant photosynthetic pigment, consequently its concentration is often used as a proxy for phytoplankton or benthic mat biomass. Usually expressed as µg l^{-1} or mg m^3.

clastic composed of fragments of pre-existing rock that have been transported some distance from their points of origin.

clathrate a chemical compound which includes a lattice that traps other molecules. An example is methane clathrate (often also termed methane hydrate), which is a solid compound in which a methane molecule is enclosed within a crystal structure of water molecules and has an ice-like appearance.

CLIMAP (climate: long range investigation, mapping, and prediction). CLIMAP was developed in the 1970s and 1980s by a consortium of American universities to create a climatological map for 18,000 years ago, based on the analysis of marine sediments.

colluvial a loose deposit of rock debris accumulated through the action of gravity at the base of a cliff or slope.

compensation point in the euphotic zone; the point where carbon fixed during photosynthesis equals carbon lost through respiration, i.e. there is no net photosynthesis.

conductivity a measure of the electrical conductance per unit distance in an aqueous solution, which is related to the concentration of dissolved salts in the water. Usually expressed as milli Siemens (mS) or micro Siemens (µS) cm^{-1}.

Copepoda a subclass in the subphylum Crustacea (phylum Arthropoda) containing micro-crustaceans found in aquatic environments: the calanoid copepods, the cyclopoid copepods, and the harpacticoid copepods.

Crustacea a subphylum in the phylum Arthropoda whose members are characterized by an exoskeleton and jointed limbs.

cryoconite the sediment found on glacier surfaces and in cryolakes and cryoconite holes.

cryoconite hole mini-lakes on glaciers and ice sheet surfaces, usually less than 0.5 m deep. They contain cryoconite sediment and an overlying water layer. In the Antarctic they are usually covered by a lid of ice.

cryptophytes a group of phytoflagellates that are very common in the plankton of Antarctic lakes. They are capable of mixotrophy (mixed nutrition).

cryolake also known as a supraglacial lake, found on glaciers and ice sheets.

Cyanobacteria a group of photosynthetic bacteria, often misnamed the 'blue-green algae' because of their colour. Filamentous colonial genera form algal mats and are also found in the plankton. The group contains the picocyanobacteria (0.2–2.0 µm) that can be common in the plankton. Some genera are capable of fixing atmospheric nitrogen using special cells called heterocysts.

cysts a resting stage that is highly resistant to wide temperature change and desiccation, found among the protozoa. The process of forming a cyst is called encystment, and emerging from a cyst, excystment.

DCM deep chlorophyll maxima. High concentration of chlorophyll *a* found on a chemocline or thermocline in lakes or the sea.

denitrification reduction of oxidized forms of nitrogen (e.g. nitrate) to nitrogen gas via anaerobic respiration. Performed mainly by certain groups of heterotrophic bacteria, e.g. pseudomonads.

desmid a group of green algae belonging to the order Desmidiales. Mostly freshwater and common in lake plankton in lakes, but not Antarctic lakes.

diatom a group of algae belonging to the Bacillariophyceae. They have a distinctive siliceous cell wall and are common in the plankton of temperate lakes and in the benthos of lakes worldwide.

dimictic a form of thermal stratification in lakes where the water column is stratified twice in an annual cycle.

dinoflagellate a group of phytoflagellates common in lake plankton and in the sea. Nutritionally complex group with autotrophic, heterotrophic, and mixotrophic species; around half of the extant species lack photosynthetic pigments.

DOC dissolved organic carbon, an important energy source to the heterotrophic bacterioplankton and benthic bacteria. Also used as an energy source by some heterotrophic and mixotrophic flagellates. Chemically complex, including dissolved free amino acids (DFAA), dissolved free carbohydrates (DFCHO), and low molecular weight (<200 Da) organic acids, lipids, vitamins, hydrocarbons, polyphenols, and enzymes.

DOM dead organic matter.

Dry Valleys the largest ice-free region in Antarctica in Victoria Land. Also often called the McMurdo Dry Valleys.

englacial lakes lakes that are enclosed within glaciers and are often small pockets of water that are the closed-off remains of crevasses or meltwater tunnels.

epiglacial lakes lakes that abut glaciers, usually covered by perennial ice and having turbid waters caused by glacial rock flour.

epilimnion the upper water layer in a thermally stratified lake which is mixed and oxygenated.

epishelf lake a freshwater lake that either overlies seawater adjacent to an ice shelf, or is joined to the sea by a conduit under an ice shelf or glacier. These lakes are tidal and are exclusive to the polar regions.

eukaryote organism that belongs to one of the major domains in the living world, the Eukaryota. Characterized by a nucleus bounded by a nuclear membrane and other membrane-bound organelles (mitochondria, chloroplasts).

euphotic zone the upper region of the water column in the sea or lake where light penetrates and primary production or photosynthesis can occur.

firnification the process of glacier ice crystal formation from old snow (or 'firn'). Firnification is caused by grain metamorphosis, aided by melt–refreezing cycles, pressure, and vapour migration in the ageing snow.

FLB fluorescently labelled bacteria: heat-killed bacteria that have been stained with the flourochrome DTAF (5-(4,6-dichlorotriazinyl)aminofluorescein). FLBs are used to assess grazing rates by heterotrophic flagellates and ciliates. The protozoans are incubated with the FLBs and the ingested FLBs counted under fluorescence microscopy.

fluorochrome is a fluorescent chemical compound (e.g. DAPI, acridine orange, SBYR gold) that can re-emit light upon light excitation within particular wavelengths. Used widely for staining and enumerating viruses, bacteria, and heterotrophic flagellates under epifluorescence microscopy.

Foraminifera a group of marine sarcodine protozoa that possess compartment shells usually formed of calcium carbonate. They form fossils and are widely used to date sediment strata in palaeontology studies.

genotyping the process of determining differences in the genetic make-up (genotype) of an individual by examining the individual's DNA sequence using one or more biological assays.

glacier large persistent body of ice that forms where the accumulation of snow exceeds its loss. Glaciers are systems with inputs and outputs: mass and energy enter the system in the form of snow (precipitation), rock debris, solar radiation, and geothermal heat, and are lost as water vapour, water, ice, rock debris, and heat. Glaciers are important sentinels of climate change.

glaciofluvial relates to streams fed by melting glaciers, or to the deposits and landforms produced by glacial streams.

glaciolacustrine sediments derived from glacial streams, deposited in lakes.

glaciomarine sediments derived from glaciers deposited in the sea or ocean.

Heliozoa a predominantly freshwater group of sarcodine protozoa that resemble stylized suns as a result of their radiating axiopoda (a form of pseudopodia). Heterotrophic, feeding mainly on other protozoans.

heterotroph organisms that have to ingest carbon (food) to grow and reproduce. They may be primary consumers, feeding on autotrophs, or secondary consumers feeding on other heterotrophs as predators.

heterotrophic bacteria bacteria that use organic carbon as an energy source.

heterotrophic flagellate small colourless flagellates (nanoflagellates) that are the main consumers of bacteria in aquatic environments. Some can take up DOC. Also found in soils.

HNAN an abbreviation for heterotrophic nanoflagellates.

Holocene meaning entirely recent, a geological epoch which began at the end of the Pleistocene (around 11,700 year ago) and continues to the present. The Holocene is part of the Quaternary period.

hyperaline lakes lakes that have salinities higher than sea water (35‰).

hypolimnion the lower layer of water below the thermocline in a thermally stratified lake.

hypoxic low in oxygen saturation, but not completely oxygen depleted (termed anoxic).

hyposaline lakes lakes that are below the salinity of seawater (35‰). Often also called brackish.

ice cover the layer of ice covering a lake which may be annual or perennial.

ice sheet a large body of terrestrial glacial ice, greater than 50,000 km^2. At the present day, three ice sheets are present on Earth, the Greenland Ice Sheet, the West, and the East Antarctic Ice Sheets.

ice shelf a large, thick mass of glacial ice of relatively flat surface topography that is floating. Formed when a glacier or ice sheet terminates in the ocean and expands over the ocean surface.

IGY International Geophysical Year: an international scientific project that lasted from July 1, 1957, to December 31, 1958. Very important in the development of Antarctic research.

Jökulhlaups an Icelandic term for glacial outburst flood events which often reflect the release of subglacial meltwater from a glacier bed. They are common in Iceland where geothermal heating of the glacier sole results in rapid melting, melt water accumulation, and periodic evacuation.

Isostatic uplift also known as isostatic rebound. A rise in land mass as the weight of overlying ice is removed by melting and retreat.

katabatic wind means 'going downhill', from the Greek katabatikos. These are strong winds in Antarctica that blow down off the continental ice sheet as a result of dense cold air flowing under the force of gravity.

LGM last glacial maximum, the peak of the last glaciation.

limnology the study of inland waters (lakes, streams, rivers, and wetlands). Mostly freshwater but also includes saline lakes.

littoral zone shallow region close to the shore or edge of a lake.

LTER McMurdo Long Term Ecosystem Research Program is an interdisciplinary and multidisciplinary study of the aquatic and terrestrial ecosystems in an ice-free region of Antarctica. MCM joined the National Science Foundation's LTER Network in 1993 and is funded through the Office of Polar Programs in six-year funding periods.

lysogenic cycle where viruses are passed by bacteria or other hosts to their progeny during cell division, during which viral nucleic acid recombines with the host genome. Once the virus has inserted itself into the host chromosome it is termed a prophage, and the cell harbouring the prophage is called a lysogen. The lysogenic cycle can continue until a factor (or factors) triggers the lytic cycle.

lytic cycle where the virus invades a host cell and turns it into a virus-producing factory, which terminates in the host cell lysing or bursting open, releasing the viruses into the environment where they find and infect new hosts. The lytic cycle is important in recycling carbon and effectively short-circuits the microbial food web, destroying bacteria before they can be consumed by flagellates and other bacterial predators.

Maritime Antarctic a generic term that covers the Antarctic Peninsula, the South Shetland Islands, the South Orkney Islands, the South Sandwich Islands, and Bouvet Island.

Mastigophora a subphylum in the phylum Sarcomastigophora, it contains the flagellates, both colourless (Zoomastigophora), and the coloured or phototrophic (Phytomastigophora).

MCV mean cell volume, the mean volume of bacterial, protozoan, or algal cells. Used to calculate community or population biomass.

meromictic lakes that are chemically stratified, usually permanently. They have strong physical and chemical gradients in their water columns.

metagenomics the study of metagenomes, genetic material recovered directly from environmental samples.

methanogenesis the production of methane, usually refers to methanogenic Archaea.

mid-Holocene hypsithermal also termed the Holocene Climatic Optimum, which was a warm event during the Holocene at around 9000–5000 yrs B.P.

MIS Marine Isotope Stage—alternating warm and cold period in paleoclimatic history, deduced by oxygen isotope variations, usually within marine core sediments.

mixolimnion the upper water layer in a meromictic lake that is mixed and oxygenated.

mixotrophy means mixed nutrition, undertaken by some phytoflagellates (e.g. cryptophytes) and some ciliates. In the case of the former, which are primarily photosynthetic, it involves ingesting bacteria and/or DOC. In the case of ciliates, it either involves endosymbiotic zoochlorellae or the sequestration of chloroplast derived from phytoflagellate prey.

monomictic lakes that undergo one period of thermal stratification in an annual cycle during the summer.

monomolimnion the lower layer of water in a meromictic lake. The waters are anoxic and often higher in temperature than the upper mixolimnion. Typically the domain of methanogenic and sulphur-reducing microorganisms.

moraine material sedimentary material comprising moraines, which are deposited by a glacier and revealed as the ice retreats.

moulin a quasi-cylindrical vertical shaft in a glacier or ice sheet surface via which surface meltwater enters the ice interior.

Myr B.P. millions of years before present.

naked amoebae descriptive term for amoebae that lack shells or tests. Members of the Sarcodina (Protozoa). Common in the benthos of aquatic environments and soils.

nematode so-called round worms belonging to the phylum Nematoda. Free-living forms are microscopic, common in the benthos of aquatic environments and the soil, where they mostly feed on bacteria and protozoa.

nitrification the biological oxidation of ammonia with oxygen to nitrite, followed by the oxidation of nitrites to nitrates. Nitrification is an important step in the nitrogen cycle in aquatic environments and the soil, and is mediated by bacteria.

NSF National Science Foundation; government body that funds scientific research in the United States, including the McMurdo Long Term Ecosystem Research Program.

oasis term applied to a fertile area in a desert. In Antarctica (a polar desert) it refers to ice-free coastal regions that support lakes and sparse populations of lichens and mosses.

oligochaete a subclass in the phylum Annelida. These are segmented worms that are found in soils (earthworms) and in the benthos of aquatic environments. Most feed on dead organic matter or detritus.

OTU operational taxonomic unit. An OTU is typically defined as a cluster of organisms that reads with 97% similarity; the expectation is that these correspond approximately with species.

oxic oxygenated.

oxycline the transition layer in a meromictic lake, where oxygen drops to hypoxic and anoxic.

PAR photosynthetically active radiation; solar radiation in the electromagnetic spectrum from 400 to 700 nanometres (visible light) that photosynthetic organisms use in the process of photosynthesis.

pelagic referring to the water column where pelagic or planktonic organisms live.

perennial ice cover ice cover that persists from year to year, found in lakes that are far south or north in the polar regions, for example the Dry Valleys lakes. Also seen in coastal lakes that abut glaciers or the continental ice sheet. This ice can be up to 5 metres thick and in some cases carries a high sediment load blown onto it from the surrounding exposed rock and soils.

photosynthesis the first stage in the carbon cycling where microscopic and macroscopic plants combine inorganic carbon (CO_2) with water using light energy (PAR) to create complex organic material or plant biomass.

phytoflagellate photosynthetic flagellates, members of the phytomastigophora in the phylum sarcomastigophora (protozoa).

phytoplankton the autotrophic component of the plankton. This is a generic term and includes a wide range of taxa, including Algae, phytoflagellate protozoans, and Cyanobacteria.

picocyanobacteria very small solitary cyanobacteria (0.2–2.0 µm) found in the plankton of aquatic environments.

plankton organisms that are free-swimming or floating in the water columns of aquatic environments.

Pleistocene the geological epoch this has lasted from about 2,588,000 to 11,700 years ago, spanning the recent glaciations and interglacial periods.

Pliocene a period in the geologic history that extends from 5.332 million to 2.588 million years before present. The Pliocene follows the Miocene epoch and is followed by the Pleistocene epoch.

PNAN abbreviation for phototrophic nanoflagellates.

POM particulate organic or POC matter includes both dead and living material.

primary production material or biomass produced during the process of photosynthesis by phototrophic organisms.

proglacial lake a lake formed either by the damming action of a moraine, ice dam, or ice shelf during the retreat of a melting glacier, or by meltwater trapped against an ice sheet due to isostatic depression of the crust around the ice.

prokaryote refers to the Bacteria and Archaea that lack nuclear membranes.

Protozoa a diverse group of single-celled eukaryotic microorganisms. Includes groups with no taxonomic affiliations and many parasitic taxa. The free-living Sarcomastigophora (flagellates and amoebae) and Ciliophora (ciliates) are important elements of the microbial community in aquatic environments and the soil.

PSU practical salinity unit, a measure of salinity used widely by marine scientists. PSU defines salinity in terms of a conductivity ratio, so it is dimensionless.

psychrohile an organism adapted to living at low temperatures with an optimum growth temperature below 15 °C.

psychrotroph an organism that can grow and reproduce at low temperatures but has an optimum growth temperature above 15 °C.

Quaternary period encompasses the Holocene and Pleistocene and spans 2.6 million years.

radiocarbon dating a radiometric dating method which uses the decay of carbon-14 (^{14}C) to determine the age of organic material.

redox an abbreviation used to refer to reduction and oxidation reactions.

rotifers a phylum (Rotifera) of microscopic multicellular (metazoan) organisms. Predominantly freshwater, common in the plankton and benthos where they feed on bacteria, protozoa, algae, detritus, and other rotifers.

16S ribosomal DNA (16s rDNA) a component of the 30S small subunit of prokaryotic ribosomes. This has played a pivotal role in the identification and characterization of bacteria.

salinity usually expressed as ‰ (parts per thousand), effectively grams of salt per kg of solution.

Sarcomastigophora a phylum in the sub-kingdom Protozoa. Contains the flagellates and the amoebae.

Scalar irradiance photon scalar irradiance h is defined as the total number of photons per unit time and area arriving at a point from all directions about the point when all directions are weighed evenly, whereas irradiance is the radiance field multiplied by the cosine if its zenith angle and then integrated over the field-of-view of the sensor.

SCAR Scientific Committee for Antarctic Research, established in 1958.

secondary production growth or biomass produced by heterotrophic organisms that either eat plants or are predators on other heterotrophic organisms.

stable isotope refers to the isotopes of an element that are not radiogenic (although can include the daughter isotopic products of radioactive decay), but which are affected by fractionation processes.

subglacial lake lakes which are located beneath a glacier or ice sheet, between the base of the ice mass and its underlying sub-strata, be it sediment or bedrock.

sublimation a phase change transition by which a solid is converted directly to gas without passing through the liquid phase. Within glaciers, this would refer to a change from ice to water vapour.

sulphate-reducing bacteria found in anoxic environments where they reduce sulphate to sulphide by means of a number of electron donors, including H_2, fatty acids, alcohols, and aromatic compounds

sulphur oxidizing bacteria obtain their energy by the oxidation of reduced inorganic sulphur compounds. Some use sulphide or a metal sulphide such as iron or copper sulphide, others use elemental sulphur or thiosulphate.

supraglacial lake also known as cryolakes. These lakes are found on glaciers and ice sheets.

tardigrades also known commonly as water bears. They are members of the phylum Tardigrada—small (<1 mm) metazoans with four pairs of limbs with terminal claws. Ubiquitous distribution, including the benthos of polar lakes and soils.

tectonic activity movement or distortion of the Earth's crust resulting from forces within it.

terragenic sediments derived from the erosion of landmasses.

thermal stratification the seasonal development of different temperature water layers in a lake water column. Summer stratification involves warmer water overlying cooler water, while in winter cooler water may overlie warmer water.

thermocline the boundary layer between the upper epilimnion and lower hypolimnion in a thermally stratified lake, over which there is a marked temperature gradient.

TOC total organic carbon, includes dissolved organic carbon, dead organic matter and living material.

TOP total organic phosphorus.

TransAntarctic Mountains mountain range that transects the Antarctic continent, dividing it into West and East Antarctica (see Figure 1.1).

VBR virus to bacteria ratio: the number of viruses divided by the number of bacteria in a sample. Gives an indication of potential infection.

virome the genomes of all the viruses that inhabit a particular organism or habitat.

viriophage a virus that infects other viruses and uses the infected helper virus's machinery for reproduction, while it inhibits replication of the helper virus.

virus a bit of nucleic acid, RNA or DNA, surrounded by a protein coat called the capsid. Viruses are not strictly living organisms because they cannot reproduce. They commandeer an infected host cell and make it produce more virus, effectively turning it into a virus-producing factory.

xanthophyll a type of carotenoid that contains oxygen. They play a major role in stimulating energy dissipation by non-photochemical quenching, thereby protecting against photoinhibition, in an enzyme mediated process called the xanthophyll cycle.

zoochlorellae endosymbiotic algae found in a variety of organisms including ciliated protozoans. The zoochlorellae benefit from the excretory products of the ciliate (ammonium and orthophosphate) required for photosynthesis and having a host that is highly motile and which can position itself in the best light climate. In return the ciliate acquires some of the photosynthate produced by the zoochlorellae.

zooplankton small metazoan organisms that swim or float in the water column of lakes or the sea. In lakes the zooplankton is made up of various crustaceans and rotifers.

References

Adamson, D.A., Mabin, M.C.G. and Luly, J.G. (1997). Holocene isostasy and late Cenozoic development of landforms including Beaver and Radok Lake basins in the Amery Oasis, Prince Charles Mountains. *Antarctic Science* **9**, 299–306.

Alexander, V., Stanley, D.W., Daley, R.J. and McCoy, C.P. (1980). Primary producers. In J.E. Hobbie (ed) *Limnology of Tundra Pond, Barrow, Alaska*, pp. 179–250. Dowden Hutchinson and Ross Inc., Stroudsberg, PA.

Alldredge, A.L. and Silver, M.W. (1988). Characteristics, dynamics and significance of marine snow. *Progress in Oceanography* **20**, 41–82.

Allende, L. and Izaguirre, I. (2003). The role of physical stability on the establishment of steady state in the phytoplankton community of two Maritime Antarctic lakes. *Hydrobiologia* **502**, 211–224.

Alley, R.B. and Whillans, I.M. (1991). Changes in the west Antarctic ice-sheet. *Science* **254**, 959–963.

Almada, P., Allende, L., Tell, G. and Izaguirre, I. (2004). Experimental evidence of the grazing impact of *Boeckella poppei* on phytoplankton in a maritime Antarctic lake. *Polar Biology* **28**, 39–46.

Amon, R.M.W. and Benner, R. (1996). Bacterial utilization of different size classes of dissolved organic carbon. *Limnology and Oceanography* **41**, 41–51.

Andersen, D.T., Sumner, D.Y., Hawes, I., Webster-Brown, J. and McKay, C.P. (2011). Discovery of large conical stromtolites in Lake Untersee, Antarctica. *Geobiology* **9**, 280–293.

Anderson, S.P. (2007). Biogeochemistry of glacial landscape systems. *Annual Review of Earth and Planetary Sciences* **35**, 375–399.

Andreoli, C., Scarabel, L., Spini, S. and Grassi, C. (1992). The picoplankton in Antarctic lakes of northern Victoria Land during summer 1989–1990. *Polar Biology* **11**, 575–582.

Anesio, A.M. and Laybourn-Parry, J. (2012). Glaciers and ice sheets as a biome. *Trends in Ecology and Evolution*, doi:10.1016/j.tree.2011.09.012.

Anesio, A.M., Mindl, B., Laybourn-Parry, J. and Sattler, B. (2007). Virus dynamics on a high Arctic glacier (Svalbard). *Journal of Geophysical Research Biogeosciences* **112**, G04S31. doi:10.1029/2006JG000350.

Anesio, A.M., Sattler, B., Foreman, C., Hodson, A.J., Tranter, M. and Psenner, R. (2011). Carbon fluxes through bacterial communities on glacier surfaces. *Annals of Glaciology* **51**(56), 32–40.

Ashley, G.M. (2002). Glaciolacustrine Environments. In Menzies, J. (ed) *Modern and Past Glacial Environments*, pp. 335–359. Butterworth-Heinemann, Oxford.

Azam, F., Fenchel, T., Field, J.G., Gray, J.S., Meyer-Reil, R.A. and Thingstad, F. (1983). The ecological role of water column microbes in the sea. *Marine Ecology Progress Series* **6**, 213–220.

Bagshaw, E.A., Tranter, M., Wadham, J.L., Fountain, A.G. and Basagic, H. (2010). Dynamic behaviour of supraglacial lakes on cold polar glaciers: Canada Glacier, McMurdo Dry Valleys, Antarctica. *Journal of Glaciology* **56**, 366–368.

Bagshaw, E.A., Tranter, M., Wadham, J.L., Fountain, A.G. and Mowlem, M. (2011). High-resolution monitoring reveals dissolved oxygen dynamics in an Antarctic cryoconite hole. *Hydrological Processes* **18**, 2868–2877.

Bamber, J., van den Broeke, M., Ettema, J., Lenaerts, J. and Rognot, E. (2012). Recent large increases in freshwater fluxes from Greenland into the North Atlantic. *Geophysical Research Letters* **39**, doi:10.1029/2012GL052552.

Bardin, V.I., Piskun, A.A. and Schmidelberg, N.A. (1990). Hydrological and hydrochemical characteristics of deep lake water basins in Prince Charles Mountains. *Antarktika: Doklady Komissii* **29**, 97–112. (in Russian with an English summary).

Barrett, P.J. (1996). Antarctic paleoenvironment through Cenozoic times: a review. *Terra Antartica* **3**, 103–119.

Bartholomew, I., Niewow, P., Sole, A., Mair, D., Crowton, T. et al. (2011). Supraglacial forcing of subglacial drainage in the ablation zone of the Greenland Ice Sheet. *Geophysical Research Letters* **38**, doi:L08502.

Bayliss, P., Ellis-Evans, J.C. and Laybourn-Parry, J. (1997). Temporal patterns of primary production in a large ultra-oligotrophic Antarctic freshwater lake. *Polar Biology* **18**, 363–370.

Bayliss, P.R. and Laybourn-Parry, J. (1995). Seasonal abundance and size variation in Antarctic populations of the cladoceran *Daphniopsis studeri*. *Antarctic Science* **7**, 393–394.

Bayly, I.A.E. and Burton, H.R. (1981). Harpacticoid copepods from a saline lake in the Vestfold Hills, Antarctica. *Australian Journal of Marine and Freshwater Research* **32**, 465–467.

Bayly, I.A.E. and Burton, H.R. (1993). Beaver Lake, Greater Antarctica, and its population of *Boeckella poppei* (Mrázek) (Copepoda: Calanoida). *Verhandlungen des Internationalen Verein Limnologie* **25**, 975–978.

Bayly, I.A.E. and Eslake, D. (1989). Vertical distribution of a planktonic harpacticoid and a calanoid (Copepoda) in a meromictic Antarctic lake. *Hydrobiologia* **172**, 207–214.

Bayly, I.A.E., Gibson, J.A.E., Wagner, B. and Swadling K.M. (2003). Taxonomy, ecology and zoogeography of two East Antarctic freshwater calanoid copepod species: *Boeckella poppei* and *Gladioferens antarcticus*. *Antarctic Science* **15**, 439–448.

Bayly, I.A.E. and Williams, W.D. (1966). Chemical and biological studies on some saline lakes of south-east Australia. *Australian Journal of Marine and Freshwater Research* **17**, 177–228.

Bell, E.M. (1998) Plankton dynamics in the saline lakes of the Vestfold Hills, Eastern Antarctica. Unpublished Ph.D. thesis, University of Nottingham, UK.

Bell, E.M. and Laybourn-Parry, J. (1999a). Annual plankton dynamics in an Antarctic saline lake. *Freshwater Biology* **41**, 507–519.

Bell, E.M. and Laybourn-Parry, J. (1999b). The plankton community of young, eutrophic, Antarctic saline lake. *Polar Biology* **22**, 248–253.

Bell, E.M. and Laybourn-Parry, J. (2003). Mixotrophy in the Antarctic phytoflagellate *Pyramimonas gelidicola* McFadden (Chlorophyta: Prasinophyceae). *Journal of Phycology* **39**, 644–649.

Bell, R.A.I. (1967). Lake Miers, South Victoria Land, Antarctica. *New Zealand Journal of Geology and Geophysics* **10**, 540–556.

Bell, R.E. (2008). The role of subglacial water in ice-sheet mass balance. *Nature Geoscience* **1**, 297–304.

Bell, R.E., Studinger, M., Fahnestock, A. and Shuman, C.A. (2006). Tectonically controlled subglacial lakes on the flanks of the Gamburtsev Subglacial Mountains, East Antarctica. *Geophysical Research Letters* **33**, doi:10.1029/2005GL025207.

Benn, D.I. and Evans, D.J.A. (2010). *Glaciers and Glaciation*, 2nd edn. Arnold, London, 734 pp.

Bennett, S.J., Sanders, R.W. and Porter, K.G. (1990). Heterotrophic, autotrophic and mixotrophic nanoflagellates: seasonal abundances and bacterivory in a eutrophic lake. *Limnology and Oceanography* **35**, 1821–1832.

Bentley, M.J., Hodgson, D.A., Smith, J.A., Ó Cofaigh, C., Domack, E.W. and seven others. (2009). Mechanisms of Holocene palaeoenvironmental change in the Antarctic Peninsula region. *The Holocene* **19**, 51–69.

Bentley, M.J., Hodgson, D.A., Sugden, D.D., Robert, S.J., Smith, J.A. and two others. (2005). Early Holocene retreat of the George VI Ice Shelf, Antarctic Peninsula. *Geology* **33**, 173–176.

Bergh Ø., Børsheim K.Y., Bratbak, G. and Heldal M. (1989). High abundance of viruses found in aquatic environments. *Nature* **340**, 467–468.

Bettarel, Y., Sime-Ngando, T., Amblard, T.C., and Dolan, J. (2004). Viral activity in two contrasting lake ecosystems. *Applied and Environmental Microbiology* **70**, 2941–2951.

Bielewicz, S., Bell, E., Kong, W., Friedberg, I., Priscu, J.C. and Morgan-Kiss, R.M. (2011). Protist diversity in a permanently ice-covered Antarctic Lake during the polar transition. *The ISME Journal* **5**, 1559–1564.

Bird, D.F. and Kalff, J. (1984). Empirical relationships between bacterial abundance and chlorophyll concentration in fresh and marine waters. *Canadian Journal of Fisheries and Aquatic Science* **41**, 1015–1023.

Bird, M.I., Chivas, A.R., Radnell, C.J. and Burton, H.R. (1991). Sedimentological and stable-isotope evolution of lakes in the Vestfold Hills, Antarctica. *Palaeogeography, Palaeoclimatology, Palaeoecology* **84**, 109–130.

Björck, S., Håkansson, H., Olsson, S, Barnekow, L. and Janssens, J. (1993). Palaeoclimatic studies in South Shetland Islands, Antarctica, based on numerous stratigraphic variables in lake sediments. *Journal of Paleolimnology* **8**, 233–272.

Björck, S., Håkansson, H., Zale, R., Karlén, W. and Jönsson, B.L. (1991). A late Holocene lake sediment sequence from Livingston Island, South Shetlands Islands, with palaeoclimatic implications. *Antarctic Science* **3**, 61–71.

Bjørnsen, P.K., Riemann, B., Horsted, S.J., Nielsen, T.G., Pock-Sten, J. (1988). Trophic interactions between heterotrophic flagellates and bacterioplankton in manipulated sea water enclosures. *Limnology and Oceanography* **33**, 409–420.

Bondarenko, N.A., Timoshkin, O.A., Röpstorf, P. and Melnik, N.G. (2006). The under-ice and bottom periods in the life cycle of *Aulacoseira baicalensis* (K.Meyer) Simonsen, a principal Lake Baikal alga. *Hydrobiologia* **568**, 107–109.

Bormann, P. and Fritzsche, D (eds) (1995). *The Schirmacher Oasis, Queen Maud Land, East Antarctica, and its Surroundings*. Justus Perthes Verlag, Gotha, Germany.

Borowitzka, L.J. (1981). The microflora, adaptations to life in extremely saline lakes. *Hydrobiologia* **81**, 33–46.

Børsheim, K.Y., Bratbak, G. and Heldal, M. (1990). Enumeration and biomass estimation of planktonic bacteria and viruses by transmission electron microscopy. *Applied and Environmental Microbiology* **56**, 352–356.

Borutsky, E.V. (1962). New data on the copepod *Acanthocyclops mirnyi* Borutsky et M. Vinogradov from Antarctic. In Russian with an English summary. *Zoologicheskii Zhurnal* **41**, 1106–1107.

Bottrell, S.H. and Tranter, M. (2002). Sulphide oxidation under partially anoxic conditions at the bed of the Haut Glacier d'Arolla, Switzerland. *Hydrological Processes* **16**, 2363–2368.

Bowman, J.P., McCammon, S.A. and Skerratt, J.H. (1997). *Methylosphaera hansonii* gen. nov., sp. nov.,a psychrophilic, group I methanotroph from Antarctic marine-salinity, meromictic lakes. *Microbiology* **143**, 1451–1459.

Bowman, J.P., Gosink, J.J., McCammon, S.A., Lewis, T.E., Nichols, D.S. et al. (1998). *Colwellia demingiae* sp. nov., *Colwellia hornarae* sp.nov., *Colwellia psychrotropica* sp. nov.: psychrophilic Antarctic species with the ability to synthesize docosahexaenoic acid (22:6ω3). *International Journal of Systematic Bacteriology* **48**, 1171–1180.

Bowman, J. P., Rea, S.M., McCammon S.A. and McMeekin, T.A. (2000). Diversity and community structure within anoxic sediment from marine salinity meromictic lakes and a coastal meromictic marine basin, Vestfold Hills, eastern Antarctica. *Environmental Microbiology* **2**, 227–237.

Boyd, E.S., Skidmore, M., Mitchell, A.C., Bakermans, C. and Peters, J.W. (2010). Methanogenesis in subglacial sediments. *Environmental Microbiology Reports* **2**, doi:10.1111/1758–1222.2110.0016X.

Brambilla, E., Hippe, H., Hagelstein, A., Tindal, B.J. and Stackebrandt, E. (2001). 16S rRNA diversity of cultured and uncultured prokaryotes of mat sample from Lake Fryxell, McMurdo Dry Valleys, Antarctica. *Extremophiles* **5**, 23–33.

Bronge, C. (1996). Hydrological and climatic changes influencing the proglacial Druzhby drainage system, Vestfold Hills, Antarctica. *Antarctic Science* **8**, 379–388.

Bulat, S.A., Alekhina, I.A., Blot, M., Petit, J.-R., de Angelis, M., Wagenbach, D., Lipenkov, V.Y. and four others. (2004). DNA signature of thermophilic bacteria from the aged accretion ice of Lake Vostok, Antarctica: implications for searching for life in extreme icy environments. *International Journal of Astrobiology* **3**, 1–12.

Bulat, S.A., Alekhina, I.A., Lipenkove, V.Y., Lukin, V.V. et al. (2009). Cell concentrations of miroorganisms in glacial and lake ice of the Vostok ice core, East Antarctica. *Microbiology* **78**, 808–810.

Burch, M.D. (1988). Annual cycle of phytoplankton in Ace Lake: an ice covered, saline meromictic lake. *Hydrobiologia* **165**, 59–75.

Burgis, M.J. and Morris, P. (1987). *The Natural History of Lakes*. Cambridge University Press, Cambridge, UK.

Burke, C.M. and Burton, H.R. (1988). Photosynthetic bacteria in meromictic lakes and stratified fjords of the Vestfold Hills, Antarctica. *Hydrobiologia* **165**, 12–23.

Burton, H.R. (1980). Methane in a saline antarctic lake. In Trudinger P.A. and Walter M.R. (eds) *Biogeochemistry of Ancient and Modern Environments*, pp. 243–251. 3053. Australian Academy of Science, Canberra.

Burton, H.R. and Barker, R.J. (1979). Sulfur chemistry and microbiological fractionation of sulfur isotopes in a saline Antarctic lake. *Geomicrobiology Journal* **1**, 329–340.

Burton, H.R. and Hamond, R. (1981). Harpacticoid copepods from a saline lake in the Vestfold Hills. *Australian Journal of Marine and Freshwater Research* **32**, 465–467.

Butler, H., Atkinson, A. and Gordon, M. (2005). Omnivory and predation impact of the calanoid copepod *Boeckella poppei* in a maritime Antarctic lake. *Polar Biology* **28**, 815–821.

Butler, H.G. (1999). Seasonal dynamics of the planktonic microbial community in a maritime Antarctic lake undergoing eutrophication. *Journal of Plankton Research* **21**, 2393–2419.

Butler, H.G., Edworthy, M.G. and Ellis-Evans, J.C. (2000). Temporal plankton dynamics in an oligotrophic maritime Antarctic lake. *Freshwater Biology* **43**, 215–230.

Camacho, A. (2006). Planktonic microbial assemblages and the potential effects of metazooplankton predation on the food web of lakes from the maritime Antarctica and sub-Antarctic islands. *Reviews in Environmental Science and Biotechnology* **5**, 167–185.

Campbell, P. and Torgersen, T. (1980). Maintenance of iron meromixis by iron redeposition in a rapidly flushed monimolimnion. *Canadian Journal of Fisheries and Aquatic Sciences* **37**, 1303–1313.

Canfield, D.E. and Green, W.J. (1985). The cycling of nutrients in a closed-basin Antarctic lake—Lake Vanda. *Biogeochemistry* **1**, 233–256.

Canfield, D.E., Green, W.J. and Nixon, P. (1995). Pb-210 and stable Pb through the redox transition zone of an Antarctic lake. *Geochimica et Cosmochimica Acta* **59**, 2459–2468.

Carter, S.P., Blankenship, D.D., Peters, M.E., Young, D.A., Holt, J.W. and Morse, D.L. (2007). Radar-based subglacial lake classification in Antarctica. *Geochemistry Geophysics Geosystems* **8**, doi:10.1029/2006GC001408.

Carter, S.P. and Fricker, H.A. (2012). The supply of subglacial meltwater to the grounding line of the Sip le Coast, West Antarctica. *Annals of Glaciology* **53**, 267–280.

Cathey, D.D., Parker, B.C., Simmons, G.M., Yongue, W.H. and Van Brunt, M.R. (1981). The microfauna of algal mats and artificial substrates in Southern Victoria Land lakes of Antarctica. *Hydrobiologia* **85**, 3–15.

Charlebois, R.L., Beiko, R.G. and Ragan, M.A. (2003). Microbial phylogenetics: branching out. *Nature* **421**, doi:10.1038/421217a.

Chinn, T.J. (1993). Physical hydrology of the Dry Valley Lakes. In Green, W.J. and Friedman, E.I. (eds) *Physical* and *Biogeochemical Processes in Antarctic Lakes*, pp. 1–51. Antarctic Research Series 59, American Geophysical Union, Washington, DC.

Christie, P. (1987). Nitrogen in two contrasting Antarctic bryophyte communities. *Journal of Ecology* **75**, 73–93.

Christner, B.C., Mosley-Thompson, E., Thompson, L. and Reeve, J.N. (2001). Isolation of bacteria and 16S rDNAs from Lake Vostok accretion ice. *Environmental Microbiology* **3**, 570–577.

Christner, B.C., Priscu, J.C., Achberger, A.M., Barbante, C. et al. (2014). Subglacial Lake Whillans: a microbial ecosystem beneath the West Antarctic Ice Sheet. Submitted to *Nature*.

Christner, B.C., Royston-Bishop, G., Foreman, C.M., Arnold, B.R., Tranter, M., Welch, K.A. and four others. (2006). Limnological conditions in subglacial Lake Vostok. *Limnology and Oceanography* **51**, 2485–2501.

Chróst, T.J., Münster, U., Rai, H., Albrecht, D., Witzel, K.P. and Overbeck, J. (1989). Photosynthetic production and exoenzymatic degradation of organic matter in the euphotic zone of a eutrophic lake. *Journal of Plankton Research* **11**, 223–242.

Clayton-Greene, J.M., Hendy, C.H. and Hogg, A.G. (1988). Chronology of a Wisconsin age proglacial lake in the Miers Valley, Antarctica. *New Zealand Journal of Geology and Geophysics* **31**, 353–361.

CLIMAP (1981). Geological Society of America, Map and Chart Series, C36.

Cochran, P.K. and Paul, J.H. (1998). Seasonal abundance of lysogenic bacteria in a subtropical estuary. *Applied and Environmental Microbiology* **64**, 2308–2312.

Codispoti, L.A., Brandes, J.A, Christensen, J.P, Devol, A.H., Naqvi, S.W.A, Paerl, H.W. and Yoshinari, T. 2001. The oceanic fixed nitrogen and nitrous oxide budgets: moving targets as we enter the anthropocene? *Scienta Marina* **65** supplement (2), 85–105.

Cofaigh, C.Ó., Dowdeswell, J.A., Evans, J. and Larter, R.D. (2008). Geological constraints on antarctic palaeo-ice-stream retreat. *Earth Surface Processes and Landforms* **33**, 513–525.

Coffin, R.B. (1989). Bacterial uptake of dissolved free and combined amino acids in estuarine waters. *Limnology and Oceanography* **34**, 531–542.

Coker, J.A., Sheridan, P.P., Loveland-Curtze, J., Gutshall, K.R., Auman, A.J., and Brenchley, J.E. (2003). Biochemical characterization of beta-galactosidase with a low temperature optimum obtained from an Antarctic *Arthrobacter* isolate. *Journal of Bacteriology* **185**, 5473–5482.

Cole, J.J., Carpenter, S.R., Kitchell, J.F. and Pace, M.L. (2002). Pathways of organic carbon utilization in small lakes: results from a whole lake ^{13}C addition and coupled model. *Limnology and Oceanography* **47**, 1664–1675.

Cole, J.J., Findlay, S. and Pace, M.L. (1988). Bacterial production in fresh and saltwater ecosystems: a cross-system overview. *Marine Ecology Progress Series* **43**, 1–10.

Colombet, J., Charpin, M., Robin, A., Portelli, C., Amblard, C., Cauchie, H.M. and Sime-Ngando, T. (2009). Seasonal depth-related gradients in virioplankton: standing stock and relationships with microbial communities in Lake Pavin (France). *Microbial Ecology* **58**, 728–736.

Conovitz, P.A., McKnight, D.M., MacDonald, L.H., Fountain, A.G. and House, H.R. (1998). Hydrologic processes influencing streamflow variation in Fryxell Basin, Antarctica. In Priscu J.C. (ed) *Ecosystem Dynamics in a Polar Desert: The McMurdo Dry Valleys*, pp. 93–108. Antarctic Research Series 72, American Geophysical Union, Washington, DC.

Cook, A.J., Fox, A.J., Vaughan, D.G. and Ferrigno, J.G. (2005). Retreating glacier fronts on the Antarctic Peninsula over the past half-century. *Science* **308**, 541–544.

Cremer, H., Gore, D., Hultzsch, N., Melles, M. and Wagner, B. (2004). The diatom flora and limnology of lakes in the Amery Oasis, East Antarctica. *Polar Biology* **27**, 513–531.

Cromer, L., Gibson, J.A.E., Swadling, K.M. and Ritz, D.A. (2005a). Faunal microfossils: indicators of Holocene ecological change in a saline Antarctic lake. *Palaeogeography, Palaeoclimology, Palaeoecology* **221**, 83–97.

Cromer, L., Williams, R. and Gibson, J.A.E. (2005b). *Trematomus scotti* in Beaver Lake: the first record of a fish from a non-marine Antarctic habitat. *Journal of Fish Biology* **66**, 1493–1497.

Crump, B.C., Kling, G.W., Bahr, M. and Hobbie, J.E. (2003). Bacterioplankton community shifts in an Arctic lake correlate with seasonal changes in organic matter source. *Applied and Environmental Microbiology* **69**, 253–2268.

Dartnall, H.J.G. (1980). Freshwater Biology at Rothera Point, Adelaide Island: 1. General description of the pool and the fauna. *British Antarctic Survey Bulletin* **50**, 51–54.

Dartnall, H.J.G. (1995). Rotifers, and other aquatic invertebrates, from the Larsemann Hills, Antarctica. *Papers and Proceedings of the Royal Society of Tasmania* **129**, 17–23.

Dartnall, H.J.G. (2000). A limnological reconnaissance of the Vestfold Hills. *Australian Antarctic Research Expeditions Reports* **141**, 1–55.

De Angelis, M., Petit, J.-R., Savarino, L., Souchez, R. and Thiemens, M.H. (2004). Contributions of ancient evaporitic-type reservoir to subglacial Lake Vostok chemistry. *Earth and Planetary Science Letters* **222**, 751–765.

DeConto, R. M., Galeotti, S., Pagani, M., Tracy, D., Schaefer, K. Zhang, T.J., Pollard, D. and Beerling, D.J. (2012). Past extreme warming events linked to massive carbon release from thawing permafrost. *Nature* **484**, 87–91.

Del Giorgio, P.A. and Cole, J.J. (1998). Bacterial growth efficiency in natural aquatic systems. *Annual Review of Ecological Systems* **29**, 503–541.

De los Rios, A., AScaso, C., Wierzchos, J., Fernández-Valiente, E. and Quesada, A. (2004). Microstructure characterization of cyanobacterial mats from the

McMurdo Ice Shelf, Antarctica. *Applied and Environmental Microbiology* **70**, 569–580.

DeMaere, M.Z., Williams, T.J., Allen, M.A., Brown, M.V., Gibson, J.A.E. and 8 others (2013). High Level of intergenera gene transfer exchange shapes the evolution of Haloarchaea in an isolated Antarctic lake. *Proceedings of the National Academy of Sciences*, doi:10.1073/pnas.1307090110.

Dhaked, R.K., Alam, S.I., Dixit, A. and Singh, L. (2005). Purification and characterization of thermo-labile alkaline phosphatase from an Antarctic psychotolerant Bacillus sp. P9. *Enzyme Microbiological Technology* **36**, 855–861.

Dieser, M., Foreman, C.M., Jaros, C., Lisle, J.T., Greenwood, M., Laybourn-Parry, J., Miller, P.L., Chin, Y.-P. and McKnight, D.M. (2012). Physicochemical dynamics in a coastal Antarctic lake as it transitions from frozen to open water. *Antarctic Science* **25**, 663–675.

Domack, E.W., Jacobson, E.A., Shipp, S. and Anderson, J.B. (1999). Late Pleistocene-Holocene retreat of the west Antarctic ice-sheet system in the Ross Sea: Part 2—sedimentologic and stratigraphic signature. *Geological Society of America Bulletin* **111**, 1517–1536.

Doran, P.T., Fritsen, C.H., McKay, C.P., Priscu, J.C. and Adams, E.E. (2003). Formation and character of an ancient 19-m ice cover and underlying trapped brine in an 'ice–sealed' east Antarctic lake. *Proceedings of the National Academy of Sciences* **100**, 26–31.

Doran, P.T., McKay, C.P., Clow, G.D., Dana, G.L., Fountain, A.G., Nylen, T. and Lyons, W.B. (2002a). Valley floor climate observations from the McMurdo Dry Valleys, Antarctica 1986–2000. *Journal of Geophysical Research* **107**, doi:10.1029/2001JD002045.

Doran, P.T., McKay, C.P., Fountain, A.G., Nylen, T., McKnight, D.M., Jaros, C. and Barrett, J.E. (2008). Hydrological response to extreme warm and cold summers in the McMurdo Dry Valleys, East Antarctic. *Antarctic Science* **20**, 499–509.

Doran, P.T., Priscu, J.C., Lyons, W.B., Walsh, J.E., Fountain, A.G. et al. (2002b). Antarctic climate cooling and terrestrial ecosystem response. *Nature* **415**, 517–520.

Doran, P.T., Wharton, R.A., Lyons, W.B., Des Marais, D.J. and Andersen, D.T. (2000). Sedimentology and geochemistry of a perennially ice-covered epishelf lake in Bunger Hills Oasis, East Antarctica. *Antarctic Science* **12**, 131–140.

Dore, J.E. and Priscu, J.C. (2001). Phytopkankton phosphorus deficiency and alkaline phosphatase activity in the McMurdo Dry Valley lakes, Antarctica. *Limnology and Oceanography* **46**, 1331–1346.

Dowdeswell, J.A. and Siegert, M.J. (1999). The dimensions and topographic setting of Antarctic subglacial lakes and implications for large-scale water storage beneath continental ice sheets. *Geological Society of America Bulletin* **111**, 254–263.

Dowdeswell, J.A. and Siegert, M.J. (2003). The physiography of modern Antarctic subglacial lakes. *Global and Planetary Change* **35**, 221–236.

Edmondson, W.T. (1955). The seasonal life history of *Daphnia* in an Arctic lake. *Ecology* **36**, 439–455.

Elliott, J.A. (2012). Predicting the impact of changing nutrient load and temperature on the phytopkankton of England's largest lake, Windermere. *Freshwater Biology* **57**, 400–413.

Ellis-Evans, J.C. (1981). Freshwater microbiology in the Antarctic: 1. Microbial numbers and activity in oligotrophic Moss Lake, Signy Island. *British Antarctic Survey Bulletin* **54**, 85–104.

Ellis-Evans, J.C. (1984). Methane in maritime Antarctic freshwater lakes. *Polar Biology* **3**, 63–71.

Ellis-Evans, J.C. (1996). Microbial diversity and function in Antarctic freshwater ecosystems. *Biodiversity and Conservation* **5**, 1395–1431.

Ellis-Evans, J.C., Laybourn-Parry, J. and Bayliss, P.R. (1997). Human impact on an oligotrophic lake in the Larsemann Hills. In Battaglia, B., Valencia, J. and Walton D.W.H. (eds) *Antarctic Communities, Species, Structure and Survival*, pp. 396–404. Cambridge University Press, Cambridge, UK.

Ellis-Evans, J.C., Laybourn-Parry, J., Bayliss, P. and Perriss S.J. (1998). Physical, chemical and microbial community characteristics of lakes of the Larsemann Hills, continental Antarctica. *Archiv für Hydrobiologie* **141**, 29–230.

Ellis-Evans, J. C. and Lemon, E.C.G. (1989). Some aspects of iron cycling in maritime Antarctic lakes. *Hydrobiologia* **172**, 149–164.

Emiliani, C. (1955). Pleistocene temperatures. *The Journal of Geology* **63**, 538–578.

Evans, J., Dowdeswell, J.A., Cofaigh, C.Ó., Benham, T.J. and Anderson, J.B. (2006). Extent and dynamics of the West Antarctic ice sheet on the outer continental shelf of pine island bay during the last glaciation. *Marine Geology* **230**, 53–72.

Fahnestock, M., Abdalat, W., Joughin, I., Brazena, J. and Prasad, G. (2001). High geothermal heat flow, basal melt and the origin of rapid ice flow in central Greenland. *Science* **294**, 2338–2342.

Felip, M. Camarero, L. and Catalan, J. (1999). Temporal changes of microbial assemblages in the ice and snow cover of a high mountain lake. *Limnology and Oceanography* **44**, 973–987.

Felip, M., Sattler, B., Psenner, R. and Catalan, J. (1995). Highly active microbial communities in the ice and snow cover of high mountain lakes. *Applied and Environmental Microbiology* **61**, 2394–2401.

Fernández-Valiente, E., Camacho, A., Rochera, C., Rico, E., Vincent, W.F. and Quesada, A. (2007). Community structure and physiological characterization of microbial mats in Byers Peninsula, Livingston Islands (South

Shetland Islands, Antarctica). *FEMS Microbiology Ecology* **59**, 377–385.

Fernández-Valiente, E., Quesada, A., Howard-Williams, C. and Hawes, I. (2001). N_2-fixation in cyanobacterial mats from ponds on the McMurdo Ice Shelf, Antarctica. *Microbial Ecology* **42**, 338–349.

Ferraccioli, F., Armadillo, E., Jordan, T., Bozzo E. and Corr, H. (2009). Aeromagnetic exploration over the east Antarctic ice sheet: A new view of the Wilkes subglacial basin. *Tectonophysics* **478**, 62–77.

Ferris, J.M. and Burton, H.R. (1988). The annual cycle of heat content and mechanical stability of hypersaline Deep Lake, Vestfold Hills, Antarctica. *Hydrobiologia* **165**, 115–128.

Filina, I.Y., Balnkenship, D.D., Thoma, M., Lukin, V.V., Masolove, V.N. et al. (2008). New 3D bathymetry and sediment distribution in Lake Vostok: Implications for pre-glacial origin and numerical modeling of the internal processes within the lake. *Earth and Planetary Science Letters* **276**, 106–114.

Flower, B.P. (1999). Cenozoic deep-sea temperatures and polar glaciation: The oxygen isotope record. In Barrett, P. and Orombelli, G. (eds) Proceedings of the Workshop: Geological Records of Global and Planetary Changes. *Terra Antarctica Reports* **3**, 27–42.

Foreman, C.M., Sattler, B., Mikucki, J.A., Porazinska, D.L. and Priscu, J.C. (2007). Metabolic activity and diversity of cryoconites in the Taylor Valley, Antarctica. *Journal of Geophysical Research* **112**, G04S32.

Foreman, C.M., Wolf, C.F. and Priscu, J.C. (2004). Impact of episodic warming events on the physical, chemical and biological relationships of lakes in the McMurdo Dry Valleys, Antarctica. *Aquatic Geochemistry* **10**, 239–268.

Francis, C.A., Beman, J.M. and Kuypers, M.M. (2007). New processes and players in the nitrogen cycle: The microbial ecology of anaerobic and archael ammonia oxidation. *The ISME Journal* **1**, 19–27.

Franzmann, P.D., Liu, Y., Balkwill, D.L., Aldrich, H.C., Conway, E. et al. (1997). *Methanogenium frigidum* sp. Nov., a psychrophilic, H_2-using methanogen from Ace Lake. *International Journal of Systematic Bacteriology* **47**, 1068–1072.

Franzmann, P.D., Roberts, N.J., Mancuso, C.A., Burton, H.R. and McMeekin, T.A. (1991). Methane production in meromictic Ace Lake, Antarctica. *Hydrobiologia* **210**, 191–201.

Franzmann, P.D. and Rohde, M. (1991). An obligately anaerobic, coiled bacterium from Ace Lake, Antarctica. *Journal of General Microbiology* **137**, 2191–2196.

Franzmann, P.D. and Rohde, M. (1992). Characteristics of a novel, anaerobic, mycoplasma-like bacterium from Ace Lake, Antarctica. *Antarctic Science* **4**, 155–162.

Franzmann, P.D., Skyring, G.W., Burton, H.R. and Deprez, P.P. (1988). Sulfate reduction rates and some aspects of the limnology of 4 lakes and a fjord in the Vestfold Hills, Antarctica. *Hydrobiologia* **165**, 25–33.

Fricker, H.A., Scambos, T., Bindschadler, R. and Padman, L. (2007). An active subglacial water system in West Antarctica mapped from space. *Science* **315**, 1544–1548.

Fritsen, C.H. and Priscu, J.C. (1998). Cyanobacterial assemblages in permanent ice covers on Antarctic lakes: distribution, growth rate and temperature response of photosynthesis. *Journal of Phycology* **34**, 587–597.

Fuhrman, J. (1987). Close coupling between release and uptake of dissolved free amino acids in seawater studied isotope dilution approach. *Marine Ecology Progress Series* **37**, 45–52.

Fuhrman, J.A. and Noble, R.T. (1995). Viruses and protists cause similar bacterial mortality in coastal seawater. *Limnology and Oceanography* **40**, 1236–1242.

Fukui, F., Torii, T. and Okabe, S. (1985). Vertical distribution of nutrients and DOC in lake waters near Syowa Station, Antarctica. *Antarctic Record* **86**, 28–35.

Fulford-Smith, S.P. and Sikes, E.L. (1996). The evolution of Ace Lake, Antarctica, determined from sedimentary diatom assemblages. *Palaeogeography, Paleoclimatology, Palaeoecology* **124**, 73–86.

Galchenko, V. (1994). Sulfate reduction, methane production, and methane oxidation in various water bodies of Bunger Hills oasis of Antarctica. *Microbiology and Molecular Biology Reviews* **63**, 388–396.

Galton-Fenzi, B.K., Hunter, J.R., Coleman, R. and Young, N. (2012). A decade of change in the hydraulic connection between an Antarctic epishelf lake and the ocean. *Journal of Glaciology* **58**, 223–228.

Gibson, J.A.E. (1999). The meromictic lakes and stratified marine basins of the Vestfold Hills, East Antarctica. *Antarctic Science* **11**, 172–189.

Gibson, J.A.E. and Andersen, D.T. (2002). Physical structure of the epishelf lakes of the southern Bunger Hills, East Antarctica. *Antarctic Science* **14**, 253–261.

Gibson, J.A.E., Dartnall, H.J.G. and Swadling, K.M. (1998). On the occurrence of males and production of ephippial eggs in populations of *Daphniopsis studeri* (Cladocera) in lakes of the Vestfold Hills and Larsemann Hill, East Antarctica. *Polar Biology* **19**, 148–150.

Gibson, J.A.E., Gore, D.B. and Kaup, E. (2002a). Algae River: an extensive drainage system in the Bunger Hills, East Antarctica. *Polar Record* **38**, 141–152.

Gibson, J.A.E., Paterson, K.S., White, C.A. and Swadling, K.M. (2009). Evidence for the continued existence of Abraxas Lake, Vestfold Hill, East Antarctica during the Last Glacial Maximum. *Antarctic Science* **21**, 269–278.

Gibson, J.A.E., Vincent, W.F., Van Hove, P., Belzile, C., Wang, X. and Muir, D. (2002b). Geochemistry of ice-covered, meromictic Lake A in the Canadian High Arctic. *Aquatic Geochemistry* **8**, 97–119.

Gilbert, J.A., Davies, P.L. and Laybourn-Parry, J. (2005). A hyperactive Ca^{2+} - dependent antifreeze protein in an Antarctic lake. *FEMS Microbiology Letters* **245**, 67–72.

Gilllieson, D., Burgess, J. Spate, A. and Cochrane, A. (1990). *An Atlas of the Lakes of the Larsemann Hills, Princess Elizabeth Land, Antarctica*. Australian Antarctic Research Expeditions, Antarctic Division, Kingston, Australia.

Gillieson, D.S. (1991). An environmental history of two freshwater lakes in the Larsemann Hills. *Hydrobiologia* **214**, 327–331.

Glatz, R.E., Lepp, P.W., Ward, B.B. and Francis, C.A. (2006). Planktonic microbial community composition across steep physical/chemical gradients in permanently ice-covered Lake Bonney, Antarctica. *Geobiology* **4**, 53–67.

Goldman, C.R., Mason, D.T. and Hobbie, J.E.(1967). Two Antarctic desert lakes. *Limnology and Oceanography* **12**, 295–310.

Goodchild, A., Saunders, N.F.W., Ertan, H., Raftery, M., Guilhaus, M. et al. (2004). A proteomic determination of cold adaptation in the Antarctic archaeon, Methanococcoides burtonii. *Molecular Microbiology* **53**, 309–321.

Goodwin, I.D. (1988). The nature and origin of a jokulhlaup near Casey Station, Antarctica. *Journal of Glaciology* **34**, 95–101.

Gore, D.B. (1997). Blanketing snow and ice may have prevented abundant biogenic sedimentation in the Eastern Antarctic oases flowing retreat of the ice sheet. *Antarctic Science* **9**, 336–346.

Gore, D.B., Pickard, J., Baird, A.S. and Webb, J.A. (1996). Glacial Crooked Lake, Vestfold Hills, East Antarctica. *Polar Record* **32**, 19–24.

Gore, D.B., Rhodes, E.J., Augustinus, P.C., Leishman, M.R., Colhourn, E.A. and Rees-Jones, J. (2001). Bunger Hills, East Antarctica: Ice free at the Last Glacial Maximum. *Geology* **29**, 1103–1106.

Gray, L., Joughin, I., Tulaczyk, S., Spikes, B., Bindschaller, R. and Jezek, K. (2005). Evidence for subglacial water transport in the West Antarctic ice sheet through three-dimensional satellite radar interferometry. *Geophysical Research Letters* **32**, doi:10.1029/2004GLO21387.

Green, W.J. and Lyons, W.B. (2009). The saline lakes of the McMurdo Dry Valleys, Antarctica. *Aquatic Geochemistry* **15**, 321–348.

Grey, J., Laybourn-Parry, J., Leakey, R.J.G. and McMinn, A. (1997). Temporal patterns of protozooplankton abundance and their food in Ellis Fjord, Princess Elizabeth Land, Eastern Antarctica. *Estuarine, Coastal and Shelf Science* **45**, 17–25.

Grossart, H.-P. and Simon, M. (1993). Limnetic macroscopic organic aggregates (lake snow): occurrence, characteristics, and microbial dynamics in Lake Constance. *Limnology and Oceanography* **38**, 532–546.

Gustafson, D.E., Stoecker, D.K., Johnson, M.D., Van Heukelem, W.F. and Sneider, K. (2000). Cryptophyte algae robbed of their organelles by the marine ciliate Mesodinium rubrum. *Nature* **405**, 1049–1052.

Hall, B. and Denton, G. (2000). Radiocarbon chronology of Ross Sea Drift, Eastern Taylor Valley, Antarctica: evidence for a grounded ice sheet in the Ross Sea at the Last Glacial Maximum. *Geografiska Annaler: Series A* **82**, 305–336.

Hall, B.L., Denton, G.H. and Oveturf, B. (2001). Glacial Lake Wright, a high-level Antarctic lake during the LGM and early Holocene. *Antarctic Science* **13**, 53–60.

Hand, R.M. and Burton, H.R. (1981). Microbial ecology of an Antarctic saline meromictic lake. *Hydrobiologia* **82**, 363–374.

Handelsman, J. (2004). Metagenomics: application of genomics to uncultured microorganisms. *Microbiological and Molecular Biology Reviews* **68**, 669–685.

Hansson, L.-A, Hylander, S., Dartnall, H.J.G., Lidström, S. and Svensson, J.-E. (2011). High zooplankton diversity in the extreme environments of the McMurdo Dry Valley lakes, Antarctica. *Antarctic Science*, doi:10.1017/S095410201100071X.

Hawes, I. (1983). Nutrients and their effects on phytoplankton populations on lakes on Signy Island, Antarctica. *Polar Biology* **2**, 115–126.

Hawes, I. (1985a). Light climate and phytoplankton photosynthesis in maritime Antarctic lakes. *Hydrobiologia* **123**, 69–79.

Hawes, I. (1985b). Factors controlling phytoplankton populations in maritime Antarctic lakes. In Siegfried, W.R., Condy, P.R. and Laws, R.M. (eds) *Antarctic Nutrient Cycles and Food Webs*, pp. 245–252. Springer-Verlag, Berlin and Heidelberg.

Hawes, I. (1990). The effects of light and temperature on photosynthate partitioning in Antarctic freshwater phytoplankton. *Journal of Plankton Research* **12**, 513–518.

Hawes, I., Howard-Williams, C. and Fountain, A.G. (2008b). Ice-based freshwater ecosystems. In Vincent, W.F. and Laybourn-Parry, J. (eds) *Polar Lakes and Rivers, Limnology of Arctic and Antarctic Ecosystems*, pp. 103–118. Oxford University Press, Oxford, UK.

Hawes, I., Howard-Williams, C. and Pridmore, R.D. (1993). Environmental control of microbial biomass in ponds of the McMurdo Ice Shelf, Antarctica. *Archiv für Hydrobiologie* **127**, 271–287.

Hawes, I., Howard-Williams, C. and Vincent, W.F. (1992).Desiccation and recovery of Antarctic cyanobacterial mats. *Polar Biology* **12**, 587–594.

Hawes, I., Moorhead, D., Sutherland, D., Schmeling, J. and Schwarz, A.-M. (2001). Benthic primary production in two perennially ice-covered Antarctic lakes: patterns of biomass accumulation with a model of community metabolism. *Antarctic Science* **13**, 18–27.

Hawes, I., Smith, R., Howard-Williams, C. and Schwarz, A.-M. (1999). Environmental conditions during freezing, and response of microbial mats in ponds of the McMurdo Ice Shelf, Antarctica. *Antarctic Science* **11**, 198–208.

Hawes, I., Safi, K., Sorrell, B., Webster-Brown, J. and Arscott, D. (2011b). Summer-winter transitions in Antarctic ponds I: the physical environment. *Antarctic Science* **23**, 235–242.

Hawes, I., Safi, K., Webster-Brown, J., Sorrell, B. and Arscott, D. (2011c). Summer-winter transitions in Antarctic ponds II: biological responses. *Antarctic Science* **23**, 243–254.

Hawes, I. and Schwarz, A.-M. (1999) Photosynthesis in an extreme shade environment: benthic microbial mats from Lake Hoare, a permanently ice-covered Antarctic lake. *Journal of Phycology* **35**, 448–459.

Hawes, I. and Schwarz, A.-M. (2000). Absorption and utilization of irradiance by cyanobacterial mats in two ice-covered Antarctic lakes with contrasting light climates. *Journal of Phycology* **37**, 5–15.

Hawes, I., Sumner, D.Y., Andersen, D.T. and MacKey, T.J. (2011a). Legacies of recent environmental change in the benthic communities of Lake Joyce, a perennially ice-covered Antarctic lake. *Geobiology* **9**, 394–410.

Hawes, T.C., Worland, M.R. and Bale, J.S. (2008a). Physiological constraints on the life cycle and distribution of the Antarctic fairy shrimp *Branchinecta gaini*. *Polar Biology* **31**, 1531–1538.

Healy, M., Webster-Brown, J.G., Brown, K.L. and Lane, V. (2006). Chemistry and stratification of Antarctic meltwater ponds II: inland ponds in the McMurdo Dry Valleys, Victoria Land. *Antarctic Science* **18**, 525–533.

Heath, C.W. (1988). Annual primary productivity of an Antarctic continental lake: phytoplankton and benthic algal mat production strategies. *Hydrobiologia* **165**, 77–87.

Hebert, P.D.N. (1981). Obligate asexuality in *Daphnia*. *American Naturalist* **117**, 784–780.

Hendy, C.H. (2000). Late Quaternary lakes in the McMurdo Sound region of Antarctica. *Geografiska Annaler* **82A**, 411–432.

Henshaw, T. and Laybourn-Parry, J. (2002). The annual patterns of photosynthesis in two large freshwater, ultra-oligotrophic Antarctic lakes. *Polar Biology* **25**, 744–752.

Herbei, R., Lyons, W.B., Laybourn-Parry, J., Gardner, C., Priscu, J.C. and McKnight, D.M. (2010). Physiochemical properties influencing biomass abundance and primary production in Lake Hoare, Antarctica. *Ecological Modelling* **221**, 1184–1193.

Hewson, I., O'Neil, J.M., Fuhrman, J.A. and Dennison, W.C. (2001). Virus-like particle distribution and abundance in sediments and overlying waters along eutrophication gradients in two subtropical estuaries. *Limnology and Oceanography* **46**, 1737–1746.

Heywood, R.B. (1972). Antarctic limnology: a review. *British Antarctic Survey Bulletin* **29**, 35–65.

Heywood, R.B. (1977). A limnological survey of the Ablation Point area, Alexander Island, Antarctica. *Philosophical Transactions of the Royal Society, London B* **279**, 39–54.

Heywood, R.B., Dartnall, H.J.G. and Priddle, J. (1979) The freshwater lakes of Signy Island, South Orkney Islands Antarctic: data sheets. Volume 3, British Antarctic Survey, Cambridge UK.

Hill, D.J., Heywood, A.M., Hindmarsh, R. and P. Valdes, (2007). Characterizing ice sheets during the Pliocene: evidence from data and models. In Williams, M., Haywood, A.M., Gregory, J., and Schmidt, D.N. (eds) *Deep Time Perspectives on Climate Change: Marrying the Signals from Computer Models and Biological Proxies, Micropalaeont. Soc. Spec. Pub.*, pp. 517–538. 3185. Geological Society, London.

Hodgson, D.A., Convey, P., Verleyen, E., Vyverman, W., McInnes, S.J., Sands, C.J. and 6 others (2010). The limnology and biology of the Dufek Massif, Transantarctic Mountains 82° South. *Polar Science* **4**, 197–214.

Hodgson, D.A., Dyson, C.L., Jones, V.J. and Smellie, J.L. (1998). Tephra analysis of sediments from Midge lake (South Shetland Islands) and Sombre Lake (South Orkney Islands), Antarctica. *Antarctic Science* **10**, 13–20.

Hodgson, D.A., Johnston, N.M., Caulkett, A.P. and Jones, V.J. (1998). Palaeolimnology of Antarctic fur seal *Arctocephalus gazelle* populations and implications for Antarctic managment. *Biological Conservation* **83**, 145–154.

Hodgson, D.A., Noon, P.E., Vyverman, W., Bryant, C.L., Gore, D.B. et al. (2001a). Were the Larsemann Hills ice-free through the last Glacial Maximum? *Antarctic Science* **13**, 440–454.

Hodgson, D.A., Roberts, D., McMinn, A., Verleyen, E. and three others (2006b). Recent rapid salinity rise in three East Antarctic lakes. *Journal of Paleolimnology* **36**, 385–406.

Hodgson, D.A., Roberts, S.J., Bentley, M.J., Smith, J.A., Johnson, J.S. et al. (2009a). Exploring former subglacial Hodgson Lake, Antarctica Paper 1: site description, geomorphology and limnology. *Quaternary Science Reviews* **28**, 2295–2309.

Hodgson, D.A., Roberts, S.J., Bentley, M.J., Carmichael, E.L., Smith, J.A. et al. (2009b). Exploring former subglacial Hodgson Lake. Paper II: Palaeolimnology. *Quaternary Science Reviews* **28**, 2310–2325.

Hodgson, D.A., Verleyen, E., Squier, A.H., Sabbe, K., Keely, B.J. et al. (2006a). Interglacial environments of coastal east Antarctica: comparison of MIS 1 (Holocene and

MIS 5e (Last Interglacial) lake-sediment records. *Quaternary Science Reviews* **25**, 179–197.

Hodgson, D.A., Vyvermann, W. and Sabbe, K. (2001b). Limnology and biology of saline lakes in the Rauer Islands, eastern Antarctica. *Antarctic Science* **13**, 255–270.

Hodgson, D.A., Vyvermann, W., Veleyen, E., Sabbe, K., Leavitt, P.R. et al. (2004). Environmental factors influencing the pigment composition of *in situ* benthic microbial communities in east Antarctic lakes. *Aquatic Microbial Ecology* **37**, 347–263.

Hodson, A.J., Anesio, A.M. and Tranter, M. (2007). A glacier respires: quantifying the distribution and respiration CO_2 flux of cryoconite across an entire supraglacial ecosystem. *Journal of Geophysical Research* **112** (G04S36), doi:10.1029/2007JG000452.

Hodson, A.J., Mumford, P.N., Kohler, J. and Wynn, P.M. (2005). The High Arctic glacial ecosystem: new insights from nutrient budgets. *Biogeochemistry* **72**, 233–256.

Hodson, A.J., Paterson, H., Westwood, K., Cameron, K. and Laybourn-Parry, J. (2013). A blue ice ecosystem on the margins of the East Antarctic Ice Sheet. *Journal of Glaciology* **59**, doi:10.3189/2013JoG12J052.

Hofer, J.S. and Sommaruga, R. (2001). Seasonal dynamics of viruses in an alpine lake: importance of filamentous forms. *Aquatic Microbial Ecology* **26**, 1–11.

Howard-Williams, C., Pridmore, R., Downes, M.T. and Vincent, W.F. (1989). Microbial biomass, photosynthesis and chlorophyll a related pigments in the ponds of the McMurdo Ice Shelf, Antarctica. *Antarctic Science* **1**, 125–131.

Howard-Williams, C., Pridmore, R.D., Broady, P.A. and Vincent, W.F. (1990). Environmental and biological variability in the McMurdo Ice Shelf ecosystem. In Kerry, K.R. and Hempel, G. (eds) *Antarctic Ecosystems: Ecological Change and Conservation*, pp. 23–31. Springer-Verlag, Berlin and Heidelberg.

Howard-Williams, C., Schwarz, A-M. and Hawes, I. (1998). Optical Properties of the McMurdo Dry Valley lakes; Antarctica. In J.C. Priscu (ed) *Ecosystem Dynamics in a Polar Desert: The McMurdo Dry Valleys, Antarctica*, pp. 189–204. Antarctic Research Series Volume 72, American Geophysical Union, Washington, DC.

Hughes, T. (1998). *Ice Sheets*. Oxford University Press, New York, 343 pp.

Huybers, P. and Denton, G. (2008). Antarctic temperature at orbital timescales controlled by local summer duration. *Nature Geoscience* **1**, 787–792.

Imura, S., Bando, T., Saito, S., Set, K. and Kanda, H. (1999). Benthic moss pillars in Antarctic lakes. *Polar Biology* **22**, 137–140.

Ingólfsson, Ó., Hjort, C., Berkman, P.A., Björck, S., Colhoun, E. and six others (1998). Antarctic glacial history since the Last Glacial Maximum: an overview of the record on land. *Antarctic Science* **10**, 326–344.

Izaguirre, I., Allende, L. and Marinone, M.C. (2003). Comparative study of the planktonic communities of three lakes of contrasting trophic status at Hope Bay (Antarctic Peninsula). *Journal of Plankton Research* **25**, 1079–1097.

Izaguirre, I., Mataloni, G., Vinocur, A. and Tell, G. (1993). Temporal and spatial variations of phytoplankton from Boeckella Lake (Hope Bay, Antarctic Peninsula). *Antarctic Science* **5**, 137–141.

James, M.R., Hall, J.A. and Laybourn-Parry, J. (1998). Protozooplankton of the Dry Valley Lakes of Southern Victoria Land. In J.C. Priscu (ed) *Ecosystem Dynamics in a Polar Desert*, pp. 225–268, Antarctic Research Series Academic Press 72, American Geophysical Union, Washington, DC.

James, M.R., Pridmore, R.D. and Cummings, V.J. (1995). Planktonic communities of melt ponds on the McMurdo Ice Shelf, Antarctica. *Polar Biology* **15**, 555–567.

James, S.R., Burton, H.R., McMeekin, T.A. and Mancuso, C.A. (1994). Seasonal abundance of Halomonas meridian, Halomonas subglaciescola, Flavobacerium gondwanense and Flavobacterium salegens in four Antarctic lakes. *Antarctic Science* **6**, 325–332.

Janiec, K. (1996). The comparison of freshwater invertebrates of Spitzbergen (Arctic) and King George Island (Antarctica). *Polish Polar Research* **17**, 173–202.

Jones, V.J. (1996). The diversity, distribution and ecology of diatoms from Antarctic inland waters. *Biodiversity and Conservation* **5**, 1433–1449.

Jones, V.J., Hodgson, D.A. and Chepstow-Lusty, A. (2000). Palaeolimnological evidence for marked Holocene environmental changes on Signy Island, Antarctica. *The Holocene* **10**, 43–60.

Jones, V.J., Juggins, S. and Ellis-Evans, J.C. (1993). The relationship between water chemistry and surface sediment diatom assemblages in maritime Antarctic lakes. *Antarctic Science* **5**, 339–348.

Joughin, I., Tulaczyk, S., MacAyeal, D.R. and Engelhardt, H. (2004). Melting and freezing beneath the Ross ice streams, Antarctica. *Journal of Glaciology* **50**, 96–108.

Jouzel, J., Lorius, C., Petit, J.R., Genthon, C., Barkov, N.I., Kotlyakov, V.M. and Petrov, V.M. (1987). Vostok ice core: a continuous isotope temperature record over the last climatic cycle (160,000 years). *Nature* **329**, 403–408.

Jouzel, J., Masson-Delmotte, V., Cattani, O., Dreyfus, G., Falourd, S. et al. (2007). Orbital and millennial antarctic climate variability over the past 800,000 years. *Science* **317**, 793–796.

Jouzel, J., Petit, J.R., Souchez, R., Barkov, N.I., Lipenkov, V.Y. et al. (1999). More than 200 meters of lake ice above subglacial Lake Vostok. *Science* **286**, 2138–2141.

Jungblut, A.-D., Hawes, I., Mountfort, D., Hitzfeld, B., Dietrich, D.R. et al. (2005). Diversity within cyanobacterial mat communities in variable salinity meltwater ponds of McMurdo Ice Shelf, Antarctica. *Environmental Microbiology* **7**, 519–529.

Jungblut, A.-D., Lovejoy, C. and Vincent, W.F. (2010). Global distribution of cyanobacterial ecotypes in the cold biosphere. *ISME Journal* **4**, 191–202.

Karl, D.M., Bird, D.F., Björkman, K., Houlihan, T., Shackelford, R. and Tupas, L. (1999). Microorganisms in the accreted ice of Lake Vostok, Antarctica. *Science* **286**, 2144–2147.

Karr, E.A., Ng, J.M., Belchik, S.M., Sattley, W.M., Madigan, M.T. and Achenbach, L.A. (2006). Biodiversity of methanogenic and other Archaea in the permanently frozen Lake Fryxell, Antarctica. *Applied and Environmental Microbiology* **72**, 1663–1666.

Karr, E.A., Sattley, M.W., Jung, D.O., Madigan, M. and Achenbach, L.A. (2003). Remarkable diversity of phototrophic purple bacteria in a permanently frozen Antarctic lake. *Applied and Environmental Microbiology* **69**, 4910–4914.

Karr, E.A., Sattley, W.M., Rice, M.R., Jung, D.O., Madigan, M.T. and Achenbach, L.A. (2005). Diversity and distribution of sulfate-reducing bacteria in permanently frozen Lake Fryxell, McMurdo Dry Valleys, Antarctica. *Applied and Environmental Microbiology* **71**, 6353–6359.

Kaspar, M., Simmons, G.M., Parker, B.C., Seaburg, K.G. and Wharton, R.A. (1982). Bryum Hedw. collected from Lake Vanda, Antarctica. *Bryologist* **85**, 424–430.

Kato, K., Arikawa, T., Imura, S. and Kanda, H. (2013). Molecular identification and phylogeny of an aquatic moss species in Antarctic lakes. *Polar Biology* **36**, 1557–1568.

Kaup, E. (1994). Annual primary production of phytoplankton in Lake Verkhneye, Schirmacher Oasis, Antarctica. *Polar Biology* **14**, 433–439.

Kaup, E. (1995). Chlorophyll and Primary Production. In Bormann, P. and Fritzsche, D. (eds) *The Schirmacher Oasis, Queen Maud Land, East Antarctica, and its Surroundings*, pp. 316–319. Justus Perthes Verlag, Gotha, Germany.

Kaup, E., Haendel, D. and Vaikmäe, R. (1993). Limnological features of the saline lakes of the Bunger Hills (Wilkes land, Antarctica). *Antarctic Science* **5**, 41–50.

Kawecka, B., Olech, M., Nowogrodzka-Zagórska, M. and Wojtun, B. (1998). Diatom communities in small water bodies at H. Arctowski Polish Antarctic Station (King George Island, South Shetland Islands, Antarctica). *Polar Biology* **19**, 183–192.

Kepner, R.L., Wharton, R.A. and Coats, D.W. (1999). Ciliated protozoa in two Antarctic lakes: analysis by quantitative protargol staining and examination of artificial substrates. *Polar Biology* **21**, 285–294.

Kepner, R.L., Wharton, R.A. and Suttle, C.A. (1998). Viruses in Antarctic lakes. *Limnology and Oceanography* **43**, 1754–1761.

Kilman, P. (1981). Pelagic bacteria: extreme abundances in African saline lakes. *Naturwissenschaften* **68**, 380–381.

Kimura, S., Ban, S., Imura, S., Kudoh, S. and Matsuzaki, M. (2010). Limnological characteristics of vertical structure in lakes of Syowa Oasis, East Antarctica. *Polar Science* **3**, 262–271.

Klokov, V., Kaup, E., Zierath, R. and Haendel, D. (1990). Lakes of the Bunger Hill (East Antarctica): chemical and ecological properties. *Polish Polar Research* **11**, 47–159.

Komárek, J. and Komárek, O. (2003). Diverity of cyanobacteria in seepages of King George Island, maritime Antarctica. In Huiskes, A.H.L., Giekes, W.W.C., Rozema, J., Schorno, R.M.L., van der Vies, S.M. and Wolf, W.J. (eds) *Antarctic Biology in a Global Context*, pp. 244–250. Backhuys Publishers, Leiden, the Netherlands.

Koonin, E.V. and Wolf, Y.I. (2008). Genomics of Bacteria and Archaea: emerging dynamic view of the prokaryote world. *Nucleic Acids Research* **36**, 6688–6719.

Kudoh, S., Tanabe, Y., Matsuzaki, M. and Imura, S. (2009). *In situ* photochemical activity of the phytobenthic communities in two Antarctic lakes. *Polar Biology* **32**, 1617–1627.

Labrenz, M. and Hirsch, P. (2001). Physiological diversity and adaptations of aerobic heterotrophic bacteria from different depth of hypersaline, heliothermal and meromictic Ekho Lake (East Antarctica). *Polar Biology* **24**, 320–327.

La Scola, B., Desnues, C., Pagnier, I., Robert, C., Barrassi, L. and 6 others (2008). The virophage as a unique parasite of the giant mimivirus. *Nature* **455**, 100–105.

Latifovie, R. and Pouliot, D. (2007). Analysis of climate change impacts in lake phenology in Canada using historical satellite data record. *Remote Sensing of Environment* **106**, 492–507.

Lauro, F.M., DeMaere, M.Z., Yay, S., Brown, M.V., Ng, C. and 8 others (2011). An integrated study of a meromictic lake ecosystem in Antarctia. *The ISME Journal* **5**, 879–895.

Lawrence, M. J. F. and Hendy, C.H. (1985). Water column and sediment characteristics of Lake Fryxell, Taylor Valley, Antarctica. *New Zealand Journal of Geology and Geophysics* **28**, 543–552.

Lawson, J., Doran, P.T., Kenig, F., Des Marais, D.J. and Priscu, J.C. (2004). Stable carbon and nitrogen isotopic composition of benthic and pelagic organic matter of the McMurdo Dry Valleys, Antarctica. *Aquatic Geochemistry* **10**, 269–301.

Laybourn-Parry, J. (1992). *Protozoan Plankton Ecology*. Chapman and Hall, London, UK, 231 pp.

Laybourn-Parry, J., Anesio, A.M., Madan, N. and Säwström, C. (2013). Viral dynamics in a large epishelf lake (Beaver Lake, Antarctica). *Freshwater Biology* doi:10.1111/fwb.12146.

Laybourn-Parry, J. and Bayliss, P. (1996). Seasonal dynamics of the planktonic community of Lake Druzhby,

Princess Elizabeth Land, Eastern Antarctica. *Freshwater Biology* **35**, 57–67.

Laybourn-Parry, J., Bayliss, P. and Ellis-Evans, J.C. (1995). The dynamics of heterotrophic nanoflagellates and bacterioplankton in a large ultra-oligotrophic Antarctic lake. *Journal of Plankton Research* **17**, 1835–1850.

Laybourn-Parry, J., Bell, E.M. and Roberts, E.C. (2000). Protozoan growth rates in Antarctic lakes. *Polar Biology* **23**, 445–451.

Laybourn-Parry, J., Ellis-Evans, J.C. and Butler, H. (1996). Microbial dynamics during the summer ice-loss phase in maritime Antarctic lakes. *Journal of Plankton Research* **18**, 495–511.

Laybourn-Parry, J., Henshaw, T., Jones, D.J. and Quayle, W. (2004). Bacterioplankton production in freshwater Antarctic lakes. *Freshwater Biology* **49**, 735–744.

Laybourn-Parry, J., Hofer, J.S. and Sommaruga, R. (2001a). Viruses in the plankton of freshwater and saline Antarctic lakes. *Freshwater Biology* **46**, 1279–1287.

Laybourn-Parry, J., James, M.R., McKnight, D.M., Priscu, J.C., Spaulding, S.A. and Shiel, R. (1997). The microbial plankton of Lake Fryxell, southern Victoria Land, Antarctica during the summers of 1992 and 1994. *Polar Biology* **17**, 54–61.

Laybourn-Parry, J., Madan, N.J., Marshall, W.A., Marchant, H.J. and Wright, S.W. (2006). Carbon dynamics in an ultra-oligotrophic epishelf lake (Beaver Lake, Antarctica) in summer. *Freshwater Biology* **51**, 1116–1130.

Laybourn-Parry, J. and Marchant, H.J. (1992). *Daphniopsis studeri* (Crustacea: Cladocera) in lakes of the Vestfold Hills, Antarctica. *Polar Biology* **11**, 631–635.

Laybourn-Parry, J., Marchant, H.J. and Brown, P.E. (1992). Seasonal cycle of the microbial plankton in Crooked Lake, Antarctica. *Polar Biology* **12**, 411–416.

Laybourn-Parry, J. and Marshall, W.A. (2003). Photosynthesis, mixotrophy and microbial plankton dynamics in two high Arctic lakes during summer. *Polar Biology* **26**, 517–524.

Laybourn-Parry, J., Marshall, W.A. and Madan, N.J. (2007). Viral dynamics and patterns of lysogeny in saline Antarctic lakes. *Polar Biology* **30**, 351–358.

Laybourn-Parry, J., Marshall, W.A. and Marchant, H.J. (2005). Flagellate nutritional versatility as a key to survival in two contrasting Antarctic saline lakes. *Freshwater Biology* **50**, 830–838.

Laybourn-Parry, J. and Perriss, S.J. (1995). The role and distribution of the autotrophic ciliate *Mesodinium rubrum (Myrionecta rubra)* in three Antarctic saline lakes. *Archiv für Hydrobiologie* **135**, 179–194.

Laybourn-Parry, J., Quayle, W. and Henshaw, T. (2002). The biology and evolution of Antarctic saline lakes in relation to salinity and trophy. *Polar Biology* **25**, 542–552.

Laybourn-Parry, J. Quayle, W.C., Henshaw, T., Ruddell, A. and Marchant, H.J. (2001b). Life on the edge: the plankton and chemistry of Beaver Lake, an ultra-oligotrophic epishelf lake, Antarctica. *Freshwater Biology* **46**, 1205–1217.

Laybourn-Parry, J., Tranter, M. and Hodson, H.J. (2008). *The Ecology of Snow and Ice Environments*, Oxford Unversity Press, Oxford.

Laybourn-Parry, J. and Walton, M. (1998). Seasonal heterotrophic flagellate and bacterial plankton dynamics in a large oligotrophic lake- Loch Ness. *Freshwater Biology* **39**, 1–8.

Laybourn-Parry, J., Walton, M., Young, J., Jones, R.I. and Shine, A. (1994). Protozooplankton and bacterioplankton in a large oligotrophic lake–Loch Ness, Scotland. *Journal of Plankton Research* **16**, 1655–1670.

Le Brocq, A.M., Ross, N., Griggs, J.A., Bingham, R.G. et al. (2013). Evidence from ice shelves for channelized meltwater flow beneath the Antarctic Ice Sheet. *Nature Geoscience* **6**, 945–948.

Lee, S.Y. and Holder, G.D. (2000). A generalized model for calculating equilibrium states of gas hydrates: Par II. *Annals of the Academy of the New York Academy of Sciences* **912**, 614–622.

Legrand, M., Petit, J.R. and Lorius, C. (1988). Vostok (Antarctica) Ice Core- atmospheric chemistry changes over the last climatic cycle (160,000 years). *Chemical Geology* **70**, 101.

Leppäranta, M., Jarvinen, O. and Matilla, O.-P. (2012) Structure and life cycle of supraglacial lakes in Dronning Maud Land. *Antarctic Science* **24**, doi:10.1017/S0954 102012001009.

Liang, Y.-L., Colgan, W., Lv, Q., Steffan, K., Abdalati, W. et al. (2012). A decadal investigation of supraglacial lakes in West Greenland using a fully automatic detection and tracking logarithm. *Remote Sensing of Environment* **123**, 127–138.

Light, J.J., Ellis-Evans, J.C. and Priddle, J. (1981). Phytoplankton ecology in an Antarctic lake. *Freshwater Biology* **11**, 11–26.

Lindholm, T. (1985). *Mesodinium rubrum*–a unique photosynthetic ciliate. *Advances in Aquatic Microbiology* **3**, 1–48.

Lisiecki, L.E. and Raymo, M.E. (2005). A Pliocene-Pleistocene stack of 57 globally distributed benthic $\delta O18$ records. Paleoceanography **20** (1), PA1003, doi:10.1029/2004PA001071.

Lisle, J.T. and Priscu, J.C. (2004). The occurrence of lysogenic bacteria and microbial aggregates in the lakes of the McMurdo Dry Valleys, Antarctica. *Microbial Ecology* **47**, 427–439.

Lizotte, M.P. (2008). Phytoplankton and primary production. In Vincent, W.F. and Laybourn-Parry, J. (eds) *Polar*

Lakes and Rivers: Limnology of Arctic and Antarctic Aquatic Ecosystems, pp. 157–178. Oxford University Press, Oxford.

Lizotte, M.P. and Priscu, J.C. (1994). Natural fluorescence and quantum yields in vertically stationary phytoplankton from perennially ice-covered lakes. *Limnology and Oceanography* **39**, 1399–1410.

Lizotte, M.P., Sharp, T.R. and Priscu, J.C. (1996). Phytoplankton dynamics in the stratified water column of Lake Bonney, Antarctica. I. Biomass and productivity during the winter-spring transition. *Polar Biology* **16**, 155–162.

Llames, M.E. and Vinocur, A. (2007). Phytoplankton structure and dynamics in a volcanic lake in Deception Island (South Shetland Islands, Antarctica). *Polar Biology* **30**, 849–857.

Logares, R., Lindström, E.S., Langenheder, S., Logue, J.B., Paterson, H. and four others (2012). Biogeography of bacterial communities exposed to progressive long-term environmental change. *The ISME Journal*, doi:10.1038/ismej.2012.168.

Loopman, A. and Klokov, V. (1988). The formation of water run-off from lake catchments of the Schirmacher Oasis in East Antarctica during summer season 1983–1984. In Martin, J. (ed) *Limnological Studies in Queen Maud Land, East Antarctica*, pp. 57–65. Valgus, Tallinn.

López-Bueno, A., Tamares, J., Velázquez, D., Moya, A. et al. (2009). High diversity of the viral community from an Antarctic lake. *Science* **326**, 858–861.

Lowe, A. and Anderson, J.B. (2002). Reconstruction of the West Antarctic ice sheet in Pine Island Bay during the Last Glacial Maximum and its subsequent retreat history. *Quaternary Science Reviews* **21**, 1879–1897.

Lowe, A. and Anderson, J.B. (2003). Evidence for abundant subglacial meltwater beneath the paleo-ice sheet in Pine Island Bay, Antarctica. *Journal of Glaciology* **49**, 125–138.

Ludlam, S.D. (1996). The comparative limnology of high arctic, coastal, meromictic lakes. *Journal of Paleolimnology* **16**, 111–131.

Lymer, D. Logue, J.B., Brussaard, C.P.D., Baudoux, A.-C, Vrede, K. and Lindström E.S. (2008). Temporal variation in freshwater viral and bacterial community composition. *Freshwater Biology* **53**, 1163–1175.

Lyons, W.B. and Finlay, J.C. (2008). Biogeohemical processes in high-latitude lakes and rivers. In Vincent, W.F. and Laybourn-Parry, J. (eds) *Polar Lakes and Rivers: Limnology of Arctic and Antarctic Aquatic Ecosystems*, pp. 137–156. Oxford University Press, Oxford.

Lyons, W.B., Fountain, A., Doran, P., Priscu, J.C., Neumann, K. and Welch, K.A. (2000). Importance of landscape position and legacy: the evolution of the lakes in the Taylor Valley. *Freshwater Biology* **43**, 355–367.

Lyons, W.B., Frape, S.K. and Welch, K.A. (1999). History of the McMurdo Dry Valley lakes, Antarctica, from stable chlorine isotope data. *Geology* **27**, 527–530.

Lyons, W.B., Tyler, S.W., Wharton, R.A., McKnight, D.M. and Vaughn, B.H. (1998a). A late Holocene desiccation of Lake Hoare and Lake Fryxell, McMurdo Dry Valley, Antarctica. *Antarctic Science* **10**, 247–256.

Lyons, W.B., Welch, K.A. and Sharma, P. (1998b). Chloride-36 in the waters of the McMurdo Dry Valley lakes, southern Victoria Land, Antarctica: revisited. *Geochimica et Cosmochmica Acta* **62**, 185–191.

Lyons, W.B., Welch, K.A., Gardner, C.B., Jaros, C., Moorhead, D.L., Knoepfle, J.L. and Doran, P.T. (2012). The geochemistry of upland ponds, Taylor Valley, Antarctica. *Antarctic Science* **24**, 3–14.

Lyons, W.B., Welch, K.A., Snyder, G. et al. (2005). Halogen geochemistry of the McMurdo Dry Valley lakes, Antarctica: clues to the origin of solutes and lake evolution. *Geochimica et Cosmochimica Acta* **69**, 305–325.

Macayeal, D.R. (1992). Irregular oscillations of the west Antarctic ice-sheet. *Nature* **359**, 29–32.

Madan, N.J., Marshall, W.A. and Laybourn-Parry, J. (2005). Virus and microbial loop dynamics over an annual cyle in three contrasting Antarctc lakes. *Freshwater Biology* **50**, 1291–1300.

Mancuso, C.A., Franzmann, P.D., Burton, H.R. and Nichols, P.D. (1990). Microbial community structure and biomass estimates of a methanogenic Antarctic lake ecosystem as determined by phospholipid analysis. *Microbial Ecology* **19**, 73–95.

Maranger, R. and Bird, D.F. (1995). Viral abundance in aquatic systems: a comparison between marine and fresh waters. *Marine Ecology Progress Series* **121**, 217–226.

Marchant, H.J. (1985). Choanoflagellates in the Antarctic marine food chain. In Siegfried, W.R., Condy, P.R. and Law, R.M. (eds) *Antarctic Nutrient Cycles and Food Webs*, pp. 271–276. Springer, Berlin and Heidelberg.

Margulis, L. (1974). Five Kingdom classification and origin and evolution of cell. In Dobzansky, T., Hecht, M.K. and Steere, W.C. (eds) *Evolutionary Biology Volume Seven*, Plenum Press, New York.

Marion, G.M. (1997). A theoretical evaluation of mineral stability in Don Juan Pond, Wright Valley, Victoria Land. *Antarctic Science* **9**, 92–99.

Markager, S., Vincent, W.F. and Tang, E.P.Y. (1999). Carbon fixation by phytoplankton in high Arctic lakes: implications of low temperature for photosynthesis. *Limnology and Oceanography* **44**, 597–607.

Marshall, W.A. (1996). Biological particles over Antarctica. *Nature* **383**, 680.

Marshall, W.A. and Laybourn-Parry, J. (2002). The balance between photosynthesis and grazing in Antarctic mixotrophic cryptophytes during summer. *Freshwater Biology* **47**, 2060–2070.

Mataloni, G., Tesolin, G. and Tell, G. (1998). Characterization of a small eutrophic Antarctic lake (Otero Lake, Cierva Point) on the basis of algal assemblages and water chemistry. *Polar Biology* **19**, 107–114.

Matsubaya, O., Torii, T., Burton, H.R., Kerry, K. and Sakai, H. (1979). Antarctic saline lakes—stable isotopic ratios, chemical compositions and evolution. *Geochimica et Cosmochimica Acta* **43**, 7–25.

Matsumoto, G.I. (1989). Biogeochemical study of organic-substances in Antarctic lakes. *Hydrobiologia* **172**, 265–289.

Matsumoto, G. I., (1993). Geochemical features of the McMurdo Dry Valley lakes, Antarctica. In Green, W.J. and Friedmann, E.I. (eds) *Physical and Biogeochemical Processes in Antarctic Lakes, Antarctic Research Series*, pp. 95–118. American Geophysical Union, Washington, DC.

McCammon, S.A. and Bowman, J.P. (2000). Taxonomy of Antarctic Flavobacterium species: description of *Flavobacterium gillisiae* sp. nov., *Flavobacterium tegetincola* sp.nov. and *Flavobacterium xanthum* sp. nov. rev and reclassification of (*Flavobacterium*) *salegens* as *Salegentibacter salegens* gen. nov., comb. nov. *International Journal of Systematic and Evolutionary Microbiology* **50**, 1055–1063.

McCammon, S.A., Innes, B.H., Bowman, J.P., Franzmann, P.D. and five others (1998). *Flavobacterium hibernum* sp. nov., a lactose-utilizing bacterium from a freshwater Antarctic lake. *International Journal of Systematic Bacteriology* **48**, 1405–1412.

McGenity, T.J. and Oren, A. (2012). Hypersaline Environments. In Bell, E.M. (ed) *Life at Extremes, Environments, Organsims and Strategies for Survival*, pp. 402–437. CABI Wallingford, UK.

McGowan, S., Barker, P., Haworth, E.Y., Leavitt, P.R., Maberly, S.C. and Pates, J. (2012). Humans and climate as drivers of algal community change in Windermere since 1850. *Freshwater Biology* **57**, 260–277.

McKay, C.P., Hand, K.P., Doran, P.T., Andersen, D.T. and Priscu, J.C. (2003). Clathrate formation and the fate of noble and biologically useful gases in Lake Vostok. *Geophysical Research Letters* **30(13)**, 35 doi:10.1029/2003GL017490.

McKenna, K.C., Moorhead, D.L., Roberts, E.C. and Laybourn-Parry, J. (2005). Simulated patterns of carbon flow in the pelagic food web of Lake Fryxell, Antarctica: little evidence of top-down control. *Ecological Modelling* **192**, 457–472.

McKnight, D.M., Aiken, G.R., Andrews, E.D., Bowles, E.C. and Harnish, R.A. (1993). Dissolved organic material in Dry Valley lakes: A comparison of Lake Frxyell, Lake Hoare and Lake Vanda. In Green, W.J and Friedmann, E.I. (eds) *Physical and Biogeochemical Processes in Antarctic Lakes, Antarctic Research Series*, pp. 119–133. American Geophysical Union, Washington, DC.

McKnight, D.M., Aiken, G.R. and Smith, R.L. (1991). Aquatic fulvic-acids in microbially based ecosystems—results from two desert lakes in Antarctica. *Limnology and Oceanography* **36**, 998–1006.

McKnight, D.M., Gooseff, M.N., Vincent, W.F. and Peterson, B.J. (2008). High latitude rivers and streams. In Vincent, W.F. and Laybourn-Parry, J. (eds) *Polar Lakes and Rivers: Limnology of Arctic and Antarctic Aquatic Ecosystems*, pp. 83–102. Oxford University Press, Oxford, UK.

McKnight, D.M., Howes, B.L., Taylor, C.D. and Goehringer, D.D. (2000). Phytoplankton dynamics in stably stratified Antarctic lake during winter darkness. *Journal of Phycology* **36**, 852–861.

McMeekin, T.A. and Franzmann, P.D. (1988). Effect of temperature on the growth rates of halotolerant and halophilic bacteria isolated from Antarctic saline lakes. *Polar Biology* **8**, 281–285.

McMillan, M., Corr, H., Shepard, A., Ridout, A., Laxon, S. and Cullen, R. (2013). Three dimensional mapping by CryoSat-2 of subglacial lake volume changes. *Geophysical Research Letters* **30**, doi:10.1002/grl.50689.

Mercer, J.H. (1978). West Antarctic ice sheet and CO_2 greenhouse effect—threat of disaster. *Nature* **271**, 321–325.

Meyer, J.L., Edwards, R.T. and Risley, R. (1997). Bacterial growth on dissolved organic carbon from a blackwater river. *Microbial Ecology* **13**, 13–29.

Michaud, L., Caruso, C., Mangano, S., Interdonato, F., Bruni, V. and Lo Guidice, A. (2012). Predominance of *Flavobacterium, Psuedomonas* and *Polaromonas* within the prokaryotic community of freshwater shallow lakes in northern Victoria Land, East Antarctica. *FEMS Microbiology Ecology* doi:10.1111/j.1574–6941.2012.01394.x.

Mikucki, J., Foreman, C.M., Sattler, B., Lyons, W.B. and Priscu, J.C. (2004). Geomicrobiology of Blood Falls: an iron rich saline discharge at the terminus of the Taylor Glacier, Antarctica. *Aquatic Geochemistry* **10**, 199–220.

Mikucki, J., Pearson, A., Johnston, D.T., Turchyn, A.V., Farquhar, J., Schrag, D.P., Anbar, A.D., Priscu, J.C. and Lee, P.A. (2009). A contemporary microbially maintained subglacial ferrous 'ocean'. *Science* **324**, 397–399.

Mindl, B., Anesio, A.M., Meirer, K., Hodson, A.J., Laybourn-Parry, J., Sommaruga, R. and Sattler, B. (2007). Factors influencing bacterial dynamics along a transect from supraglacial runoff to pro-glacial lakes of high Arctic glaciers. *FEMS Microbiology Ecology* **59**, 762–771.

Miteva, V.I. and Brenchley, J.E. (2005). Detection and isolation of ultrasmall microorganisms from a 120,000-year-old Greenland ice core. *Applied and Environmental Microbiology* **71**, 7806–7818.

Miteva, V.I., Sheridan, P.P. and Brenchley, J.E. (2004). Phylogenetic and physiological diversity of microorganisms isolated from a deep Greenland glacier ice core. *Applied and Environmental Microbiology* **70**, 202–213.

Montecino, V., Pizarro, G., Cabrera, S. and Contreras, M. (1991). Spatial and temporal photosynthetic compartments during summer in Antarctic Lake Kitiesh. *Polar Biology* **11**, 371–377.

Moorhead, D.L. (2007). Mesoscale dynamics of ephemeral wetlands in the Antarctic Dry Valleys: implications to production and distribution of organic matter. *Ecosystems* **10**, 86–94.

Moorhead, D.L., Barrett, J.E., Virginia, R.A., Wall, D.H. and Porazinska, D. (2003). Organic matter an soil biota of upland wetlands in Taylor Valley, Antarctica. *Polar Biology* **26**, 567–576.

Moorhead, D.L., Schmeling, J. and Hawes, I. (2005). Modelling the contribution of benthic microbial mats to net primary production in Lake Hoare, McMurdo Dry Valleys. *Antarctic Science* **17**, 33–45.

Morgan, R.M., Ivanov, A.G., Priscu, J.C., Maxwell, D.P. and Hutner, N.P.A. (1998). Structure and composition of the photochemical apparatus of the Antarctic green alga, *Chlamydomonas subcaudata*. *Photosynthesis Research* **56**, 303–314.

Morgan-Kiss, R.M., Priscu, J.C., Pocok, T., Gudynaite-Savitch, L. and Huner, N.P.A. (2006). Adaptation and acclimation of photosynthetic microorganism to permanently cold environments. *Microbiology and Molecular Biology Reviews* **70**, 222–252.

Mosier, A.C., Murray, A.E. and Fritsen, C.H. (2007). Microbiota within the perennial ice cover of Lake Vida, Antarctica. *FEMS Microbiology Ecology* **59**, 274–288.

Moss, B. (1980). *Ecology of Freshwaters*. Blackwell Scientific Publications, Oxford.

Mountford, D.O., Kaspar, H.F., Asher, R.A. and Sutherland, D. (2003). Influences of pond geochemistry, temperature, and freeze-thaw on terminal anerobic processes occurring in sediments of six ponds of the McMurdo Ice Shelf, near Bratina Island, Antarctica. *Applied and Environmental Microbiology* **69**, 583–592.

Mueller, D.R., Vincent, W.F., Bonilla, S. and Laurion, I. (2005). Extremeophiles and broadband pigmentation strategies in a high arctic ice shelf ecosystem. *FEMS Microbiology Ecology* **53**, 73–87.

Mueller, D.R., Vincent, W.F. and Jeffries, M.O. (2003). Break-up of the largest Arctic ice shelf and the associated loss of an epishelf lake. *Geophysical Research Letters* **30**, doi:10.1029/2003GL017931.

Murray, A.E., Kenig, F., Fritsen, C.H., McKay, C.P., Cawley, K.M. and 13 others (2012). Microbial life at –13 °C in the brine of an ice-sealed Antarctic lake. *Proceedings of the National Academy of Sciences* **109**, doi:10.1073/pnas.1208607109.

Murtagh, G.J., Dyer, P.S., Rogerson, A., Nash, G.V. and Laybourn-Parry, J. (2001). A new species of *Tetramitus* in the bentos of a saline lake. *European Journal of Protistology* **37**, 437–443.

Naish, T., Powell, R., Levy, R., Wilson, G., Scherer, R. et al. (2009). Obliquity-paced Pliocene West Antarctic ice sheet oscillations. *Nature* **458**, 322–384.

Nakai, R., Abe, T., Baba, T., Imura, S. Kagoshima, H. et al. (2012). Microflora of aquatic moss pillars in a freshwater lake, East Antarctica, based on fatty acid and 16S rRNA gene analysis. *Polar Biology* **35**, 425–433.

Neale, P.J. and Priscu, J.C. (1995). The photosynthetic apparatus of phytoplankton from a perennially ice-covered lake: acclimation to an extreme shade environment. *Plant Cell Physiology* **36**, 253–263.

Neale, P.J. and Priscu, J.C. (1998). Fluorescence quenching in phytoplankton of the McMurdo Dry Valley lakes (Antarctica): implications for the structure and function of the photosynthetic apparatus. In J.C. Priscu (ed) *Ecosystem Dynamics in a Polar Desert: The McMurdo Dry Valleys, Antarctica*, pp. 241–253. Antarctic Research Series Volume 72, American Geophysical Union, Washington, DC.

Nedzarek, A. and Pociecha, A. (2010). Limnological characterization of freshwater systems of the Thomas Point Oasis (Admiralty Bay, King George Island, West Antarctica). *Polar Science* **4**, 457–467.

Neumann, K., Lyons, W.B., Priscu, J.C. and Donahoe, R.J. (2001). CO_2 concentrations in perennially ice-covered lakes of the Taylor Valley, Antarctica. *Biogeochemistry* **56**, 27–50.

Nichols, D.S., Miller, M.R., Davies, N.W., Goodchild, A, Raftery, M and Cavicchioli, R. (2004). Cold adaptation in the Antarctic Archeaon *Methanococcoides burtonii* involves membrane lipid unsaturation. *Journal of Bacteriology* **186**, 8508–8515.

Nichols, D.S., Nichols, P.D. and McMeekin, T.A. (1993). Polyunsaturated fatty acids in Antarctic bacteria. *Antarctic Science* **5**, 149–160.

Noon, P.E., Leng, M.J. and Jones, V.J. (2003). Oxygen-isotope ($\delta^{18}O$) evidence of Holocene hydrological changes at Signy Island, maritime Antarctica. *The Holocene* **13**, 251–263.

O'Brien, P. E., Goodwin, I., Forsberg, C.F., Cooper, A.K. and Whitehead, J. (2007). Late Neogene ice drainage changes in Prydz Bay, East Antarctica and the interaction of Antarctic ice sheet evolution and climate. *Palaeogeography Palaeoclimatology Palaeoecology* **245**, 390–410.

O'Brien, W.J., Bahr, M., Hershey, A.E., Hobbie, J.E., Kipphut, G.W. and six others (1997). The limnology of Toolik Lake. In Milner, A.M. and Oswood, M.W. (eds) *Freshwaters of Alaska*, pp. 61–106. Springer-Verlag, New York.

Ollivier, B., Caumette, P., Garcia, J-L. and Mah, R.A. (1994). Anaerobic bacteria from hypersaline environments. *Microbiological Reviews* **58**, 27–38.

Oppenheim, D.R. (1990). A Preliminary study of benthic diatoms in contrasting lake environments. In Kerry, K.R. and Hempel, G. (eds) *Antarctic Ecosystems, Ecological Change and Conservation*, pp. 91–99. Springer-Verlag, Berlin and Heidelberg.

Oswald, G.K.A. and Robin, G. de Q. (1973). Lakes beneath the Antarctic Ice Sheet. *Nature* **245**, 251–254.

Ouellet, M., Bisson, M., Page, P. and Dickman, M. (1987). Physicochemical limnology of meromictic saline Lake Sophia, Canadian Arctic Archipeligo. *Arctic and Alpine Research* **19**, 305–312.

Ouellet, M., Dickman, M., Bisson, M. and Page, P. (1989). Physico-chemical characteristics and origin of hypersaline meromictic Lake Garrow in the Canadian High Arctic. *Hydrobiologia* **172**, 215–234.

Pace, M.L. and Cole, J.J. (1994). Primary and bacterial production in lakes: are they coupled over depth? *Journal of Plankton Research* **16**, 661–672.

Paerl, H.W. and Priscu, J.C. (1998). Microbial phototrophic, heterotrophic, and diazotrophic activities associated with aggregates in the permanent ice cover of Lake Bonney, Antarctica. *Microbial Ecology*, **36**(3), 221–230.

Paggi, J.C. (1996). Feeding ecology of *Branchinecta gaini* (Crustacea: Anostraca) in ponds of South Shetland Islands, Antarctica. *Polar Biology* **16**, 13–18.

Palethorpe, B., Hayes-Gill, B., Crowe, J., Sumner, M., Crout, N. et al. (2004). Real time physical data acquisition through a remote sensing platform on a polar lake. *Limnology and Oceanography Methods* **2**, 191–201.

Parker, B.C., Simmons, G.M., Seaburg, K.G., Cathey, D.D. and Allnutt, F.C.T. (1982). Comparative ecology of plankton communities in seven Antarctic oasis lakes. *Journal of Plankton Research* **4**, 271–286.

Patterson, H. and Laybourn-Parry, J. (2012). Sea-ice microbial dynamics over an annual ice cycle in Prydz Bay (Antarctica). *Polar Biology* doi 10.1007/s00300-011-1146-3.

Pattyn, F. (2010). Antarctic subglacial conditions inferred from a hybrid ice sheet/ice stream model. *Earth and Planetary Science Letters* **295**, 451–461.

Pearce, D.A. (2003). Bacterioplankton community structure in a maritime Antarctic oligotrophic lake during a period of holomixis by denatured gradient gel electrophoresis (DGGE) and Fluorescence *in Situ* Hybridization (FISH). *Microbial Ecology* **46**, 92–105.

Pearce, D.A. (2005). The structure and stability of the bacterioplankton community in Antarctic freshwater lakes, subject to extremely rapid environmental change. *FEMS Microbiology Ecology* **53**, 61–72.

Pearce, D.A., Cockell, C.S., Lindström, E.S. and Tranvik, L.J. (2007). First evidence for a bipolar distribution of dominant freshwater lake bacterioplankton. *Antarctic Science* **19**, 245–252.

Pearce, D.A., Hodgson, D.A., Thorne, M.A.S., Burns, G. and Cockell, C.S. (2013). Preliminary analysis of life within a former subglacial lake sediment in Antarctica. *Diversity* **5**, 680–702.

Pearce, D.A., van der Gast, C.J., Lawley, B. and Ellis-Evans, J.C. (2003). Bacterioplankton community diversity in a maritime Antarctic lake, determined by culture-dependent and culture-independent techniques. *FEMS Microbiology Ecology* **45**, 59–70.

Peeters, K., Hodgson, D.A., Convey, P. and Willems, A. (2011). Culturable diversity of heterotrophic bacteria in Forlidas Pond (Pensacola Mountains) and Lundström Lake (Shackleton Range), Antarctica. *Microbial Ecology* **62**, 399–413.

Peeters, K., Verleyen, E., Hodgson, D.A., Convey, P. et al. (2012). Heterotrophic bacterial diversity in aquatic microbial mat communities from Antarctica. *Polar Biology* **35**, 543–554.

Perriss, S.J. and Laybourn-Parry, J. (1997). Microbial communities in saline lakes of the Vestfold Hills (Eastern Antarctica). *Polar Biology* **18**, 135–144.

Perriss, S.J., Laybourn-Parry, J. and Marchant, H.J. (1995). The widespread occurrence of the autotrophic ciliate *Mesodinium rubrum* (Ciliophora: Haptorida) in the freshwater and brackish lakes of the Vestfold Hills, Eastern Antarctica. *Polar Biology* **15**, 423–428.

Perry, J.J., Staley, J.T. and Lory, S. (2002). *Microbial Life*, Sinauer Associates, Publishers, Sunderland, MA.

Petit, J.R., Basil, I., Leruyuet, A., Raynaud, D., Lorius, C. et al. (1997). Four climate cycles in Vostok ice core. *Nature* **387**, 359–360.

Petit, J.R., Jouzel, J., Raynaud, D., Barkov, N.I., Barnola, J-M. et al. (1999). Climate and atmosphere history of the past 420,000 years from the Vostok ice core, Antarctica. *Nature* **399**, 429–436.

Petz, W., Valbonesi, A., Schiftner, U., Quesada, A. and Ellis-Evans, J.C. (2007). Ciliate biogeography in Antarctic and Arctic freshwater ecosystems: endemism of global distribution of species? *FEMS Microbiology Ecology* **59**, 396–408.

Pick, F.R. and Caron, D.A. (1987). Picoplankton and nanoplankton biomass in Lake Ontario: relative contribution of phototrophic and heterotrophic communities. *Canadian Journal of Fisheries and Aquatic Sciences* **44**, 2164–2172.

Pienitz, R., Doran, P.T. and Lamoureux, S.F. (2008). Origin and geomorphology of lakes in the polar regions. In Vincent, W.F. and Laybourn-Parry, J. (eds) *Polar Lake* and Rivers Limnology of Arctic and Antarctic Aquatic Ecosystems, pp. 25–41. Oxford University, Press Oxford, UK.

Pociecha, A. (2008). Density dynamics of *Notholca squamula* saline Focke (Rotifera) in Lake Wujka, a freshwater Antarctic lake. *Polar Biology* **31**, 275–279.

Pociecha, A. and Dumont, H.J. (2008). Life cycle of *Boeckella poppei* Mrazek and *Branchinecta gaini* Daday (King George Island, South Shetlands). *Polar Biology* **31**, 245–248.

Pocock, T., Lachance, M.-A., Pröschold, T., Priscu, J.C., Kim, S.S. and Huner, N.P.A. (2004). Identfcation of a psychrophilic green alga from Lake Bonney Antarctica: *Chlamydomonas raudensis* Ettl. (UWO 241) Chlorophyceae. *Journal of Phycology* **40**, 1138–1148.

Pollard, D. and DeConto, R.M. (2009). Modelling West Antarctic ice sheet growth and collapse through the past five million years. *Nature* **458**, 329–332, doi:10.1038/Nature07809.

Pomeroy, L.R. (1974). The ocean's food web, a changing paradigm. *Bioscience* **9**, 499–504.

Porter, J., Arzberger, P., Braun, H-W., Bryant, P. et al. (2005). Wireless sensor networks for ecology. *Bioscience* **55**, 561–572.

Powell, L.M., Bowman, J.P., Skerratt, J.H., Franzmann, P.D. and Burton, H.R. (2005). Ecology of a novel *Synechococcus* clade occurring in dense populations in saline Antarctic lakes. *Marine Ecology Progress Series* **291**, 65–80.

Prairie, Y.T., Bird, D.F. and Cole, J.J. (2002). The summer metabolic balance in the epilimnion of southeastern Quebec lakes. *Limnology and Oceanogrphy* **47**, 316–321.

Priddle, J. (1980a). The production ecology of benthic plants in some Antarctic lakes. I. *In situ* production studies. *Journal of Ecology* **68**, 141–153.

Priscu, J.C. (1992). Particulate organic matter decomposition in the water column of Lake Bonney, Taylor Valley, Antarctica. *Antarctic Journal of the United States* **27**, 260–262.

Priscu, J.C. (1995). Phytoplankton nutrient deficiency in lakes of the McMurdo Dry Valleys, Antarctica. *Freshwater Biology* **34**, 215–227.

Priscu, J.C. (1997). The biogeochemistry of nitrous oxide in permanently ice-covered lakes of the McMurdo Dry Valleys, Antarctica. *Global Change Biology* **3**, 301–315.

Priscu, J.C. (1998). Preface. In Priscu J.C. (ed) *Ecosystem Dynamics in a Polar Desert: The McMurdo Dry Valleys, Antarctica*, pp xi. Antarctic Research Series 72, American Geophysical Union, Washington, DC.

Priscu, J.C., Adams, E.E., Lyons, W.B., Voytek, M.A., Mogk, D.W., Brown, R.L., McKay, C.P., Takacs, C.D., Welch, K.A., Wolf, C.F., Kirshtein, C.D. and Avci, R. (1999a). Geomicrobiology of subglacial ice above Lake Vostok, Antarctica. *Science* **286**, 2141–2144.

Priscu, J.C., Achberger, A.M., Cahoon, J.E., Christner, B.C., Edwards, R.L. and 8 others (2013a). A microbiologically clean strategy for access to the Whillans Ice Stream subglacial environment. *Antarctic Science* doi:10.1017/S0954102013000035.

Priscu, J.C., Adams, E.E., Paerl, H.W., Fritsen, C.H., Dore, J.E., Wolf, F. and Mikuchi, J. (2005a). Perennial Antarctic ice: a refuge for cyanobacteria in an extreme environment. In Castello, J.D. and Rogers, S.O. (eds) *Life in Ancient Ice*, pp. 22–49. Princeton Press.

Priscu, J.C., Christner,B., Skidmore,M., Mikucki, J. et al. (2013b). Geomicrobiology of Subglacial Lake Whillans, Antarctica, AGU Abstracts B22B-08, American Geophysical Union Meeting 2013, San Francisco, USA.

Priscu, J.C., Downes M.T. and McKay, C.P. (1996). Extreme supersaturation of nitrous oxide in a poorly ventilated antarctic lake. *Limnology and Oceanography* **41**, 1544–1551.

Priscu, J.C., Fritsen, C.H., Adams, E.E., Giovannoni, S.J., Paerl, H.W., McKay, C.P., Doran, P.T., Gordon, D.A., Lanoil, B.D. and Pinckney, J.L. (1998). Perennial Antarctic lake ice: an oasis for life in a polar desert. *Science* **280**, 2095–2098.

Priscu, J.C., Kennicutt, M.C., Bell, R.E., Bulat, S.A., Ellis-Evans, J.C. and Lukin, V.V. (2005b). Exploring subglacial Antarctic lake environments. *EOS* **86**, 193–200.

Priscu, J.C., Priscu, L.R., Vincent, W.F. and Howard-Williams, C. (1987). Photosynthate distribution by microplankton in permanently ice-covered Antarctic desert lakes. *Limnology and Oceanography* **32**, 260–270.

Priscu, J.C., Tulczyk, S., Studinger, M., Kennicutt, M.C., Christner, B.C. and Forman, C.M. (2008). Antarctic subglacial water: origin, evolution and ecology. In Vincent, W.F. and Laybourn-Parry, J. (eds) *Polar Lakes and Rivers: Limnology of Arctic and Antarctic Aquatic Systems*, pp. 119–136. Oxford University Press, Oxford.

Priscu, J.C., Vincent, W.F. and Howard-Williams, C. (1989). Inorganic nitrogen uptake and regeneration in perennially ice covered Lakes Fryxell and Vanda, Antarctica. *Journal of Plankton Research* **11**, 335–351.

Priscu, J. C., Wolf, C.F., Takacs, C.D., Fritsen, C.H., Laybourn-Parry, J. et al. (1999b). Carbon transformations in a perennially ice-covered antarctic lake. *Bioscience* **49**, 997–1008.

Pross, J., Contreras, L., Bijl, P.K., Greenwood, D.R., Bohaty, S.M. et al. (2012). Persistent near-tropical warmth on the Antarctic continent during the early Eocene epoch. *Nature* **488**, 73–77.

Quayle, W.C., Peck, L.S., Peat, H., Ellis-Evans, J.C. and Harrigan, P.R. (2002). Extreme responses to climate change in Antarctic lakes. *Science* **295**, 645.

Quesada, A., Fernández-Valiente, E., Hawes, I. and Howard-Williams, C. (2008). Benthic primary production in polar lakes and rivers. In Vincent, W.F. and Laybourn-Parry, J. (eds) *Polar Lakes and Rivers: Limnology of Arctic and Antarctic Aquatic Ecosystems*, pp. 179–196. Oxford University Press, Oxford.

Rankin, L.M., Franzmann, P.D., McMeekin, T.A. and Burton H.R. (1997). Seasonal distribution of picocyanobacteria in Ace Lake, a marine derived Antarctic lake. In Battaglia, B., Valencia, J. and Walton, D.W.H. (eds) *Antarctic Communities, Species, Structure and Survival*, pp. 178–184. Cambridge University Press, Cambridge.

Rankin, L.M., Gibson, J.A.E., Franzmann, P.D. and Burton, H.R. (1999). The chemical stratification and microbial communities of Ace Lake: a review of the characteristics of a marine derived meromictic lake. *Polarforschung* **66**, 33–52.

Rautio, M., Bayly, I.A.E., Gibson, J.A.E. and Nyman, M. (2008). Zooplankton and zoobenthos in high-latitude water bodies. In Vincent, W.F. and Laybourn-Parry, J. (eds) *Polar Lakes and Rivers: Limnology of Arctic and Antarctic Aquatic Ecosystems*, pp. 231–247. Oxford University Press, Oxford.

Ravelo, A.C., Andreasen, D.H., Lyle, M., Lyle, A.O. and Wara, M.W. (2004). Regional climate shifts caused by gradual global cooling in the Pliocene epoch. *Nature* **429**, 263–267.

Raymond, J., Zhaxybayeva, O., Gogarten, J.P., Gerdes, S.Y. and Blankenship, R.E. (2002). Whole-genome analysis of photosynthetic prokaryotes. *Science* **298**, 1616–1620.

Rengefors, K., Laybourn-Parry, J., Logares, R., Marshall, W.A. and Hansen, G. (2008). Marine-derived dinoflagellates in Antarctic saline lakes: community composition and annual dynamics. *Journal of Phycology* **44**, 592–604.

Rengefors, K., Logares, R. and Laybourn-Parry, J. (2012). Polar lakes may act as ecological islands to aquatic protists. *Molecular Ecology*, doi:10.1111/j.1365–1294X.2012.05596.x.

Reynolds, C.S. (1984). *The Ecology of Freshwater Phytoplankton*, 384 pages. Cambridge University Press, Cambridge.

Reynolds, J.M. (1981). Lakes on George VI Ice Shelf, Antarctica. *Polar Record* **20**, 425–432.

Richter, W. (1995). Biology. In Bormann, P. and Fritzsche, D (eds) *The Schirmacher Oasis, Queen Maud Land, East Antarctica, and its Surroundings*, pp. 321–347. Justus Perthes Verlag, Gotha, Germany.

Richter, W. and Bormann, P. (1995). Hydrology. In Bormann, P. and Fritzsche, D (eds) *The Schirmacher Oasis, Queen Maud Land, East Antarctica, and its Surroundings*, pp. 259–319. Justus Perthes Verlag, Gotha, Germany.

Riddle, M.J. and Muir, D.C.G. (2008). Direct human impacts on high-latitude lakes and rivers. In Vincent, W.F. and Laybourn-Parry, J. (eds) *Polar Lakes and Rivers: Limnology of Arctic and Antarctic Aquatic Ecosystems*, pp. 291–306. Oxford University Press, Oxford, UK.

Roberts, D. and McMinn, A. (1996). Relationships between surface sediment diatom assemblages and water chemistry gradients in saline lakes of the Vestfold Hills, Antarctica. *Antarctic Science* **8**, 331–341.

Roberts, D. and McMinn, A. (1998). A weighted-averaging regression and calibration model for inferring lake water salinity from fossil diatom assemblages in saline lakes of the Vestfold Hills: a new tool for interpreting Holocene lake histories in Antarctica. *Journal of Paleolimnology* **19**, 99–113.

Roberts, D. and McMinn, A. (1999). A diatom-based palaeosalinity history of Ace Lake, Vestfold Hills, Antarctica. *The Holocene* **9**, 401–408.

Roberts, D., McMinn, A., Johnston, N., Gore, D.B., Melles, M. and Cremer, H. (2001). An analysis of the limnology and sedimentology diatom flora of fourteen lakes and ponds from the Windmill Islands, East Antarctica. *Antarctic Science* **13**, 140–419.

Roberts, D., McMinn, A. and Zwartz, D. (2000). An initial palaeolimnology history of Jaw Lake, Bunger Hills based on a diatom-salinity transfer function applied to sediment cores. *Antarctic Science* **12**, 172–176.

Roberts, E.C. and Laybourn-Parry, J. (1999). Mixotrophic cryptophytes and their predators in the Dry Valley lakes of Antarctica. *Freshwater Biology* **41**, 737–746.

Roberts, E.C., Laybourn-Parry, J., McKnight, D.M. and Novarino, G. (2000). Stratification and dynamics of microbial loop communities in Lake Fryxell, Antarctica. *Freshwater Biology* **44**, 649–661.

Roberts, E.C., Priscu, J.C. and Laybourn-Parry, J. (2004b). Microplankton dynamics in a perennially ice-covered Antarctic lake–Lake Hoare. *Freshwater Biology* **49**, 853–869.

Roberts, E.C., Priscu, J.C., Wolf, C., Lyons, W.B. and Laybourn-Parry, J. (2004a). The distribution of microplankton in the McMurdo Dry Valley lakes, Antarctica: response to ecosystem legacy or present-day climatic controls? *Polar Biology* **27**, 238–249.

Robin, G.D., Swithinbank, C.W.M. and Smith, B.M.E. (1970). Radio echo exploration of the Antarctic ice sheet. *International Association of Scientific Hydrology Publication* **86**, 97–115.

Rochera, C., Justel, A., Fernández-Valiente, E., Bañón, M. et al. (2010). Inter-annual meteorological variability and its effects on a lake from maritime Antarctica. *Polar Biology* **33**, 1615–1628.

Rogerson, A. and Laybourn-Parry, J. (1992). Bacterioplankton abundance and production in the Clyde, Scotland. *Archiv für Hydrobiologie* **126**, 1–14.

Rojas, J.L., Martin, J., Tormo, J.R., Vincente, F., Brunati, M. and 7 others (2009). Bacterial diversity from benthic algal mat of Antarctic lakes as a source of new bioactive metabolites. *Marine Genomics* **2**, 33–41.

Ross, N., Siegert, M., Rivera, A., Bentley, M.J. et al. (2011). Ellsworth Subglacial Lake, West Antarctica: a review of its History and Recent Field Campaigns. *Geophysical Monographs* **192**, *Antarctic Subglacial Aquatic Environments*, 221–233. American Geophysical Union.

Royston-Bishop, G., Priscu, J.C., Tranter, M., Christner, M., Siegert, M.J. and Lee, V. (2005). Incorporation of particulates into accreted ice above subglacial Lake Vostok and other Antarctic subglacial lakes. *Annual Review of Earth and Planetary Sciences* **33**, 215–245.

Sabacká, M. and Elster, J. (2006). Response of cyanobacteria and algae from Antarctic wetland habitats to freezing and desiccation stress. *Polar Biology* **30**, 31–37.

Sabbe, K., Hodgson, D.A., Verleyen, E., Taton, A., Wilmotte, A. et al. (2004). Salinity, depth and the structure and composition of microbial mats in continental Antarctic lakes. *Freshwater Biology* **49**, 296–319.

Sabbe, K., Verleyen, E., Hosgson, D.A., Vanhoutte, K. and Vyverman, W. (2003). Benthic diatom flora of freshwater and saline lakes in the Larsemann Hill and Rauer Islands, East Antarctica. *Antarctic Science* **15**, 227–248.

Safi, K., Hawes, I. and Sorrell, B. (2012). Microbial population responses in three stratified Antarctic meltwater ponds during the autumn freeze. *Antarctic Science* **24**, 571–588.

Samarkin, V.A., Madigan, M.T., Bowles, M.W., Casciotti, K.L., Priscu, J.C. and two others. (2010). Abiotic nitrous oxide emission from the hypersaline Don Juan Pond in Antarctica. *Nature Geoscience* **3**, 341–344.

Sattley, W.M. and Madigan, M.T. (2006). Isolation, characterization, and ecology of cold-active, chemolithotrophic, sulfur-oxidizing bacteria from perennially ice-covered Lake Fryxell, Antarctica. *Applied and Environmental Microbiology* **72**, 5562–5568.

Säwström, C., Anesio, A.M., Granéli, W. and Laybourn-Parry, J. (2007d). Seasonal viral loop dynamics in two large ultra-oligotrophic Antarctic freshwater lakes. *Microbial Ecology* **53**, 1–11.

Säwström, C., Granéli, W., Laybourn-Parry, J. and Anesio, A.M. (2007a). High viral infection rates in Antarctic and Arctic bacterioplankton. *Environmental Microbiology* **9**, 250–255.

Säwström, C., Karlson, J., Laybourn-Parry, J. and Granéli, W. (2009). Zooplankton feeding on algae and bacteria under ice in Lake Druzhby, East Antarctica. *Polar Biology* **32**, 1195–1202.

Säwström, C., Laybourn-Parry, J., Anesio, A.M., Priscu, J.C. and Lisle, J. (2008). Bacteriophage in polar inland waters. *Extremophiles* **12**, 167–175.

Säwström, C., Laybourn-Parry, J., Granéli, W. and Anesio, A.M. (2007c). Heterotrophic bacteria and viral dynamics in Arctic freshwaters: results from a field study and nutrient-temperature manipulation experiments. *Polar Biology* **30**, 1407–1415.

Säwström, C., Pearce., I., Davidson, A.T., Rosén, P. and Laybourn-Parry, J. (2007b). Influence of environmental conditions, bacterial activity and viability on the viral component in 10 Antarctic lakes. *FEMS Microbiology Ecology* **63**, 12–22.

Scherer, R. P. (1991). Quaternary and Tertiary microfossils from beneath Ice Stream-b—evidence for a dynamic West Antarctic ice-sheet history. *Global and Planetary Change* **90**, 395–412.

Scherer, R.P., Aldahan, A., Tulaczyk, S., Possnert, G., Engelhardt, H. and Kamb, B. (1998). Pleistocene collapse of the West Antarctic Ice Sheet. *Science* **281**, 82–85.

Scherer, R.P., Bohaty, S.M., Dunbar, R.B., Esper, O., Flores, J.A. et al. (2008). Antarctic records of precession-paced insolation-driven warming during early Pleistocene marine isotope stage 31. *Geophysical Research Letters* **35** (3), L03505. doi:10.1029/2007GL032254.

Schiaffino, M.R., Unrein, F., Gasol, J.M., Farias, M.E., Estevez, C., Balagué, V. and Izaguirre, I. (2009). Comparative analysis of bacterioplankton assemblages from maritime Antarctic freshwater lakes with contrasting trophic status. *Polar Biology* **32**, 923–936.

Shackleton, N. J. (1987). Oxygen isotopes, ice volume and sea level. *Quaternary Science Reviews* **6**, 183–190.

Sherr, E.B. and Sherr, B.F. (1993). Protistan grazing rates via uptake of fluorescently labelled prey. In Kemp, P., Sherr, B.F. and Cole, J.J. (eds) *Handbook of Methods in Aquatic Microbial Ecology*, pp. 695–701. Lewis Publishing, Ann Abor, MI.

Siegal, B.Z., McMurty, G., Siegal, S.M. et al. (1979). Life in the calcium chloride environment of Don Juan Pond, Antarctica. *Nature* **280**, 828–829.

Siegert, M.J. (2001). *Ice Sheets and Late Quaternary Environmental Changes*, John Wiley & Sons Ltd, New York.

Siegert, M.J. (2005). Lakes beneath the ice sheet: the occurrence, analysis and future exploration of Lake Vostok and other Antarctic subglacial lakes. *Annual Review of Earth and Planetary Science* **33**, 215–245.

Siegert, M.J., Clarke, R.J., Mowlem, M., Ross, N., Hill, C.S. et al. (2012). Clean access, measurement and sampling of Ellsworth Subglacial Lake: a method for exploring deep Antarctic subglacial lake environments. *Reviews of Geophysics* **50**, doi:10.1029/2011 RG000361.

Siegert, M.J., Dowdeswell, J.A., Gorman, M.R. and McIntyre, N.F. (1996). An inventory of Antarctic subglacial lakes. *Antarctic Science* **8**, 281–286.

Siegert, M.J., Ellis-Evans, J.C., Tranter, M., Mayer, C., Petit, J-R, Salamatin, A. and Priscu, J.C. (2001). Physical, chemical and biological processes in Lake Vostok and other Antarctic subglacial lakes. *Nature* **414**, 603–609.

Siegert, M.J., Hindmarsh, R., Corr, R., Smith, A., Woodward, J. et al. (2004). Subglacial Lake Ellsworth: a candidate for *in situ* exploration in West Antarctica. *Geophysical Research Letters* **31**, doi:10.1029/2004GL021477.

Siegert, M.J., Kwok, R., Mayer, C. and Hubbard, B. (2000). Water exchange between the subglacial Lake Vostok and the overlying ice sheet. *Nature* **403**, 643–646.

Siegert, M.J., Le Brocq, A. and Payne, A.J. (2007). Hydrological connections between Antarctic subglacial lakes, the flow of water beneath the East Antarctic ice sheet and implications for sedimentary processes. In Hambrey, H.J., Christoffersen, P., Galsser, N.F. and Hubbard, B. (eds) *Glacial Sedimentary Processes and Products*, pp. 3–10. Wiley-Blackwell, Hoboken, NJ.

Siegert, M.J., Tranter, M., Ellis-Evans, J.C., Priscu, J.C. and Lyons, W.B. (2003). The hydrochemistry of Lake Vostok and the potential for life in Antarctic subglacial lakes. *Hydrological Processes* **17**, 795–814.

Simek, K. and Straskrabová, V. (1992). Bacterioplankton production and protozoan bacterivory in a mesotrophic reservoir. *Journal of Plankton Research* **14**, 773–787.

Simoes, J.C., Petit, J.R., Sochez, R., Lipenkov, V.Y., de Angelis, M. et al. (2002). Evidence of glacial flour in the deepest 89 m of the Vostok ice core. *Annals of Glaciology* **35**, 340–346.

Skidmore, M., Tranter, M., Tulaczyk, S. and Lanoil, B.D. (2010). Hydrochemistry of ice stream beds–evaporitic or microbial effects? *Hydrological Processes* **24**, 517–523.

Smith, B.E., Fricker, H.A., Joughin, I.R. and Tulaczyk, S. (2009). An inventory of active subglacial lakes in Antarctica detected by Icesat (2003–2008). *Journal of Glaciology* **55**, 573–595.

Smith, J.A., Hodgson, D.A., Bentley M.J., Verleyen, E., Leng, M.J. and Roberts, S.J. (2006). Limnology of two Antarctic epishelf lakes and their potential to record periods of ice shelf loss. *Journal of Paleolimnology* **35**, 373–394.

Smith, R., Miller, J. and Howes, B. (1993). The geochemistry of methane in Lake Fryxell, an amictic, permanently ice-covered, Antarctic lake. *Biogeochemistry* **21**, 95–115.

Smith, R.I.L. (1988). Destruction of Antarctic terrestrial ecosystems by a rapidly increasing fur seal population. *Biological Conservation* **45**, 55–72.

Sobek, S., Tranvik, L.J., Prairie, Y.T., Kortelainen, P. and Cole, J.J. (2007). Patterns and regulation of dissolved organic carbon: an analysis of 7500 widely distributed lakes. *Limnology and Oceanography* **52**, 1208–1219.

Sommaruga, R., Psenner, R., Schafferer, E., Koinig, K.A. and Sommaruga-Wögrath, S. (1999). Dissolved organic carbon concentration and phytoplankton biomass in high-mountain lakes of the Austrian Alps: potential effect of climatic warming on UV underwater attenuation. *Arctic, Antarctic and Alpine Research* **31**, 247–253.

Souchez, R., Jean-Baptiste, P., Petit, J.R., Lipenkov, V.Y. et al. (2003). What is the deepest part of the Vostok ice core telling us? *Earth-Science Reviews* **60**, 131–146.

Souchez, R., Petit, J.R., Tison, J.-L., Jouzel, J. and Verbeke, V. (2000). Ice formation in suglacial Lake Vostok, central Anatarctica. *Earth and Planetary Science Letters* **181**, 529–538.

Spaulding, S.A., McKnight, D.M., Smith, R.L. and Dufford, R. (1994). Phytoplankton population dynamics in perenially ice-covered Lake Fryxell, Antarctica. *Journal of Plankton Research* **16**, 527–541.

Spigel, R.H. and Priscu, J.C. (1996). Evolution of temperature and salt structure of Lake Bonney, a chemically stratified Antarctic lake. *Hydrobiologia* **321**, 177–190.

Spigel, R.H. and Priscu, J.C. (1998). Physical limnology of the McMurdo Dry Valley lakes. In Priscu, J.C. (ed) *Ecosystem Dynamics in a Polar Desert: The McMurdo Dry Valleys, Antarctica*, pp. 152–188. Antarctic Research Series 72, American Geophysical Union, Washington, DC.

Squier, A.H., Hodgson, D.A. and Keely, B.J. (2005). Evidence of late quaternary environmental change in a continental east Antarctic lake from lacustrine sedimentary pigment distributions. *Antarctic Science* **17**, 361–376.

Steward, G.F., Smith, D.C. and Azam, F. (1996). Abundance and production of bacterial and viruses in the Bering and Chukchi seas. *Marine Ecology Progress Series* **131**, 287–300.

Steward, G.F., Wikner, J., Cochlan, W.P. and Azam, F. (1992). Estimation of virus production in the sea, I. Method development. *Marine Microbial Food Webs* **6**, 57–78.

Stibal, M., Wadham, J.L., Lis, G.P., Telling, J., Pancost, R.D. et al. (2012). Methanogenic potential of arctic and Antarctic subglacial environments with contrasting organic carbon sources. *Global Change Biology*, doi:10.1111/j.1365–2486.2012.02763.xn/a-n/a.

Stingl, U., Foo, W., Vergin, K.L., Lanoil, B. and Giovannoni, S.J., (2008). Dilution-top-extinction culturing of psychrotolerant planktonic bacteria from permanently ice-covered lakes in the McMurdo Dry Valleys, Antarctica. *Microbial Ecology* **55**, 395–405.

Studinger, M., Bell, R.E., Karner, G.D., Tikku, A.A., Holt, J.W. et al. (2003). Ice cover, landscape setting and geological framework of Lake Vostok, East Antarctic. *Earth and Planetary Science Letters* **205**, 195–210.

Stuiver, M., Denton, G.H., Hughes, T.J. and Fastook, J.L. (1981). History of the marine ice sheet in West Antarctica during the last glaciation, a working hypothesis. In Denton, G.H. and Hughes, T.H. (eds) *The Last Great Ice Sheets*, pp. 319–336. Wiley-Interscience, New York.

Sugden, D.E., Bentley, M.J. and O'Cofaigh, C. (2006). Geological and geomorphological insights into Antarctic ice sheet evolution. *Philosophical Transactions of the Royal Society A* **364**, 1607–1625.

Sundal, A.V., Shepherd, A., Nienow, P., Hanna, E., Palmer, S. and Huybrechts, P. (2009). Evolution of supra-glacial lakes across the Greenland Ice Sheet. *Remote Sensing of Environment* **113**, 2164–2171.

Suren, A. (1990). Microfauna associated with algal mats in melt ponds of the Ross Ice Shelf. *Polar Biology* **10**, 329–335.

Sutherland, D.L. (2009). Microbial mat communities in response to recent changes in the physiochemical environment of the meltwater ponds on the McMurdo Ice Shelf, Antarctica. *Polar Biology* **32**, 1023–1032.

Suttle C.A., Chan A.M. and Cottrell M.T. (1990). Infection of phytoplankton by viruses and reduction of primary productivity. *Nature* **347**, 467–469.

Swadling, K.M. and Gibson, J.A.E. (2000). Grazing rates of a calanoid copepod (*Paralabidocera antarctica*) in a continental Antarctic lake. *Polar Biology* **23**, 301–308.

Swadling, K.M., Nichols, P.D., Gibson, J.A.E. and Ritz, D.A. (2000). Role of lipid in the life-cycles of ice-dependent and ice independent populations of the copepod *Paralabidocera antarctica*. *Marine Ecology Progress Series* **208**, 171–182.

Tabacco, I.E., Clanferra, P., Forier, A., Salvini, F. and Zirizotte, A. (2006). Physiography and tectonic setting of the subglacial lake district between Vostok and Belgica subglcial highlands (Antarctica). *Geophysical Journal International* **165**, 1029–1040.

Takacs, C.D. and Priscu, J.C. (1998). Bacterioplankton dynamics in the McMurdo Dry Valley lakes, Antarctica: production and biomass over four seasons. *Microbial Ecology* **36**, 239–250.

Takacs, C.D., Priscu, J.C. and McKnight D.M. (2001). Bacterial dissolved organic carbon demand in McMurdo Dry Valley lakes, Antarctica. *Limnology and Oceanography* **46**, 1189–1194.

Takii, S., Konda, T., Hiraishi, A. et al. (1986). Vertical distribution in and isolation of bacteria from Lake Vanda: an Antarctic lake. *Hydrobiologia* **135**, 15–21.

Tang, E.P.Y., Tremblay, R. and Vincent, W.F. (1997). Cyanobacterial dominance of polar freshwater ecosystems: are high latitude mat-formers adapted to low temperature? *Journal of Phycology* **33**, 171–181.

Tapper, M.A. and Hicks, R.E. (1998). Temperate viruses and lysogeny in Lake Superior bacterioplankton. *Limnology and Oceanography* **43**, 95–103.

Taton, A., Grubisic, S., Balthasart, P., Hodgson, D.A., Laybourn-Parry, J. and Wilmotte, A. (2006a). Biogeographical distribution and ecological ranges of benthic cyanobacteria in East Antarctic lakes. *FEMS Microbiology Ecology* **57**, 272–289.

Taton, A., Grubisic, S., Ertz, D., Hodgson, D.A., Piccardi, R. et al. (2006b). Polyphasic study of Antarctic cyanobacterial strains. *Journal of Phycology* **42**, 1257–1270.

Taylor, G. (1922). *The Physiography of the McMurdo Sound and Granite Harbour Region: British Antarctic 'Terra Nova' Expedition, 1910–1913*. Harrison & Sons, London.

Thoma, M., Mayer, C. and Grosfeld, K. (2008). Sensitivity of subglacial Lake Vostok's flow regime on environmental parameters. *Earth and Planetary Science Letters* **269**, 242–247.

Thomas, D.N., Fogg, G.E., Convey, P., Fritsen, C.H., Gili, J.-M. and 4 others (2008). *The Biology of the Polar Regions*, pp. 116–142. Oxford University Press, Oxford.

Thomson, P.G., McMinn, A., Kiessling, I., Watson, M. and Goldsworthy. P.M. (2006). Composition and succession of dinoflagellates and chrysophytes in the upper fast ice of Davis Station, East Antarctica. *Polar Biology* **29**, 337–345.

Thurman, J., Parry, J.D., Hill, P.J. and Laybourn-Parry, J. (2010). The filter-feeding ciliates *Colpidium striatum* and *Tetrahymena pyriformis* display selective feeding behaviours in the presence of mixed, equally sized, bacterial prey. *Protist* **161**, 577–588.

Thurman, J., Parry, J., Hill, P.J., Priscu, J.C., Vick, T.J., Chiuchiolo, A. and Laybourn-Parry, J. (2012). Microbial dynamics and flagellate grazing rates during transition to winter in Lakes Hoare and Bonney, Antarctica. *FEMS Microbiology Ecology* **82**, 449–458.

Tominaga, H. and Fukui, F. (1981). Saline lakes at Syowa Oasis, Antarctica. *Hydrobiologia* **82**, 375–389.

Tomiyama, C. and Kitano, Y. (1985). Salt origin in the Wright Valley, Antarctica. *Antarctic Record* **86**, 17–26.

Tong, S., Vors, N. and Patterson, D.J. (1997). Heterotrophic flagellates, centrohelid heliozoan and filose amoebae from marine and freshwater sites in the Antarctic. *Polar Biology* **18**, 91–106.

Torella, F. and Morita, R.Y. (1979). Evidence by electron micrographs for a high incidence of bacteriophage particles in the waters of Yaguina Bay, Oregon: ecological and taxonomic implications. *Applied and Environmental Microbiology* **37**, 774–778.

Torii, T., Matsumoto, G.I. and Nakaya, S. (1988). The chemical characteristics of Antarctic lakes and ponds, with special emphasis on the distribution of nutrients. *Polarforschung* **58**, 219–230.

Toro, M., Camacho, A., Rochera, C., Rico, E., Bañón, M. et al. (2007). Limnological characteristics of the freshwater ecosystems of Byers Peninsula, Livingston Island, in maritime Antarctica. *Polar Biology* **30**, 635–649.

Tranter, M., Fountain, A.G., Fritsen, C.H., Lyons, W.B., Priscu, J.C., Statham, P.J. and Welch, K.A. (2004). Extreme hydrochemical conditions in natural microcosms entombed within Antarctic ice. *Hydrological Processes* **18**, 379–387.

Tranter, M., Sharp, M.J., Lamb, H.R., Brown, G.H., Hubbard, B.P. and Williss, I.C. (2002). Geochemical weathering at the bed of Haut Glacier d'Arolla, Switzerland–a new model. *Hydrological Processes* **16**, 959–993.

Turner, J., Colwell, S.R., Marshall, G.J., Lachlan-Cope, T.A., Carleton, A.M. et al. (2005). Antarctic climate change during the last 50 years. *International Journal of Climatology* **25**, 279–294.

Uemura, T., Taniguchi, M. and Shibuya, K. (2011). Submarine groundwater discharge in Lützow-Holm Bay, Antarctica. *Geophysical Research Letters* **38**, 959–993.

Unrein, F., Izaguirre, I., Massana, R., Balagué, V. and Gasol, J.M. (2005). Nanoplankton assemblages in maritime

Antarctic lakes: characterization and molecular fingerprinting comparison. *Aquatic Microbial Ecology* **40**, 269–282.

Unrein, F. and Vinocur, A. (1999). Phytoplankton structure and dynamics in a turbid Antarctic lake (Potter Peninsula, King George Island). *Polar Biology* **22**, 93–101.

Van den Hoff, J. and Franzmann, P.D. (1986). A choanoflagellate in a hypersaline Antarctic lake. *Polar Biology* **6**, 71–73.

Van Trappen, S., Mergaert, J., Van Eygen, S., Dawyndt, P., Cnockaert, M.C. and Swings, J. (2002). Diversity of 746 bacteria isolated from microbial mats from ten Antarctic lakes. *Systematic and Applied Microbiology* **25**, 603–610.

Vaqué, D. and Pace, M.L. (1992). Grazing on bacteria by flagellates and cladocerans in lakes of contrasting food web structure. *Journal of Plankton Research* **14**, 307–321.

Vaughan, D.G. and Doake, C.S.M. (1996). Recent atmospheric warming and retreat of ice shelves on the Antarctic Peninsula. *Nature*, **379**, 328–331.

Vaughan, D.G., Rivera, A., Woodward, J., Corr, H.F.G., Wendt, J. and Zamora, R. (2007). Topographic and hydrological controls on subglacial Lake Ellsworth, West Antarctica. *Geophysical Research Letters* **34**, (18), L18501. doi:10.1029/2007GL030769.

Veillette, J., Mueller, D.R., Antoniades, D. and Vincent. W.F. (2008). Arctic epishelf lakes as sentinel ecosystems: past, present and future. *Journal of Geophysical Research* **113**, G04014. doi:10.1029/2008JG000730.

Verkulich, S.R., Melles, M., Hubberon, H.-W. and Pushina, Z.V. (2002). Holocene environmental changes and development of Figurnoye Lake in the southern Bunger Hills. *Journal of Paleolimnology* **28**, 253–267.

Vincent, W.F. (1981). Production strategies in Antarctic inland waters: phytoplankton ecophysiology in a permanently ice-covered lake. *Ecology* **62**, 1215–1224.

Vincent, W.F. (2000). Cyanobacteria dominance in the Polar regions. In Whitton, A. and Potts, M. (eds) *The Ecology of Cyanobacteria*, pp. 321–340. Kluwer Academic Publishers, Dordrecht, the Netherlands.

Vincent, W.F., Castenholz, R.W., Downes, M.T. and Howard-Williams, C. (1993). Antarctic Cyanobacteria: light, nutrients and photosynthesis in the microbial mat environment. *Journal of Phycology* **29**, 745–755.

Vincent, W.F., Gibson, J.A.E. and Jeffries, M.O. (2001). Ice-shelf collapse, climate change, and habitat loss in the Canadian high Arctic. *Polar Record* **37**, 133–142.

Vincent, W.F., Hobbie, J.E. and Laybourn-Parry, J. (2008). Introduction to the limnology of high-latitude lake and river ecosystems. In Vincent. W.F. and Laybourn-Parry, J. (eds) *Polar Lakes and Rivers: Limnology of Arctic and Antarctic Aquatic Ecosystems*, pp. 1–18. Oxford University Press, Oxford.

Vincent, W.F. and James, M.R. (1996). Biodiversity in extreme aquatic environments: lakes, pond and steams of the Ross Sea sector, Antarctica. *Biodiversity and Conservation* **5**, 1451–1471.

Vincent, W.F. and Vincent, C.L. (1982). Factors controlling phytoplankton production in Lake Vanda (77°S). *Canadian Journal of Fisheries and Aquatic Science* **39**, 1602–1609.

Vinocur, A. and Pizarro, H. (2000). Microbial mats of twenty-six lake from Potter Peninsula, King George Island, Antarctica. *Hydrobiologia* **437**, 171–185.

Vopel, K. and Hawes, I. (2006). Photosynthetic performance of benthic microbial mats in Lake Hoare, Antarctica. *Limnology and Oceanography* **51**, 1801–1812.

Voytek, M.A., Ward, B.B. and Priscu, J.C. (1998). The adundance of ammonium-oxidizing bacteria in Lake Bonney, Antarctica determined by immunofluorescence, PCR and in situ hybridization. In Priscu, J.C. (ed) *Ecosystem Dynamics in a Polar Desert, the McMurdo Dry Valleys, Antarctica* pp. 217–228. Antarctic Research Series, American Geophysical Union, Washington DC, Volume **72**.

Wadham, J.L. Arndt, S., Tulaczyk, S., Stibal, M., Tranter, M. et al. (2012). Potential methane reservoirs beneath Antarctica. *Nature* **488**, 633–737.

Wadham, J.L., Bottrell, S., Tranter, M. and Raiswell, R. (2004). Stable isotope evidence for microbial sulphate reduction at the bed of a polythermal high Arctic glacier. *Earth and Planetary Science Letters* **219**, 341–355.

Wadham, J.L., Death, R., Monteiro, F., Tranter, M. et al. (2013). The potential role of the Antarctic Ice Sheet in global geochemical cycles. *Earth and Environmental Science Transactions of the Royal Society of Edinburgh* **104**, 55–67.

Wadham, J.L., Tranter, M., Hodson, A.J., Hodgkins, R., Bottrell, S. et al. (2010a) Hydro-biogeochemcial coupling beneath a large polythermal Arctic glacier: implications for subice sheet biogeochemistry. *Journal of Geophysical Research–Earth Surface* **15**, doi:10.1029/2009JF001602.

Wadham, J.L., Tranter, M., Skidmore, M., Hodson, A.J., Priscu, J.C., Lyons, W.B., Sharp, M., Wynn, P. and Jackson, M. (2010b). Biogeochemcial weathering under ice: size matters. *Global Biogeochemcial Cycles* **24** (3), GB3025.

Wagner, B. and Cremer, H. (2006). Limnology and sedimentology record of Radok Lake, Amery Oasis, East Antarctica. In Fütterer, D.K., Damaske, D., Kleinschmidt, G., Miller, H. and Tessensohn, F. (eds) *Antartica: Contributions to Global Earth Sciences*, pp. 447–545. Springer, Berlin, Heidelberg and New York.

Wagner, B., Cremer, H., Hultzsch, N., Gore, D.B. and Melles, M. (2004). Late Pleistocene and Holocene history of Lake Terrasovoje, Amery Oasis, East Antarctica, and its climatic and environmental implications. *Journal of Paleolimnology* **32**, 321–339.

Wagner, B., Hultzsch, N., Melles, M. and Gore, D.B. (2007). Indications of Holocene sea-level rise in Beaver Lake, East Antarctica. *Antarctic Science* **19**, 125–128.

Wagner, B., Melles, M., Doran, P.T., Kenig, F., Forman, S.L., Pierau, R. and Allen, P. (2006). Glacial and postglacial sedimentation in the Fryxell basin, Taylor Valley, southern Victoria Land, Antarctica. *Palaeogeography, Palaeoclimatology, Palaoecology* **241**, 320–337.

Wagner, B., Ortlepp, S., Doran, P.T., Kenig, F., Melles, M. and Burkemper, A. (2011). The Holocene environmental history of Lake Hoare, Taylor Valley, Antarctica, reconstructed from sediment cores. *Antarctic Science*, doi:10.1017/S0954102011000125.

Wagner, B. and Seppelt, R. (2006). Deep-water occurrence of the moss *Bryum pseudotriquetrum* in Radok Lake, Amery Oasis, East Antarctica. *Polar Biology* **29**, 791–795.

Walder, J.S. and Fowler, A. (1994). Channelized subglacial drainage over a deformable bed. *Journal of Glaciology* **40**, 3–15.

Wand, U. (1995). Hydrogeochemistry. In Bormann, P. and Fritzsche, D. (eds) *The Schirmacher Oasis, Queen Maud Land, East Antarctica, and its Surroundings*, pp. 309–312. Justus Perthes Verlag, Gotha, Germany.

Wand, U., Hermichen, W.-D., Brüggemann, E., Zierath, R. and Klokov, V.D. (2011). Stable isotope and hydrogeochemcial studies of Beaver Lake and Radok Lake, MacRobertson Land, East Antarctica. *Isotopes in Environmental and Health Studies* **47**, 407–414.

Wand, U. and Perlt, J. (1999). Glacial boulders 'floating' on the ice cover of Lake Untersee, East Antarctica. *Antarctic Science* **11**, 256–260.

Wand, U., Samarkin, V.A., Nitzsche, H.M. and Hubberten, H.W. (2006). Biogeochemistry of methane in the permanently ice-covered Lake Untersee, Central Dronning Maud Land, East Antarctica. *Limnology and Oceanography* **51**, 1180–1194.

Wand, U., Schwartz, G., Brüggemann, E. and Bräuer, K. (1997). Evidence for physical and chemical stratification in Lake Untersee (central Dronning Maud Land, East Antarctica). *Antarctic Science* **9**, 43–45.

Ward, B.B., Granger, J., Maldonado, M.T. and Wells, M.L. (2003). What limits bacterial production in the suboxic region of permanently ice-covered Lake Bonney, Antarctica? *Aquatic Microbial Ecology* **31**, 33–47.

Watanabe, O., Jouzel, J., Johnsen, S., Parrenin, F., Shoji, H. and Yoshida, N. (2003). Homogeneous climate variability across East Antarctica over the past three glacial cycles. *Nature* **422**, 509–512.

Weand, B.L., Hoehn, R.C. and Parker, B.C. (1977). Nutrient fluxes in Lake Bonney–meromictic Antarctic lake. *Archiv für Hydrobiologie* **80**, 519–530.

Webster, J.G., Brown, K.L. and Vincent, W.F. (1994). Geochemical processes affecting meltwater chemistry and the formation of saline ponds in the Victoria Valley and Bull Pass region, Antarctica. *Hydrobiologia* **281**, 171–186.

Webster, J., Hawes, I., Downes, M., Timperley, M. and Howard-Williams, C. (1996). Evidence for regional climate change in the recent evolution of a high altitude pro-glacial lake. *Antarctic Science* **8**, 49–59.

Webster-Brown, J., Gall., M., Gibson, J., Wood, S. and Hawes, I. (2010). The biogeochemistry of meltwater habitats in the Darwin Glacier region (80ºS). Victoria Land, Antarctica. *Antarctic Science* **22**, 646–661.

Weinbauer M.G. (2004). Ecology of prokaryotic viruses. *FEMS Microbiology Ecology* **28**, 127–181.

Weinbauer, M.G., Brettar, I. and Höfle, M. (2003). Lysogeny and virus-induced mortality of bacterioplankton in surface, deep and anoxic waters. *Limnology and Oceanography* **48**, 1457–1465.

Weinbauer, M.G. and Suttle, C.A. (1996). Potential significance of lysogeny to bacteriophage production and bacterial mortality in coastal water of the Gulf of Mexico. *Applied and Environmental Microbiology* **62**, 4374–4380.

Weinbauer, M.G. and Suttle, C.A. (1999). Lysogeny and prophage induction in coastal and offshore bacterial communities. *Aquatic Microbial Ecology* **18**, 217–225.

Weisse, T., Muller, H., Pinto-Coelho, R.M., Schweizer, A, Springman, D. and Baldringer, G. (1990). Response of the microbial loop to the phytoplankton spring bloom in a large prealpine lake. *Limnology and Oceanography* **35**, 781–794.

Wetzel, R.G. (2001). *Limnology: Lake and River Ecosystems*, 3rd edn. Academic Press, San Diego, CA.

Whalen, S.C. and Alexander, V. (1984). Influence of temperature and light on rates of inorganic nitrogen transport by algae in an Arctic lake. *Canadian Journal of Fisheries and Aquatic Sciences* **41**, 1310–1318.

Whittaker, R.H. (1969). New concepts of kingdoms of organisms. *Science* **163**, 150–160.

Wilson, A.T. (1964). Evidence from chemical diffusion of a climatic change in the McMurdo Dry Valleys 1200 years ago. *Nature* **210**, 176–177.

Wilson, A.T., Holdsworth, R. and Hendy, C.H. (1974). Lake Vanda: source of heating. *Antarctic Journal of the United States* **9**, 137–138.

Wilson, W.H., Lane, D., Pearce, D.A. and Ellis-Evans, J.C. (2000). Transmission electron microscope analysis of virus-like particles in the freshwater lakes of Signy Island, Antarctica. *Polar Biology* **23**, 657–660.

Wingham, D. J., Siegert, M.J., Shepherd, A. and Muir, A.S. (2006). Rapid discharge connects antarctic subglacial lakes. *Nature* **440**, 1033–1036.

Wommack K.E. and Colwell R.R. (2000). Virioplankton: viruses in aquatic systems. *Microbiology and Molecular Biology Reviews* **64**, 69–114.

Woodward, J., Smith, A.M., Ross, N., Thoma, M., Corr, J. et al. (2010). Location for direct access to subglacial

Lake Ellsworth: an assessment of geophysical data and modeling. *Geophysical Research Letters* **37**, doi:10.1029/2010GL042884.

Wright, A. and Siegert, M.J. (2011). The identification and physiographical setting of Antarctic subglacial lakes: an update based on recent discoveries. *Antarctic Subglacial Aquatic Environments* **192**, 9–26.

Wright, A. and Siegert, M.J. (2012). A fourth inventory of Antarctic subglacial lakes. *Antarctic Science* **24**, 659–664.

Wright, S.W. and Burton H.R. (1981). The biology of Antarctic saline lakes. *Hydrobiologia* **82**, 319–338.

Wynn, P.M., Hodson, A.J. and Heaton, T. (2006). Chemical and isotopic switching within the subglacial environments of a high Arctic glacier. *Biogeochemistry* **78**, 173–193.

Yang, X.X., Lin, X.Z., Bian, J., Sun, X.Q. and Hunag, X.H. (2004). Identification of five strains of Antarctic bacteria producing low temperature lipase. *Acta Oceanologica Sinica* **23**, 717–723.

Yau, S., Lauro, F.M., DeMaere, M.Z., Brown, M.V., Thomas, T. and 6 others. (2011). Virophage control of Antarctic alagal host-virus dynamics. *Proceedings of the National Academy of Sciences* **108**, 6163–6168.

Young, D.A., Wright, A.P., Roberts, J.L., Warner, R.C., Young, N.W. et al. (2011). A dynamic early East Antarctic ice sheet suggested by ice-covered fjord landscapes. *Nature* **474**, 72–75.

Zachos, J., Pagani, M., Sloan, L., Thomas, E. and Billups, K. (2001). Trends, rhythms, and aberrations in global climate 65 ma to present. *Science* **292**, 686–693.

Zachos, J. C., Breza, J.R. and Wise, S.W. (1992). Early Oligocene ice-sheet expansion on Antarctica—stable isotope and sedimentological evidence from Kerguelen Plateau, Southern Indian Ocean. *Geology* **20**, 569–573.

Zachos, J.C., Dickens, G.R. and Zeebe, R.E. (2008) An early Cenozoic perspective on greenhouse warming and carbon-cycle dynamics, *Nature* **451**, 279–283.

Zinabu, G.M. and Taylor, W.D. (1997). Bacteria-chlorophyll relationships in Ethiopian lakes of varying salinity: are soda lakes different? *Journal of Plankton Research* **19**, 647–654.

Zwartz, D., Bird, M., Stone, J. and Lambeck, K. (1998). Holocene sea-level change and ice-sheet history in the Vestfold Hills, East Antarctica. *Earth and Planetary Science Letters* **155**, 131–145.

Index

A
ablation zone ponds 147
Acanthoamoeba polypahga 112
Acanthocyclops mirnyi 74, 88
Acanthoecopsis unguiculata 119
accretion ice 160, 165–166, 169–172
Achnanthes 72
Achnanthes brevipes 121
Achnanthes pinnata 87
Achnanthes subatomoides 87
Actinobacteria 64, 88, 107, 131, 171, 173
air temperatures 5–7
airborne radio echo sounding 14, 156, 158, 179
Algae 34–35
  in freshwater lakes 72
  in saline lakes 121–122
algal mats 40, 83–87
Alona 37
Amery Ice Shelf 52
Amery Oasis 52–54
ammonium 21, 62–63, 102–104, 139, 150
amoebae 31–32, 71–72, 131
Amphiascoides 36, 122
Amphora venata 87
Anabaena 29, 153, 155
Ankistrodesmus 72
Ankistrodesmus falcatus 72, 83
Antarctic Peninsula 6, 58
Antarctic Treaty 1–2
Aphanocapsa 130
Archaea 29–30, 106, 109, 131
Arthrobacter agilis 88
Arthrobacter flavus 130
Askenasia 69–70, 115
Aspidisca cicada 69
assimilation number 129, 143
atmospheric CO_2 9
autochthonous organic carbon 20, 27
autotrophic flagellates
  in epishelf lakes 140–142
  in freshwater lakes 65, 70–71
  in saline lakes 108, 116–119
autotrophic production 126

B
Bacillariophyceae *see* diatoms
Bacteria
  in algal mats 87–88
  diversity in algal mats 130–131
  Gram positive 88
  green sulphur 112
  heterotrophic 29–30
  methylotrophic 171
  purple non-sulphur 111–112
  purple sulphur 50, 112
  sulphur oxidizing 110
  sulphate reducing 110, 153
bacterial chlorophylls 111–112
bacterial concentrations
  in cryoconite holes 152
  in cryolakes 150
  in epishelf lakes 141
  in freshwater lakes 65–66
  in ice shelf ponds 154
  in saline lakes 108
  in subglacial lakes 171–172
bacterial production (growth)
  in cryoconite holes 152
  in epishelf lakes 143–144
  in freshwater lakes 48–49, 78–80
  in saline lakes 95, 125, 130–131
  in subglacial lakes 172
Bacteriodetes 64, 171
bacterioplankton 78–80
  biomass 80
basal melting 157
benthic algal mats 83–85, 88, 130–131
benthic mosses 86–88
biomass values for Ace Lake 128
Blepharisma 69
Blood Falls 97–98
blue ice region 149–150
Bodo 131
Boeckella poppei 37, 39, 74, 82–83, 122, 142–143

grazing rates 82–83
Bølingen Islands 85
Brachinecta gainii 37, 39–40, 74
Brachionus 74
Broknes Peninsula
Bryum pseudotriquetrum 86
Bunger Hills 1, 4, 44, 47, 51–52, 96, 101

C
Calliergon sarmentosum 87
carbon budget 79
carbon cycling
  in benthos 88–90, 131–132
  models 125–126
  in plankton 75–78, 125–129, 143–144
carbon dating *see* radio carbon dating
carbon dioxide 4, 7, 105
Carnobacterium 107
Cenozoic 8–10
chemoautotrophy 172
chemolithic primary production 172
chemocline 101–102, 106, 109
choanoflagellates 32
Chilodonella algivora 69
Chilodonella sp. 69, 115
Chlamydomonas 71, 123, 131
Chlamydomonas raudensis 116
Chlamydomonas reinhardtii 77
Chlamydomonas subcaudata 77, 117, 123
Chlorella 83
chloride 91, 97–98
Chlorobium 112
chlorophyll *a* 41, 45, 48–49, 85, 96, 106, 139, 149, 150
chlorophytes
  in paleolimnology 50
Chroococcus 150
Chroomonas 71, 153
chrysophytes 71

ciliated Protozoa 33–34, 68–70, 114–118, 142, 153
  diversity 70
Cinetochilum margaritaceum 69
Cladocera 36–37, 122
climate history 8–10
climate trends 5–7
climatic conditions in Antarctica 4–7
Clostridium estherheticum 130
Colwellia 107
community respiration 79
conductivity *see* salinity
copepods
  calanoid 36
  cyclopoid 144
  harpacticoid 36
Cosmarium 35, 72
Craticula 87
Crustacea 36–37
cryoconite holes 147–148, 151–152
cryolakes 11
  development 148
  drainage 148–149
cryptophytes 32–33, 71, 77–78, 116, 121, 127, 153
Cryptomonas 32, 71, 116
Cryptomonas undulate 116
Cyanobacteria 29–30
  in freshwater lakes 50, 72–73, 88
  in ice shelf ponds and lakes 153–155
  in paleolimnology 50
  in perennial ice cover 38, 129–130
  in saline lakes 110–111
  in saline lakes benthic mats 130–131
  in saline lakes ice cover 129–130
cyanobacterial mats 83–85, 130–132
  diversity 84
Cyanomonas 71
Cyanothece 150
Cyclidium glaucoma 69
Cyclotella 72, 121
Cytophaga-Flavobacterium-Bacteridetes 88, 107, 131

D
Dactylobiotus ambiguous 88
Daphniopsis studeri 37, 73–74, 82, 105, 122
  grazing rates 82
deep chlorophyll maximum (DCM) 103, 106, 123
δD 22–23
$\delta^{18}O$ 22–23, 172
denitrification 18, 21
Desmidiales-see desmids

desmids 34–35
Desmarella 131
Desulfobacter 110
Desulfovibrio 110
deuterium 6, 22–23
Diadesmis 87
Diaphanoeca grandis 32
diatoms 34–35, 72, 86–87, 100–101, 121, 130, 140, 144–145
diatom distribution 87
diatom diversity 87
Didinium 33, 69, 115
dimictic lakes 60
dinitrification 18–19
dinoflagellates 32–33, 71, 117–119
dissolved inorganic carbon 105, 140
dissolved organic carbon (DOC) 20, 22. 28, 63–64, 79–80, 103–104, 125, 127, 139–140, 150, 166
dissolved organic carbon budget 79
dissolved oxygen 101
Drepanocladus 87
Drepanocladus longifolius 74
Dunaliella 105, 117

E
East Antarctic Ice Sheet 50
Enchelys mutans 69
endemism 64, 84, 107, 174
englacial lakes, 147
ephippial eggs 36–37, 73–74
epiglacial lakes 25
epilimnion 60–62
epishelf lakes 12–13, 45
  in Arctic 134
  chlorophyll *a* 139
  formation 135–137
  hydrology 138
  photosynthetically active radiation 138
  salinity 136
  structure 135
  temperature profiles 137
  tidal ranges 138
episodic draining of subglacial lakes 156–159
euglenoids 35
euphotic zone 41
Euplotes 34, 70, 115
eutrophication 20, 175

F
fairy shrimp 37, 39, 74
feeding selectivity in protozoans 70
Filinia 74
flagellated protozoa 70–71
Flavobacteria-Cytophaga 64

Flavobacterium 107
Flavobacterium frigidarium 88
Flavobacterium gondwanense 65
Flavobacterium hibernum 64, 130
Flavobacterium slaegens 65
food web
  for Ace Lake 128
  planktonic 27–28, 128
Foraminifera 7, 100
Fragilaria 72
Fragilaria contreuns 87
freeze-thaw patterns 154
freshwater lakes
  depths 42–45
  surface areas 42–45
  water supplies 46, 49
Frontonia acuminate 69
Fontonia angusta 69
Fragilaria 72
fur seals 60
fulvic acids 20, 104

G
Geminigera cryophila 116
geochemical conditions 15–22
geomorphology 23–26
Glacial Crooked Lake 26, 51
Glacial Lake Trowbridge 26, 56
Glacial Lake Washburn 26, 56, 70
Glacial Lake Wright 26
glaciers
  Canada 56–57, 149
  Darwin 150
  Joyce 151
  Sørsdal 12, 72, 151
  Taylor 97
  Wright 96
glaciation 23
glaciostatic uplift 25–26, 99
Gladioferens antarcticus 142
Gomphonema 72
Gomphonema angustatum 87
Grazing rates of nanoflagellates 82
Greater Lake Priscu 97
Gymnodinium 71, 83, 117–118, 131
Gyrodinium glaciale 117–118

H
Halobacteriaceae 107
Halobacterium 107
Halomonas meridian 65, 107
Halomonas subglaciescola 65, 107
Hantzschia amphioxys 85
Hantzschia sp. 87
Halteria 69, 115
heliozoans 31–32, 72
heterocysts 29

Heteromita 71
heterotrophic bacteria
  in epishelf lakes 141
  in freshwater lakes 64–66
  in saline lakes 106–110
heterotrophic flagellates
  concentrations in epishelf lakes 141
  concentrations in freshwater lakes 65–66
  concentrations in saline lakes 108, 119
  grazing rates 81–82, 120, 144
  ingestion of DOC 121
  specific growth rates 119
heterotrophic grazing 80–83, 125–127
heterotrophic production 126
history of Antarctic limnology 1–4
history-geochemical indicators 22–23
history-glaciological 7–11
Holocene 10–11, 50–53, 100–101, 164
Holophyra 69
horizontal gene transfer 30, 107
hypolimnion 60–61
Hypsibius antarcticus 131
Hypsibius papillifer 88
Hypsibius renaudi 88

I
ice cores 8
ice cover
  annual 38–39, 41
  biota 129–130
  perennial 38–39, 41–42, 105, 129–130
  rafted 13
  saline lakes 129–130
ice sheets 7, 10–11, 48, 56, 156, 179
ice shelves
  Amery 138
  King George VI 59, 136, 147
  Larsen A 6
  Larsen B 6
  McMurdo 3, 11, 147, 152–153
  Ross 26, 162
  Ward Hunt 134, 147
ice shelf ponds 152–155
Idomene scotti 36, 131
inter-annual variation 69, 78, 175–176
interglacial period 6–7
International Geophysical Year (IGY) 3–4
isotopes of oxygen 7–8, 22–23
iron reduction 19

J
Janthinobacterium lividum 130

K
Kelliocottia 35, 74
Keratella 74

L
lake diversity 11–15
lake formation 49–60, 97–101
lake trophic status 41
lakes
  Ablation 134, 136, 139–140, 143–144
  Abraxas 93, 100–101, 117, 119, 122
  Ace 17–18, 26, 92, 96, 99–100, 113–114, 117–120, 122, 125, 127–129, 176
  Anderson 93
  Beall 93, 100–101
  Beaver 13, 42, 53, 134–135, 137–144
  Bisernoye 48, 63, 65, 67
  Boeckella 42, 48, 62–63, 71
  Bonney 26, 39, 58, 92, 96–97, 115–116, 120, 124–125
  Burton 92
  Chelnok 12, 42, 48
  Chico 63
  Clear 92
  Collerson 96
  Corner 65
  Crooked 42, 48, 51, 61, 63, 65–68, 70–74, 75, 79–82, 109
  Deep 24, 93, 96, 102–103, 107
  Deprez 93
  Discussion 44, 49, 65
  Druzhby 42, 48, 51, 61, 63, 65–68, 70–71, 73–74, 79–82, 109
  Ekho 92, 96, 101
  Ellsworth 156, 158–159, 161
  Figurnoya (also known as Algae Lake) 42, 48, 51–52, 87
  Fletcher 93
  Fryxell 16–18, 26, 39, 56–58, 70, 78, 91, 96, 98, 112, 115–116, 120, 125–127
  Garrow 91
  Greater Lake Priscu 97
  Heidi 48, 65, 71
  Heywood 44–45, 48, 59–60, 62–63, 65, 67, 71, 76, 81–82
  Highway 73, 92, 96, 99, 113–114, 117–118, 120, 122, 125
  Hoare 26, 39, 41–42, 48, 56–58, 61, 63, 65, 67–71, 74, 76–78, 81–82, 88–89, 130
  Hodgson 156, 158, 163–164, 173
  Holl 93, 100
  Hunazoko 92, 102
  Joyce 92, 96, 115
  Karovoye 134, 143

Kholodnoya 134
Kirisjes Pond 43, 48, 51, 65, 71
Kitiesh 89
Laternula 92
Lichen 48, 63, 65, 67, 72
Limnopolar 44, 48, 65, 71
McCallum 93
Midge 43, 48, 60, 63, 65, 71
Miers 41–42, 48, 56, 63, 65, 67–69, 130
Moss 44, 49, 59, 63, 87
Moutonée 43, 134, 136, 139–140, 143–145
Nella 48, 62–63, 65
Northern 134
Nottingham 48, 65, 67, 72
No Worries 44, 48, 62–63, 65
Nurume 23–24, 93
Oblong 92
Organic 93, 96, 99–100, 107
Oval 92
Ozhidaniya 134
Pauk 65, 67
Pendant 92, 96, 113–114, 117–118, 121–122, 125
Pol'anskogo 134
Polest 92
Predgornoye 134
Prival'noya 43, 134, 143
Progress 42, 48, 50, 62–63, 65
Radok 42, 53–54, 135
Reid 87, 93, 101
Rookery 24, 92
Sarah Tarn 94
Shield 92
Sombre 44, 48, 59, 61, 63, 65, 67, 71, 76, 82–83
Sophia 91
Southern 134
Surabati 92
Suvivesi 149–150
Terrasovoje 42, 52–54, 136
Tranquil 44, 48, 59, 61, 65, 67, 71
Transkriptsii Gulf 42, 134, 136–139
Tres Hermanos 43, 48
Untersee 19, 42, 48, 61–63
Vanda 21, 23, 26, 39, 49, 91, 96, 98–99, 101–102, 115, 124, 131–132
Vereteno 96
Verkhneye 44, 61, 76
Vida 132–133
Vostok 14, 24, 156, 158–160
Watts 26, 43, 48, 51, 73, 76, 88–89, 132
Whillans 15, 156, 158, 162
White Smoke 134–135

lakes (cont.)
    Williams 92, 117, 125
    Wilson 91, 96, 99
    Wujka 45, 74
    Zigzag 134
    Zvezda 42, 63, 65
Larsemann Hills 1, 45, 47, 51, 69, 96, 101
Last Glacial Maximum (LGM) 26, 50–51, 99, 101, 135
Lecane 74
Lepadella 74
Leptobryum 86–87
Leptobryum wilsonii 87
Leptolyngbya 40, 130
Leptolyngbya angustissima 86
Leptolyngbya antarctica 84, 86
Leptolyngbya fridgida 84
Livingstone Island 87
Long Term Ecosystem Research Program (LTER) 3, 96
Luticola muticopsis 87, 130
Lyngbya 72, 83, 153

M
Macrothrix ciliate 74
Magnifollicullina 99
Mantoniella 140
Marine Isotope Stage 31
Maritime Antarctic 1, 12, 58–60, 72, 96
Marinomonas prioryensis 109
McMurdo Dry Valleys
    lake formation 26, 54–58
    lake types 96–97
mean cell volume (MCV) 66
meromictic lakes 16–17
    Arctic 91
    structure 101
Mesodinium 69, 142
Mesodinium rubrum 34, 114–116, 121, 127–128
methane production 7, 19, 109, 173
Methanococcoides burtonii 109
Methanoculleus palmolei 131
methanogens 19, 109–110, 153
methanogenesis 19, 153, 167
Methanogenium 107
Methanogenium frigidum 109
methanotrophs 109
Methylobacillus 171
Methylophilus 171
Methylosphaera 107
Methylosphaera hansonii 109
microbial loop 27
Microcystis 73
mixolimnion 16–17, 98–99, 101–102
mixotrophic ciliates 34, 70

mixotrophic phytoflagellates 33, 81–82, 121, 127
    grazing rates 120
mixotrophy 33–34, 77–78, 116–117
Moderate Resolution Imaging Spectroradiometer (MODIS) 156, 179–180
Monodinium 69, 115, 142
monomictic lakes 60–61
monomolimnion 16–17, 101–102
Monosiga 71
morphometry
    freshwater lakes 42–45
    saline lakes 92–94
mosses 40, 86–87, 89
moulin 147–148
mycosporine-like amino acids (MAAs) 84

N
Navicula 72, 121
Navicula cf. *cryptotenella* 100
Navicula directa 130
Navicula glaciei 100
nauplii 36–37
National Science Foundation 3, 176
nematodes 36, 88, 131
nitrogen
    ammonium 21, 62–63, 102–104, 124
    cycle 18–22
    fixation 29, 155
    nitrate 62–63, 104
Nitzschia 72, 121
Nitzschia Antarctica 153
Nodularia 29, 153, 155
Nostoc 39, 83, 129–130
Notholca 74, 99, 123
Notholca squamula salina 88
nutrient supply 19–22

O
Ochromonas 71, 117, 123, 131
oligochaetes 38
optical sensors 178
organic carbon supply 19–22, 45
Oscillatoria 39–40, 72–73, 83, 86, 110, 129–130, 153
Oscillatoria deflexa 153
Oscillatoria limosa 153
oxycline 18
oxygen dissolved 18
oxygen isotopes 22–23
oxygen saturation 18

P
palaeolimnology 26
Paralabidocera antarctica 36, 40, 122, 129

Paraphysomonas vestita 71
particulate organic carbon (POC)
    in freshwater lakes 63
    in saline lakes 104
perennial ice cover 38
Peridinium 71
Philodina 35, 74, 123, 131, 153
Philodina alata 34, 74, 123
Philiodina gregarina 75
Phormidium 39, 72–74, 83, 110, 129, 150, 153
Phormidium autumnale 153
Phormidium fragile 153
Phormidium frigidum 153
Phormidium laminosum 153
Phormidium priestleyi 84
phosphorus
    in cryolakes 150
    cycling 19–23
    in freshwater lakes 62–63
    in saline lakes 102–104, 124
photosynthetic bacteria 29–30, 110–112
photosynthetic efficiency 129, 143
photosynthesis definition 18
photosynthesis rates *see* primary production
photosynthetically active radiation (PAR) 4, 39, 41–42, 76, 131, 138–139, 151, 177–178
phototrophic nanoflagellates *see* autotrophic flagellates
phycobilin 85, 131
phycoerthrin 85, 131
phycocyanin 85, 131
phytoplankton 72–73, 116–122
picocyanobacteria 29, 73, 110–111, 153
Pinnularia borealis 87
Pinnularia cymatopleura 153
Pinnularia microstauron 130
Plagiocampa 69–70, 115, 127
Planktolyngbya 150
plasmids 30
Pleistocene 9–10, 52, 101
Podophyra 33
Polarella glacialis 117–118
ponds 15
    see also saline ponds
Pony Pond 97
predation pressure 81–83, 122, 127, 129
primary production 28
    in cryolakes 151
    in epishelf lakes
        planktonic 143–144
        rates 143
    in freshwater lakes

benthic 88–90
  in ice shelf ponds
    benthic 155
  in saline lakes
    benthic 131-132
    planktonic 95, 123–125
    rates 95
Proteobacteria 64–65, 88, 107, 171, 172–173
Protozoa 31–34, 68–72, 114–121
Psammothidium abundans 87, 101
Pseudoanabaena 150
Pseudokephyrion 71
Psychrobacter glacincola 88
psychrophiles 64–65
purple non-sulphur bacteria 111–112
purple sulphur bacteria
  in palaeolimnology 50
Pyramimonas 71
Pyramimonas gelidicola 32, 116, 121

Q
Quaternary 10, 50

R
radiocarbon dating 10–11, 99–100, 137
Rauer Islands 85, 96
redox conditions 15, 17–19
respiration 79, 126, 155, 167
Resticula gelida 88
Rhinomonas 116
Rhodopseudomonas palustris 112
Ross Ice Sheet 56–57
rotifers 35–36, 74–75, 88, 122–123, 131, 142, 153

S
saline lakes
  definition 91
  distribution 94–97
  formation 97–101
  Vestfold Hills 16
saline ponds 96–97
  Brack 154
  Don Juan 97
  Floridas 97, 130–131
  Marr Ponds 97
  Parera Ponds 97
salinity 15–17, 91, 101–102, 136, 154
Sarcodina 71–72
Sarcomastogophora 31
satellite imagery 178
Schirmacher Oasis 1, 96
Scientific Committee on Antarctic Research (SCAR) 3
Scrippsiella aff hangoei 117–119

Scytonema 39, 129
Scytonemin 84
sediment cores 26, 50–54, 56–57
seepages 15
sedimentation rates 50
shade adaptation in phytoplankton 77
Shewanella baltica 88
Signy Island 57–60, 69, 71, 76, 87
silicate dissolution 167
snow cover 76
soda lakes 91
South Orkney Islands 58–59
South Shetland islands 58, 69
Spathidium 69
Sphaerophyra 69, 115
stable isotopes 7–8, 97, 104–105
Stichococcus 78
Stichococcus bacillaris 121–122
stratification of water column
  in freshwater lakes 60–62
  in saline lakes 101–102
Strombidium 34, 69–70, 115, 142
subglacial drainage 163–166
subglacial lakes, for specific lakes *see under* lakes
subglacial lakes
  biota 170–173
  chemistry 168–170
  distribution in Antarctica 13–15, 157–159
  formation 163–166
  geochemical conditions 166–168
  hydrological conditions 163–166
  nutrient levels 166–169, 171
  physiographic settings 157–163
  sulphide oxidation 170
  weathering processes 167
suctorian protozoans 33
supraglacial drainage 149
supraglacial lakes 147–152
Synechococcus 39, 73, 111, 129, 153
Syowa Oasis 1, 23–24, 86–87, 94–96

T
tardigrades 38, 88, 131
tectonic activity 23–24
temperatures
  air 5
  cryolakes 151
  epishelf lakes 136–137
  freshwater lakes 61–62
  ice shelf ponds 154
  saline lakes 101–102
Terra Nova Bay 69
Tetramitus 32
thermal stratification 60–62, 102

thermocline 60–62
Thiobacillus thioparus 110
Thiocapsa roseopersicina 112
Trematomus bernacchii 144
Trematomus scotti 144
trophic status 41
Tropidoneis laevissima 121, 153

U
ultramicrobacteria 133
ultra violet radiation
  protective mechanisms 130
Urotricha 69

V
Vestfold Hills 1, 23–24, 45–46, 51, 69–70, 96
Viriophage 112–113
viruses
  burst sizes 31, 68
  concentrations in epishelf lakes 140, 142
  concentrations in freshwater lakes 66–67, 80
  concentrations in saline lakes 113
  definition 30
  infection 31, 68
  lytic cycle 30, 68, 114
  lysogenic cycle 30, 68, 113
  morphological diversity 67
  virus to bacterium ratios 67–68, 113, 140
Vorticella mayeri 69
Vorticella sp. 69–70, 115, 153
water chemistry
  in epishelf lakes 139–140
  in freshwater lakes 62–64
  in saline lakes 102–105
  in subglacial lakes 166–170
  in water supply to stations
    Boeckella Lake 46
    No Worries Lake 46–49
    unnamed lake Deception Island 46

W
wetlands 15
Windmill Islands 96
wireless networks 178

Z
zoochlorellae 34
zooplankton
  in freshwater lakes 73–75
  grazing 82–83
  in saline lakes 122–123